BUSINESS SERIES

CONVERSATIONAL STATISTICS

BY
Harry V. Roberts
GRADUATE SCHOOL OF BUSINESS
UNIVERSITY OF CHICAGO

EDITED BY
Christine Doerr
HEWLETT PACKARD

PUBLISHED BY
The Scientific Press
THE STANFORD BARN
PALO ALTO, CA 94304
(415) 322-5221

© *Copyright, 1974, by*
HEWLETT-PACKARD COMPANY
Cupertino, California
Printed in the U.S.A.

The Hewlett-Packard University Business Series represents the published results of a Curriculum Development project sponsored by the General Systems Division of Hewlett-Packard Company.

This material is designed to be used with a computing package called IDA, which was originally written in HP BASIC for use with the HP 2000 and 3000 series minicomputers. In 1976, IDA was converted to a FORTRAN version designed to be readily transported to any computer with a reasonable implementation of FORTRAN.

Library of Congress Catalog Card Number: 76-48778

ISBN 0-89426-000-6

The Hewlett-Packard Company makes no warranty, expressed or implied, and assumes no responsibility in connection with the operation of the program material attached hereto.

Copyright ® 1974 by Hewlett-Packard Company, Cupertino, California. All rights reserved. No part of this publication may be reproduced, stored in a retrieval system (e.g., in memory, disc or core) or be transmitted by any means, electronic, mechanical, photocopy, recording or otherwise, without prior written permission from the publisher.

PREFACE

This book is designed to introduce statistics as a tool for solving practical problems of management. The treatment of the underlying theory is almost entirely intuitive, i.e., there is very little mathematical notation and no reliance on mathematical ideas beyond those ordinarily covered in the first eight grades of school. This approach reflects the many years of experience by the author in teaching statistics with a heavy emphasis on theory: He found that few students are sufficiently intrigued by statistical theory to study it with the enthusiasm needed to tackle practical applications, and even fewer are comfortable with mathematical reasoning even when their formal background includes a fair amount of work in college-level mathematics. This book is based on the conclusion that genuine interest in statistical theory and mathematical reasoning is usually best aroused by exposing students to applications first. It has been used quite successfully by the author with students in the MBA and Executive Programs of the Graduate School of Business, University of Chicago.

The electronic computer has made possible the implementation of the concept behind the book. Only with this wonderful tool can beginners be led quickly into the heart of the action on real-life, interesting applications. Furthermore, the computer is most effective when it is used in the time-sharing mode, so that the student can specify tasks in simple English commands, get back responses almost immediately, and decide what to do next. This is called conversational or interactive computing: hence the title, *Conversational Statistics*. The particular computer package that accomplishes the tasks set by this book is called IDA (Interactive Data Analysis). IDA was developed by Robert Ling with the author's assistance, and later with the assistance of the author's wife, June, at the Graduate School of Business.

The book is conversational in another sense. It is written in the format of a question-and-answer script, a medium first used by the author during expert legal testimony. This format curbs a natural tendency to ramble, plod, and philosophize, or to spend too much time on minor details at the expense of the main points. The questioner is drawn from life: he (or she) is a composite of many students, and many of the questions came up originally in a live context in the classroom.

Three introductory chapters paint something of the background against which statistical studies are conducted, and include an introduction to randomization in experimentation and sample surveys. Then the book turns to a systematic development of the ordinary regression model, starting with univariate data and ending with an introduction to those components of modern time series analysis that can be executed by ordinary regression and autoregression. The statistical model, its interpretation, and the diagnostic checks for its appropriateness provide the unifying theme. Everything else that might seem to belong to an elementary text but that does not contribute to this development is simply omitted. There is no mention, for example, of the binomial distribution. Key terms such as *mean* and *probability* are introduced as they are needed rather than developed in traditional sequences. Some teachers will be a little shocked at these departures from tradition. The reward, however, is that beginners can, in 11 weeks of a class meeting three hours a week, acquire a working understanding of one of the most important tools of statistics, namely, regression. Since the ideas of statistics are necessarily tied up with any tool, these are acquired in the process.

With the exception of a few computer simulations, all the examples in the text and in Appendix A involve real data in serious applications. Students with access to conversational computing can enrich their understanding of the text by working on these examples, and should be required to do so. In many instances, students have found important aspects of these examples that had eluded the original analysis.

Although the author's orientation toward statistics is Bayesian, it is possible to stress those fundamentals on which Bayesians and sampling-theorists agree at the operational level. Only at a few points are discussions of statistical foundations necessary, and these are brief. For example, estimation is introduced in terms of point estimates and standard errors; the reader is free to interpret these either in the context of posterior distributions or of confidence intervals, although the author's preference is obvious. Testing is carried out almost exclusively on the role of conformity measures used in selection of an appropriate model, and certain rules-of-thumb are offered for guidance in this task. The usual two-tail tests of sharp null hypotheses are omitted.

The book has been used in a first course for business and public management students, and also by a number of students and former students for independent study (for them it has given background in statistical application, guidance in the use of IDA for conversational computing, and sometimes a supplementary perspective on a more theoretical statistics course they were taking). Some of the author's colleagues have found the book helpful for supplementary reading in other statistics courses. It could be teamed with any of a number of texts in a variety of courses stressing regression and time-series analysis of business, economic, and social science data, especially at the intermediate level. An instructor who wishes to use *Conversational Statistics* on a stand-alone basis, but who wishes to delve further into theory, can use part of the classroom sessions for systematic lecturing on theory.

At present IDA is available only on Hewlett-Packard computers, and it was designed to be system-dependent, though there is some possibility that IDA may be modified to run on certain other types of hardware. In any event, *Conversational Statistics* could probably be used in conjunction with other interactive statistical packages and possibly with some of the more flexible batch packages as well. Of course, the book can be read and followed without access to any computing at all, but it is felt that students in all but the most mathematical statistics courses benefit from substantial exposure to computing.

A NOTE ON IDA, 1977:

IDA is very much a living language, and as such it is being continually improved and extended. As a result, the printouts on runs made by IDA during different phases of the preparation of *Conversational Statistics* may differ somewhat in format from the current system of IDA. These minor variations should not cause you any trouble.

In 1976 the latest version of IDA in HP BASIC was converted into FORTRAN by Gary A. Curtis. This conversion was designed to be readily transported to any computer with a reasonable implementation of FORTRAN. It is currently being tested at a number of sites on different computing systems, and current plans are to offer FORTRAN IDA for sale in early 1977. For detailed information, write to

>Mr. Gary A. Curtis
>Graduate School of Business
>University of Chicago
>Chicago, Illinois 60637

New statistical capabilities will be added in parallel to the HP BASIC and FORTRAN versions of IDA.

Access to IDA is assumed, but not strictly required, throughout this book.

ACKNOWLEDGEMENTS

I cannot possibly acknowledge all who have contributed to *Conversational Statistics*. Professor William Kruskal read an early draft and gave me valuable criticisms. Over the years many colleagues have given me virtually tutorial instruction in statistical applications, and I am deeply grateful to them all. Deans Sidney Davidson, Gary Eppen, and Robert Graves had the foresight and courage to bring interactive computing to the Graduate School of Business at a time when academic budgets were already austere; without this computing base, the book would not have been written. Professor Ling, already mentioned as the main architect of IDA, was indispensable, as were June Roberts and William Wecker, who have substantially expanded IDA's capabilities and ease of use. My students have given me the encouragement I needed to persist and, by their questions, the key to clarification of many points that otherwise would have remained obscure.

Charles Dixon, Jean Danver, and Chris Doerr of Hewlett-Packard have made it possible for the local Chicago version of *Conversational Statistics* to become the present book.

The RAND Corporation has given consent for the reprinting of the short table of random numbers in Chapter 2, and the Macmillan Company has given permission for use of the example of sampling of castings used in Chapter 2.

ABOUT THE AUTHOR

Harry V. Roberts is Professor of Statistics in the Graduate School of Business at the University of Chicago, where he first joined the faculty in 1949. Before coming to the university, he worked for several years as a marketing research analyst at McCann-Erickson, Inc., and before that he was scout and later loader on a medium tank with the Fourth Armored Division, Army of the United States, during the Second World War.

Roberts' interests in statistics encompass most of the wide range of fields to which the subject is applied, from medical and other scientific applications to applications to management problems in business and government. He has written *Statistics: a New Approach* (with W. Allen Wallis) and *Basic Methods of Marketing Research* (with James H. Lorie), as well as a number of articles both on statistical theory and on statistical applications in marketing, finance, and other areas of business. He is a member of a number of professional societies and a fellow of the American Statistical Association. Among his avocations are local government (he is currently chairman of the zone board of appeals in Homewood, Illinois), naturalistic horticulture (his yard is the most permissive in Homewood), jogging (he holds the faculty record for the marathon run), and computing (whenever possible he will be found using one of the two terminals in his home).

His wife June has become addicted to computing as a result of his interest, and his daughter DD has become an addict on her own. Only his son, Andy, has managed to stay away from terminals; he is too busy playing the string bass.

TABLE OF CONTENTS

1	CHAPTER ONE: INTRODUCTION
11	CHAPTER TWO: EFFECTIVE USES OF STATISTICS
41	CHAPTER THREE: OBSERVATION AND MEASUREMENT
43	CHAPTER FOUR: ANALYSIS OF SEQUENCE
69	CHAPTER FIVE: DISTRIBUTIONS
87	CHAPTER SIX: NORMAL REGRESSION
111	CHAPTER SEVEN: INFERENCE
125	CHAPTER EIGHT: SIMPLE REGRESSION
153	CHAPTER NINE: AUTOREGRESSION
169	CHAPTER TEN: MULTIPLE REGRESSION
189	CHAPTER ELEVEN: TIME SERIES
217	APPENDIX A: APPLICATIONS
267	APPENDIX B: COMPUTER NOTES
277	SUMMARY AND GLOSSARY

CHAPTER ONE: INTRODUCTION

STATISTICS AND COMMON SENSE.

What is statistics?

I like to think of statistics as common sense disciplined by calculation. (This was suggested to me by Laplace's famous definition of probability: "La probabilité n'est que le bon sens réduit au calcul.")

My impression of statistics has more to do with calculation than with common sense. In fact, I'm surprised to hear you mention common sense.

In our writing and teaching, we statisticians often place the main emphasis on statistical calculation and the formal mathematical theory that underlies calculation.

Why does the "common sense" aspect get lost?

Partly because statisticians, like other technical people, tend to become absorbed in the technical side of their discipline; partly because it is hard to teach common sense.

Isn't your definition too broad?

I'll have to agree that the definition includes things not ordinarily thought of as statistics. A paraphrase of my definition, for example, might serve for mathematical economics or mathematical physics, and the definition is adapted, as I mentioned, from a definition of probability. But it's so important to combat the notion that statistics is totally unrelated to the thought patterns of a normal, healthy human being, that I like to start with the short definition.

If it won't carry us too far afield, could you comment on the qualifications necessary to distinguish statistics from other disciplines, such as those you have mentioned?

The most important distinguishing feature of statistics is its concern with applications in which *uncertainty* is present.

Would that suffice to distinguish statistics from probability?

In my opinion, statistics and probability are so closely intertwined that it is unfruitful to spend time trying to delineate jurisdictional boundaries, if any. Some statisticians, however, tend to stress that probability is a deductive subject and statistics is an inductive one.

Is untrained common sense dangerous in statistics?

I've stressed the need for discipline by calculation. Still, it is better to rely on your own common sense than to accept conclusions secondhand from someone else's statistical analysis, unless you have very good reason to be confident of the other person's judgment, competence, fairness, and honesty. As you will soon see, if you hadn't already been aware of it, other people often have a vested interest in their results. Moreover, people tend to do statistics as a mechanical exercise in calculation according to cookbook rules, and they often fall into gross violations of common sense as a result.

Further, I think you will agree that most of the statistical ideas that we shall discuss are consistent with your common sense.

You said, "most". Not all?

Some important statistical ideas do contradict what appears to be common sense to most people. Example: It is commonly believed that the *percentage* of a population included in a sample is the key to the accuracy of a sample. In fact, in most applications of statistical sampling, the *absolute* sample size is relevant while the percentage sample size is virtually irrelevant. We'll see more of that point later.

Can common sense in statistics be cultivated?

To some degree. For example, it is common sense to look carefully at data. Instruction can suggest more useful ways to look at a specific set of data.

Is mathematics needed to learn statistics?

Basic arithmetic will go a long way in allowing you to study elementary statistics. You must, however, be willing and able to think carefully and patiently, to take ideas in stride without skipping or procrastinating, and to follow close reasoning. For *advanced* statistical work, mathematical skills become essential since the underlying theory is heavily mathematical. But you can go a substantial distance toward use and understanding of statistical tools without getting into the underlying theory very deeply.

Do I have to learn computer programming to learn statistics?

No. But I will show you how conversational use of the computer, which does not require programming by you, can help greatly.

Is statistics dull?

A few people--those intrigued by mathematical reasoning and the challenges of scientific inference--find statistics interesting for its own sake. Most people, however, find statistics interesting only to the extent that they can learn to use statistics for interesting and perhaps profitable applications. I shall make every effort to show you applications that will help you find statistics interesting.

What kinds of applications do you mean?

First, you need to see applications of statistics to your own work or professional interests, or at least to avocational interests.

Second, you need to see applications of statistics to a wide range of problems outside your own areas of greatest interest. In that way you can gain understanding and perspective about the statistical approach and its fundamental universality.

APPLICATIONS OF STATISTICS.

Since this book is designed primarily for students of business and public administration, I shall stress the applications arising in these fields. But I will also sprinkle in other types of applications from time to time.

Is statistics well-used in business and public administration?

Seldom. The potential for good statistical application is great but as yet largely untapped. Many organizations scarcely use statistics at all. A few years ago, a representative of a large publishing firm, well-known for its distinguished list of statistics books, asked for my recommendation on the publication of a particular statistical manuscript. I replied that the book was well-written and full of techniques that almost any company could apply with profit. His company, I suggested, could use them in its own decision-making. "Our management", he replied, "*never* uses statistics".

Surely there are exceptions to this dismal picture!

Yes, there are oases in the desert, such as Bell Laboratories or the United States Bureau of the Census. Sometimes, however, the oases may be in small parts of an organization: experiments in the R and D labs may use highly sophisticated statistical techniques, while major decisions of management are made without any reliance on statistics.

Is statistics generally well-used in other fields, such as research in science and social science?

No, not so far as I can tell. There are opportunities there, too.

What are the major divisions of the field of statistics?

Statistical inference and statistical decision theory, about which we'll say more later. Some statisticians would add a third field, statistical description; personally, I find it hard to separate this subject from decision theory. In fact, even inference can logically be regarded as a branch of decision theory.

Since you are directing your attention to those interested in business and public administration, I suppose that you will give the main emphasis to decision theory.

I'd like to, but instead I'll emphasize inference and touch only lightly on decision theory. It's hard to cover both inference and decision theory within our time limits. I give the first priority to inference because, in my experience, it is operational for a much wider range of business applications than is decision theory.

Do different statisticians take fundamentally different approaches to statistics?

At present, there are two major tribes of statisticians: the sampling-theorists or "classical" statisticians, and the Bayesians. I belong to the Bayesian tribe.

STATISTICAL INFERENCE AND DECISION THEORY.

4 □ CONVERSATIONAL STATISTICS

Will your presentation be colored by your tribal affiliation?

I think that the coloring will be very slight. We will be dealing with fundamental questions of handling data about which there is very little operational difference between the two tribes, even though the underlying philosophical gap is enormous.

Will you cover the traditional topics of introductions to statistics?

I shall use a broad body of techniques called "regression", which are very useful in application, as a vehicle for covering important statistical ideas. We shall be able to go much further with tools that you can use than would be possible otherwise.

The price paid is neglect of certain traditional topics, and a certain amount of hopping and skipping (as a colleague describes it).

I've heard it said that an introduction to statistics can't hope to enable people to produce good statistical studies; rather, the goal is often said to be that of making consumers, or intelligent critics, of statistics. You seem to disagree.

I do disagree in part.

First, I think that very few people can draw much inspiration from an attempt to inculcate "statistics-appreciation"; the notion is just too passive, however desirable it may be to cultivate skill in statistical criticism.

Second, I think that the route to enlightened statistical consumption is to be found in an orientation that stresses actually doing statistics, rather than just talking about it.

Third, I feel that there is in fact a great need to be able to apply statistics yourself, even if you are limited to very simple tools. Both professionally and in everyday life, we are constantly confronted by problems that can be illuminated by statistical reasoning.

Fourth, in my experience it is possible for people exposed only to the material we will cover to do quite good statistical studies, at least with a little guidance.

Before we get into more detailed questions, can you say anything that will be useful to me as a consumer of statistics?

As a consumer, the most important thing to learn is how to avoid being duped by misuses of statistics.

But how can I protect myself?

It is reassuring to realize that most of the statistical misuses that swarm around us from day to day are offenses against common sense, rather than technical errors, although the latter can be present, too.

But how does that help me?

You can try not to run away from statistical arguments but,

CONSUMPTION VERSUS PRODUCTION OF STATISTICS.

rather, stand your ground and try to figure out what is going on and whether or not it seems sensible. Don't be afraid to ask common sense questions, even if they sound naive.

Could you suggest some examples for practice?

A number of statistical misuses have been captured and caged in Chapter 4 of *The Nature of Statistics* (henceforth abbreviated *TNOS*) by W. Allen Wallis and Harry V. Roberts. I urge you to read and study these examples carefully so that you can begin to acquire the proper skepticism required of any intelligent statistical critic.

But let's start immediately on a more detailed example that may give you the right spirit from the start. In 1972 I was asked to serve as a rebuttal witness in a lawsuit involving an accident at a railroad grade crossing. The plaintiff had brought in an expert witness who tried to demonstrate statistically that the crossing in question was a highly dangerous one. The attorney who approached me had not followed the technical argument very well, and the information that he was able to furnish me consisted of the exhibits or extracts of exhibits reproduced on the following pages. Here's a brief guide to this material; study it for a little and then we'll discuss it.

AN EXAMPLE OF STATISTICAL MISUSE.

EXPERT LEGAL TESTIMONY ON SAFETY OF GRADE CROSSING (MARCH, 1972)

This is a summary of information on a chart that was too large for me to reproduce. The part omitted was a large graph, called a nomograph, that permitted one to carry out graphically the calculations outlined in the formula at the top of the page. Hint: Don't panic. Take your time, read everything slowly, then proceed, even if you don't fully grasp what is going on.

Table 1-1. Abstract of Essential Material From a Large Chart Presented By a "Safety Expert"

$$F = 0.412 - 0.033 x_{87} + 0.024 x_{88} + 0.007 x_{96} - 0.021 x_{100} - 0.195 x_{101} - 0.525 x_{102} + 0.013 x_{104}$$

x_{87}: Line-of-sight ratio (Scale 0 to 3)

x_{88}: Average daily traffic, ADT/1000 (Scale 0 to 16)

x_{96}: Average trains per day, TPD (Scale 0 to 80)

x_{99-102}: Type of protective device

x_{99}: Painted crossbuck

x_{100}: Reflectorized crossbuck

x_{101}: Flasher

x_{102}: Gate

x_{104}: Sum of distractions. (Number of business and advertising signs, on both sides of the roadway, along a section extending 500 feet from the crossing to 200 feet beyond the crossing from one approach direction.) (Scale 0 to 60.)

$$Pr = \begin{cases} 0, & \text{if } F < 0 \\ F, & \text{if } 0 \leq F \leq 1 \\ 1, & \text{if } 1 < F \end{cases}$$

"Pr" is described as "Potential hazard", ranging from 0 to 1. It is also called the "probability a grade crossing is accident prone".

6 □ CONVERSATIONAL STATISTICS

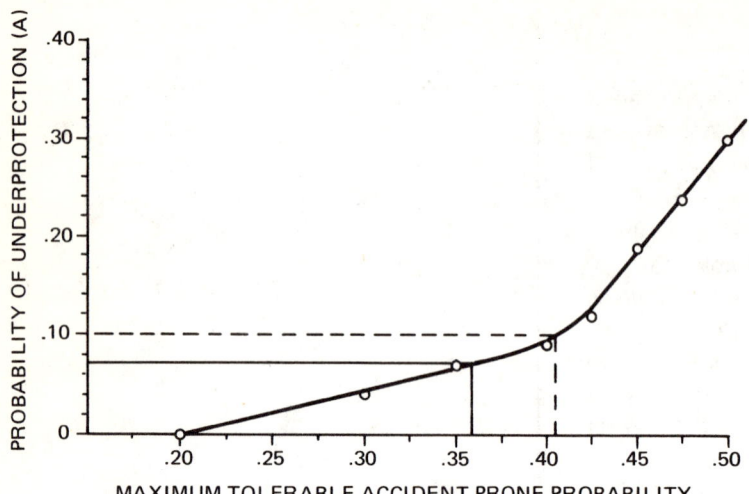

Figure 1-1. Conversion Chart for Probability of Underprotection

This chart was used to make a graphical conversion from something called "Maximum tolerable accident prone probability" to something called "Probability of underprotection". For example, it appears that a value of .41 on the horizontal scale corresponds to a value of .10 on the vertical scale, as shown by the dashed line segments. Hint: Be patient: you will be able to figure out later that "Maximum tolerable accident prone probability" means the same thing as "Potential hazard".

Table 1-2. Safety Evaluation

These are calculations placed in evidence by the safety expert. I suggest that you start with the third one and verify it as far as you can (up to "Potential hazard") by reference to Table 1-1. To go from "Potential hazard" to "Pr (underprotection)", look at Figure 1-1. Then do the same for the first two. For the first two, the final step is not obvious. I was told that the witness simply said that the "Potential hazard" was so high that it was off his chart.

Hazard ratings for urban area: V = 25 mph

		+0.412
LOS ratio	— 1.00	−0.033
TPD	— 14	+0.098
ADT	— 500	+0.012
SOD	— 5	+0.065
PD	— Painted crossbuck	0.000
	Potential hazard	0.554
	Pr (underprotection)	0.43
		+0.412
LOS ratio	— 1.00	−0.033
TPD	— 14	+0.098
ADT	— 500	+0.012
SOD	— 5	+0.065
PD	— Reflectorized crossbuck	−0.021
	Potential hazard	0.533
	Pr (underprotection)	0.38
		+0.412
LOS ratio	— 1.00	−0.033
TPD	— 14	+0.098
ADT	— 500	+0.012
SOD	— 5	+0.065
PD	— Flasher	−0.195
	Potential hazard	0.359
	Pr (underprotection)	0.07

Table 1-3. Safety Evaluation

More calculations. Can you figure out how they relate to the ones already given?

Hazard ratings for urban area: V = 30 mph

		+0.412
LOS ratio	— 0.95	−0.031
TPD	— 14	+0.098
ADT	— 500	+0.012
SOD	— 5	+0.065
PD	— Painted crossbuck	0.000
Potential hazard		0.556
Pr (underprotection)		0.43
		+0.412
LOS ratio	— 0.95	−0.031
TPD	— 14	+0.098
ADT	— 500	+0.012
SOD	— 5	+0.065
PD	— Reflectorized crossbuck	−0.021
Potential hazard		0.535
Pr (underprotection)		0.38
		+0.412
LOS ratio	— 0.95	−0.031
TPD	— 14	+0.098
ADT	— 500	+0.012
SOD	— 5	+0.065
PD	— Flasher	−0.195
Potential hazard		0.361
Pr (underprotection)		0.07

My first reaction is that this material is terribly confusing. Is that a fair criticism?

Yes, but it doesn't absolve you from trying to figure out what is going on. And remember that the jury in the case might conceivably be impressed by the fact that the material does appear confusing and hard to understand.

Where did the safety expert get his formula?

A good question. I asked the same thing, but there was no way to find out. I had to go on the evidence that I have given you.

I know a little bit about regression already, and the formula looks like something that comes out of a regression analysis.

It does indeed. You might find it instructive to ask yourself just what kind of data would be needed for such an analysis, and whether or not such data is likely to be available. But common sense should still be adequate to make you distrust the testimony.

EVALUATING THE FORMULA.

Well, isn't it fair to criticize the presentation for not furnishing a clear explanation for the basis of the formula?

That's fair--and telling. But it would be far from decisive in a courtroom. After all, the man was a safety expert.

What is "line-of-sight ratio"?

I was told only that it referred to the angle at which the road approached the railroad at the crossing. From Table 1-1 we see that it is scaled numerically from 0 to 3. Can you figure out whether higher values on this scale are supposed to be good or bad from the viewpoint of safety?

From the formula in Table 1-1, I see that we are supposed to multiply the line-of-sight ratio by a negative number, -0.033. Hence, a higher value of the line-of-sight ratio has the effect of diminishing F, which is good.

Good. You might also notice that the line-of-sight ratio is taken as 1.00 in the calculations in Table 1-2, where the approach velocity is taken as 25 m.p.h., and only 0.95 in Table 1-3, where the approach velocity is taken to be 30 m.p.h. Thus it would appear that the safety expert assumes a given line-of-sight angle to be more dangerous at higher speeds.

That seems to be the only way in which velocity affects the calculation, does it not?

Yes.

I notice on the worksheets the expression, "Hazard ratings for urban area". Was this crossing in an urban area?

From the location on the map and from photographs that I was shown, the crossing appeared to be located in open farm country in Northern Illinois. Thus, your question leads to an important reservation about the value of the statistical argument.

I tried to go through the calculation at the top of Table 1-3. It appears that the expert was substituting the numbers shown into the formula in Table 1-1, multiplying, and then adding up to get F. Is that right?

Yes.

But then what happened to some of the x's in the formula? For example, what did he do with x_{100} for "Reflectorized crossbuck"?

Here you'll be helped by noticing that there are four types of protective devices: painted crossbuck, reflectorized crossbuck, flasher, and gate. It appears that these are mutually exclusive, so only one is used in any one computation. Technically, such a variable takes a value of "1" if the condition exists (for example, if there is a reflectorized crossbuck) and "0" if the condition does not exist. (In a choice of jargon amusing to outsiders, statisticians often call such a variable a "dummy variable"!) So reflectorized crossbuck, flasher, and gate drop out of this computation, because the actual protective device was painted crossbuck.

SPOTTING DISCREPANCIES IN THE TESTIMONY.

But painted crossbuck, x_{99} in the middle of Table 1-1, does not appear in the formula at the top of that table, and it is given a 0 in the calculation in Table 1-3 that we're talking about.

Only three of the four protective devices need be shown in the equation. If these three are all absent, the fourth must be present. Thus, if you assign 0 to x_{100}, x_{101}, and x_{102}, you are left with an automatic 0 for the contribution to F of x_{99}, the painted crossbuck.

Well, I'm not at all sure that I would have been able to figure that out on my own! Isn't this example beginning to be pretty technical? I thought that you said common sense would suffice!

I did. This is the one aspect of the problem where a little specialized knowledge helps. But it's not essential: you could have gone ahead without having figured it out fully, knowing that you must be pretty much on the right track. Don't be too compulsive about all the details; some gaps have a way of closing themselves if you have the guts to push on.

All right. I see that the numbers in Table 1-3 add to 0.556, and from what you've explained about Figure 1-1, I see that we get a Pr (underprotection) that's off the upper end of the chart. It looks pretty bad for the crossing?

Yes, it does. And that was apparently the impression left in the courtroom by the safety expert.

That leaves me on dead center. Except for the quibbles about clarity of presentation and documentation of method, who am I to question a safety expert?

Here's where you must have faith in your common sense. There are no holds barred and no rules. Mull on.

Well, I notice that the contribution of average daily traffic is small-- only 0.012.

You're on the track.

Well, I guess that means that with no cars at all, the crossing would have an F of 0.556 - 0.012, or 0.544. I guess that's still off the upper end of the chart!

With that simple arithmetic, and the reasoning that goes with it, you've put the whole method of approach in severe jeopardy. The result is extremely insensitive to the number of cars crossing the railroad, so much so that the crossing would appear to be extremely hazardous even if no cars ever crossed it. Something is wrong.

A similar criticism, though not so extreme, would seem to apply to the number of trains per day.

At the other extreme, note that the formula is dominated by the type of protective device used. Any crossing without a flasher or gate is almost bound to appear hazardous according to the formula.

COMMON SENSE AGAIN.

Is your point, then, that a little arithmetic used with common sense discredits the testimony of the traffic expert?

Yes. But remember that it takes determination also. If you look away in terror, you'll never find the flaw.

Let me try to defend the traffic expert. By setting average daily traffic equal to 0, you're getting outside the range of validity of his formula.

Let's look at the problem from a positive point of view. If you let me use a technical term that has a pretty clear intuitive meaning, a good measure of the danger of a crossing might be the expected number of accidents per year. It seems to me to be common sense that the expected number of accidents, other things equal, ought to be at least roughly proportional to the number of opportunities for accidents, at least within a wide range. If the traffic volume doubles, the expected number of accidents would double, roughly speaking. Now the formula used by the expert doesn't have anything like this property: both average daily traffic and average trains per day enter in additively rather than multiplicatively.

Further, it is intuitively plausible to me that the effect of protective devices ought to be multiplicative; a particular protective device might have the effect of reducing the expected number of accidents by, say, 30% rather than by some fixed number regardless of the crossing.

Your question here has led me into a more technical discussion than I intended when I introduced the example. I hope, however, that this discussion is closely tied to common sense.

If you want to measure danger in terms of the expected number of accidents per year, why not just average the number of accidents per year for, say, the 30 years prior to this accident? Wouldn't that be a good estimate of the expected number?

Let's confine our attention to fatal accidents, since the accident at this crossing involved a fatality. There had never been a prior fatal accident.

That seems pretty good evidence that the crossing was not particularly dangerous, does it not?

Surprisingly enough, this evidence is not very strong either way. 30 years sounds like a long time, but the evidence of 30 years without a fatal accident contributes relatively little. Consider the following example. Suppose that a new type of commercial airplane has just logged its one millionth mile of use, without a fatal crash. At first you might think that this evidence suggests that the airplane is unusually safe. But if I tell you that the average rate of fatal accidents for *all* commercial aircraft is something like one in *100* million miles, you can see intuitively that one million safe miles doesn't rule out a substantial probability that this new type may be more dangerous than the average.

MULTIPLICATION VERSUS ADDITION.

CHAPTER TWO: EFFECTIVE USES OF STATISTICS

READING: TNOS, Chapters 1-2.

What is the distinction between statistical inference and decision theory?

1. Inference pertains to the process by which we learn from experience. *Knowledge* is the key word. *Statistical* inference narrows the scope to situations in which relevant parts of experience are captured in data.

2. Decision pertains to the process by which we act on the basis of knowledge. *Action* is the key word. *Statistical* decision theory narrows the scope to situations in which the decision process can be described mathematically.

Furthermore, the word "statistical" suggests that there is an element of uncertainty in the process of inference or decision. By contrast, many of the decisions treated in economics deal with deterministic applications: for example, the firm may be assumed to know both its demand curve and its marginal costs when making a decision about the rate of output of a product. Yet the close connection between statistical decision theory and economics is suggested by the subtitle of Robert Schlaifer's famous text, *Probability and Statistics for Business Decisions* [*McGraw-Hill, 1959*]: "An Introduction to Managerial Economics Under Uncertainty". Schlaifer's introductory sentences [page 2] develop the idea that statistics can be regarded as economics under uncertainty:

> When all of the facts bearing on a business decision are accurately known--when the decision is made "under certainty"--careless thinking is the only reason why the decision should turn out, after the fact, to have been wrong. But when the relevant facts are not all known--when the decision is made "under uncertainty"--it is impossible to make sure that every decision will turn out to have been right in this same sense. Under uncertainty, the businessman is forced, in effect, to gamble. His previous actions have put him in a position where he *must* place bets, hoping that he will win but knowing that he may lose. Under such circumstances, a right decision consists in the choice of the best possible bet, whether it is won or lost after the fact. . . .

Could you show me where inference and decision come in in the context of an application?

Consider a study designed to predict the machining time required to fill an order received by a steel warehousing firm. Previously, an experienced foreman, using his best judgment, had made estimates; this process took about two days, so that it was impossible to give a quick answer to a potential customer who had just phoned in an inquiry.

KNOWLEDGE AND ACTION.

PRACTICAL APPLICATION OF DECISION THEORY.

There were past records of the amount of machining time required for orders of bar stock (the product in question) with varying specifications of length, diameter, and machining tolerances. The statistical study resulted in a formula that permitted inferences about the amount of machining time required on new orders. These inferences, of course, were uncertain, but they promised good accuracy and could be made almost instantaneously. Moreover, they did not require an experienced foreman; a statistical clerk could make them.

But this application of inference was but one component of the business decision. The ultimate decision was the quotation of a price to a potential customer of bar stock who had just phoned in a request for a price quotation. The machining time on the order, as estimated by the statistical formula, was only one of many components of the final decision; cost of the bar stock itself and the degree of competitiveness of the market are examples of other considerations. The final decision on the quotation involved judgment although one of the inputs to the decision was based on statistical analysis.

Can the process of decision-making be reduced to a mathematical description?

In principle, yes. The key elements that *should* enter into any decision have been formalized by statistical decision theorists. Whether or not it is possible to implement this grand formal design in a set of particular circumstances is another question.

In some relatively simple situations, it is sometimes possible to make the formalization operational. For example, in statistical quality control, the rejection or acceptance of a manufacturing lot, or the quality surveillance of a continuing industrial process, may be made in accord with rules that can be programmed for computers. In certain types of medical diagnosis, decision procedures have been programmed, and sometimes these have led to better decisions than those previously attainable by skilled clinicians. (When this has happened, the clinicians have sometimes responded to the challenge by improving their performance to the point where they could beat the computer!) There is some evidence that for many problems involving personnel selection--students, employees, credit risks--computer decisions can improve on judgmental performance.

In today's state of the practical art of decision theory, however, decision-making is usually too complicated to be reduced entirely to computer procedures.

I want to make sure that I understand what you're saying. Even when decisions can be implemented by computers, I assume that human inputs are still necessary.

Very much so. First, the computer programs are written by people, who must formulate a mathematical description of the decision process and then put it in a form that the computer can process.

Second, and less obvious, there must be human inputs to the computer program. Decision theorists have seen that these are of two distinct kinds: (1) assessment of probabilities concerning the uncertain facts surrounding a decision, and (2) assessment of preferences for the various possible outcomes of a decision.

Is statistical inference helpful for decisions too complicated to be made by computers?

Yes, statistical inference can provide essential background for judgment in attacking tough decisions. In my opinion, this is by far the most important contribution of statistics to practical application.

Would you illustrate what you mean?

1. In the 1949-1950 recession, factory shipments of home freezers by the Maytag Company dropped so sharply that management was concerned as to whether the company should discontinue the product line. By coincidence, a statistical study of the market for home freezers, which had been started before the recession began, showed that actual consumer purchases did not drop during the recession. The decline of factory shipments reflected a reduction of inventories rather than a basic weakness of the market. The study was one of several factors that contributed to the management decision to continue the home freezers.

2. During the early 1950's, the University of Chicago was faced by a very sharp decline in enrollment. A statistical study showed that this decline was unique to the University of Chicago, not a pervasive phenomenon affecting colleges and universities generally or private universities similar to the University of Chicago. It suggested further that the major cause of the decline could be traced to the consequences of a major decision taken during the mid-1940's, in which the structure of the College was altered in order to accept students after the second year of high school rather than after graduation from high school. This study was one of the inputs to the decision of President Lawrence Kimpton to go back to a more traditional college structure.

3. Local zoning authorities must frequently make decisions about possible rezoning to permit the construction of apartments on land parcels currently zoned for single family dwellings only. It is hard to imagine a decision-theoretic formulation of this kind of decision, and even if one did, it would be unlikely to command acceptance in the political environment in which zoning decisions must be made. In one suburb of Chicago, however, one component of the decision was isolated: the economic impact on the community as reflected in property taxes. On the basis of data collected from various apartment projects of the kind usually proposed in that suburb, it was found: (1) apartment projects entailed approximately the same number of school children per acre as did single family dwellings; (2) apartment projects generated approximately three times as much in property taxes. Further, about three-fourths of property taxes went to support of public schools. It was therefore apparent that in one particular respect apartments were very desirable.

INFERENCE AS A BACKGROUND FOR DECISION.

But that wouldn't mean that all apartment proposals should be approved.

By no means, since zoning boards must take lots of other things into consideration: for example, impact on property values of surrounding property owners, preferences of citizens of the community, demand for local municipal services, effects on traffic congestion, environmental impact, plus a host of considerations listed by statute, such as unique circumstances, hardship, retention of the basic character of neighborhoods, and even altruistic motivation by the apartment developer. But the biggest tax impact, that concerning schools, has been delineated by the statistical studies.

Is statistical reasoning useful only when data are available?

No, although development of the point would require familiarity with technical statistical concepts such as weighted means, regression, and the like. But to illustrate: I would predict confidently, without any specific data, that the 10 leading groundgainers in the National Football League this year will, on average, do less well next year.

Can statistics illuminate questions of cause and effect?

My answer here must be cautious. Statistical associations or relationships do not necessarily imply causation. A company may find that as its advertising expenditures increase, so do its sales. This statistical relationship is not convincing evidence of the effect of advertising expenditures on sales. For example, many companies budget advertising as a relatively fixed percentage of anticipated sales. Thus one can see that a causal connection from sales to advertising may be partly reflected in the observed association.

But in one circumstance, the randomized experiment, unambiguous insight into causation may be obtained. Here groups that are to receive different experimental treatments are formed *at random*.

An example?

An excellent and very important example is described by Paul Meier, "The Biggest Public Health Experiment Ever: The 1954 Field Trial of the Salk Poliomyelitis Vaccine", pages 2-13 of *Statistics: A Guide to the Unknown* (henceforth abbreviated SAGTU), edited by Judith M. Tanur and by Frederick Mosteller, William H. Kruskal, Richard F. Link, Richard S. Pieters, and Gerald R. Rising. Another interesting application, which ran into some practical difficulties, is the study of Vitamins and Endurance recounted on pages 67-80 of TNOS.

What does randomization accomplish?

In my opinion, the most important thing is to form two or more groups that will be closely comparable in respects other than the experimental treatments if only the groups are moderately large.

A secondary advantage is that randomization permits precise assessment of the degree of uncertainty of comparisons of responses of the different treatment groups.

ASSOCIATION DOES NOT IMPLY CAUSATION.

Many non-Bayesian statisticians would interchange the priority of these two advantages.

What is the key idea behind randomization?

Extremely thorough shuffling.

How is randomization best achieved in practice?

By the use of random numbers.

What is a random number?

Imagine a ten-sided "die", on each side of which appears one of the digits 0 through 9:

Suppose that this die is tossed into the air spinning rapidly about its longitudinal axis and caught so that the thumb of the hand is pressing one of the numbered sides. The number on that side is recorded and the die is tossed again. The numbers recorded in this way comprise a *table of random numbers*.

On the following pages, a short table of random numbers is reproduced. (This table was produced by a much more complicated process than the die, but the difference is not important: you can think of and use the table in exactly the same way.)

Don't you have to be careful about the way you spin the die?

Yes. The die must spin rapidly, be tossed high enough to make a large number of revolutions, and there must be no sleight-of-hand to shift the die after it is caught. Under these conditions, minute differences in the starting conditions and timing of the catch will make the outcomes unpredictable.

Is there any special significance to the grouping of random digits by fives in the table?

No. This is done solely for convenience in reading the numbers.

RANDOM NUMBERS.

16 □ CONVERSATIONAL STATISTICS

Table 2-1. 10,000 Random Digits*

Line No.	1-5	6-10	11-15	16-20	21-25	26-30	31-35	36-40	41-45	46-50
0	10097	32533	76520	13586	34673	54876	80959	09117	39292	74945
1	37542	04805	64894	74296	24805	24037	20636	10402	00822	91665
2	08422	68953	19645	09303	23209	02560	15953	34764	35080	33606
3	99019	02529	09376	70715	38311	31165	88676	74397	04436	27659
4	12807	99970	80157	36147	64032	36653	98951	16877	12171	76833
5	66065	74717	34072	76850	36697	36170	65813	39885	11199	29170
6	31060	10805	45571	82406	35303	42614	86799	07439	23403	09732
7	85269	77602	02051	65692	68665	74818	73053	85247	18623	88579
8	63573	32135	05325	47048	90553	57548	28468	28709	83491	25624
9	73796	45753	03529	64778	35808	34282	60935	20344	35273	88435
10	98520	17767	14905	68607	22109	40558	60970	93433	50500	73998
11	11805	05431	39808	27732	50725	68248	29405	24201	52775	67851
12	83452	99634	06288	98083	13746	70078	18475	40610	68711	77817
13	88685	40200	86507	58401	36766	67951	90364	76493	29609	11062
14	99594	67348	87517	64969	91826	08928	93785	61368	23478	34113
15	65481	17674	17568	50950	58047	76974	73039	57186	40218	16544
16	80124	35635	17727	08015	45381	22374	21115	78253	14385	53763
17	74350	99817	77402	77214	43236	00210	45521	64237	96286	02655
18	69916	26803	66252	29148	36936	87203	76621	13990	94400	56418
19	09893	20505	14225	68514	46427	56788	96297	78822	54382	14598
20	91499	14523	68479	27686	46162	83554	94750	89923	37098	20048
21	80336	94598	26940	36858	70297	34135	53140	33340	42050	82341
22	44104	81949	85157	47954	32979	26575	57600	40881	22222	06413
23	12550	73742	11100	02040	12860	74697	96644	89439	28707	25815
24	63606	49329	16505	34484	40219	52563	43651	77082	07207	31790
25	61196	90446	26457	47774	51924	33729	65394	59593	42582	60527
26	15474	45266	95270	79953	59367	83848	82396	10118	33211	59466
27	94557	28573	67897	54387	54622	44431	91190	42592	92927	45973
28	42481	16213	97344	08721	16868	48767	03071	12059	25701	46670
29	23523	78317	73208	89837	68935	91416	26252	29663	05522	82562
30	04493	52494	75246	33824	45862	51025	61962	79335	65337	12472
31	00549	97654	64051	88159	96119	63896	54692	82391	23287	29529
32	35963	15307	26898	09354	33351	35462	77974	50024	90103	39333
33	59808	08391	45427	26842	83609	49700	13021	24892	78565	20106
34	46058	85236	01390	92286	77281	44077	93910	83647	70617	42941
35	32179	00597	87379	25241	05567	07007	86743	17157	85394	11838
36	69234	61406	20117	45204	15956	60000	18743	92423	97118	96338
37	19565	41430	01758	75379	40419	21585	66674	36806	84962	85207
38	45155	14938	19476	07246	43667	94543	59047	90033	20826	69541
39	94864	31994	36168	10851	34888	81553	01540	35456	05014	51176
40	98086	24826	45240	28404	44999	08896	39094	73407	35441	31880
41	33185	16232	41941	50949	89435	48581	88695	41994	37548	73043
42	80951	00406	96382	70774	20151	23387	25016	25298	94624	61171
43	79752	49140	71961	28296	69861	02591	74852	20539	00387	59579
44	18633	32537	98145	06571	31010	24674	05455	61427	77938	91936
45	74029	43902	77557	32270	97790	17119	52527	58021	80814	51748
46	54178	45611	80993	37143	05335	12929	56127	19255	36040	90324
47	11664	49883	52079	84827	59381	71539	09973	33440	88461	23356
48	48324	77928	31249	64710	02295	36870	32307	57546	15020	09994
49	69074	94138	87637	91976	35584	04401	10518	21615	01848	76938

*Reprinted with permission of the RAND Corporation

Table 2-1. 10,000 Random Digits (Continued)

Line No.	1-5	6-10	11-15	16-20	21-25	26-30	31-35	36-40	41-45	46-50
50	09188	20097	32825	39527	04220	86304	83389	87374	64278	58044
51	90045	85497	51981	50654	94938	81997	91870	76150	68476	64659
52	73189	50207	47677	26269	62290	64464	27124	67018	41361	82760
53	75768	76490	20971	87749	90429	12272	95375	05871	93823	43178
54	54016	44056	66281	31003	00682	27398	20714	53295	07706	17813
55	08358	69910	78542	42785	13661	58873	04618	97553	31223	06420
56	28306	03264	81333	10591	40510	07893	32604	60475	94119	01840
57	53840	86233	81594	13628	51215	90290	28466	68795	77762	20791
58	91757	53741	61613	62269	50263	90212	55781	76514	83483	47055
59	89415	92694	00397	58391	12607	17646	48949	72306	94541	37408
60	77513	03820	86864	29901	68414	82774	51908	13980	72893	55507
61	19502	37174	69979	20288	55210	29773	74287	75251	65344	67415
62	21818	59313	93278	81757	05686	73156	07082	85046	31853	38452
63	51474	66499	68107	23621	94049	91345	42836	09191	08007	45449
64	99559	68331	62535	24170	69777	12830	74819	78142	43860	72834
65	33713	48007	93584	72869	51926	64721	58303	29822	93174	93972
66	85274	86893	11303	22970	28834	34137	73515	90400	71148	43643
67	84133	89640	44035	52166	73852	70091	61222	60561	62327	18423
68	56732	16234	17395	96131	10123	91622	85496	57560	81604	18880
69	65138	56806	87648	85261	34313	65861	45875	21069	85644	47277
70	38001	02176	81719	11711	71602	92937	74219	64049	65584	49698
71	37402	96397	01304	77586	56271	10086	47324	62605	40030	37438
72	97125	40348	87083	31417	21815	39250	75237	62047	15501	29578
73	21826	41134	47143	34072	64638	85902	49139	06441	03856	54552
74	73135	42742	95719	09035	85794	74296	08789	88156	64691	19202
75	07638	77929	03061	18072	96207	44156	23821	99538	04713	66994
76	60528	83441	07954	19814	59175	20695	05533	52139	61212	06455
77	83596	35655	06958	92983	05128	09719	77433	53783	92301	50498
78	10850	62746	99599	10507	13499	06319	53075	71839	06410	19362
79	39820	98952	43622	63147	64421	80814	43800	09351	31024	73167
80	59580	06478	75569	78800	88835	54486	23768	06156	04111	08408
81	38508	07341	23793	48763	90822	97022	17719	04207	95954	49953
82	30692	70668	94688	16127	56196	80091	82067	63400	05462	69200
83	65443	95659	18288	27437	49632	24041	08337	65676	96299	90836
84	27267	50264	13192	72294	07477	44606	17985	48911	97341	30358
85	91307	06991	19072	24210	36699	53728	28825	35793	28976	66252
86	68434	94688	84473	13622	62126	98408	12843	82590	09815	93146
87	48908	15877	54745	24591	35700	04754	83824	52692	54130	55160
88	06913	45197	42672	78601	11883	09528	63011	98901	14974	40344
89	10455	16019	14210	33712	91342	37821	88325	80851	43667	70883
90	12883	97343	65027	61184	04285	01392	17974	15077	90712	26769
91	21778	30976	38807	36961	31649	42096	63281	02023	08816	47449
92	19523	59515	65122	59659	86283	68258	69572	13798	16435	91529
93	67245	52670	35583	16563	79246	86686	76463	34222	26655	90802
94	60584	47377	07500	37992	45134	26529	26760	83637	41326	44344
95	53853	41377	36066	94850	58838	73859	49364	73331	96240	43642
96	24637	38736	74384	89342	52623	07992	12369	18601	03742	83873
97	83080	12451	38992	22815	07759	51777	97377	27585	51972	37867
98	16444	24334	36151	99073	27493	70939	85130	32552	54846	54759
99	60790	18157	57178	65762	11161	78576	45819	52979	65130	04860

18 ☐ **CONVERSATIONAL STATISTICS**

Table 2-1. 10,000 Random Digits (Continued)

Line No.	1-5	6-10	11-15	16-20	Column Number 21-25	26-30	31-35	36-40	41-45	46-50
100	03991	10461	93716	16894	66083	24653	84609	58232	88618	19161
101	38555	95554	32886	59780	08355	60860	29736	47762	71299	23853
102	17546	73704	92052	46215	55121	29281	59076	07936	27954	58909
103	32643	52861	95819	06831	00911	98936	76355	93779	80863	00514
104	69572	68777	39510	35905	14060	40619	29549	69616	33564	60780
105	24122	66591	27699	06494	14845	46672	61958	77100	90899	75754
106	61196	30231	92962	61773	41839	55382	17267	70943	78038	70267
107	30532	21704	10274	12202	39685	23309	10061	68829	55986	66485
108	03788	97599	75867	20717	74416	53166	35208	33374	87539	08823
109	48228	63379	85783	47619	53152	67433	35663	52972	16818	60311
110	60365	94653	35075	33949	42614	29297	01918	28316	98953	73231
111	83799	42402	56623	34442	34994	41374	70071	14736	09958	18065
112	32960	07405	36409	83232	99385	41600	11133	07586	15917	06253
113	19322	53845	57620	52606	66497	68646	78138	66559	19640	99413
114	11220	94747	07399	37408	48509	23929	27482	45476	85244	35159
115	31751	57260	68980	05339	15470	48355	88651	22596	03152	19121
116	88492	99382	14454	04504	20094	98977	74843	93413	22109	78508
117	30934	47744	07481	83828	73788	06533	28597	20405	94205	20380
118	22888	48893	27499	98748	60530	45128	74022	84617	82037	10268
119	78212	16993	35902	91386	44372	15486	65741	14014	87481	37220
120	41849	84547	46850	52326	34677	58300	74910	64345	19325	81549
121	46352	33049	69248	93460	45305	07521	61318	31855	14413	70951
122	11087	96294	14013	31792	59747	67277	76503	34513	39663	77544
123	52701	08337	56303	87315	16520	69676	11654	99893	02181	68161
124	57275	36898	81304	48585	68652	27376	92852	55866	88448	03584
125	20857	73156	70284	24326	79375	95220	01159	63267	10622	48391
126	15633	84924	90415	93614	33521	26665	55823	47641	86225	31704
127	92694	48297	39904	02115	59589	49076	66821	41575	49767	04037
128	77613	19019	88152	00080	20554	91409	96277	48257	50816	97616
129	38688	32486	45134	63545	59404	72059	43947	51680	43852	59693
130	25163	01889	70014	15021	41290	67312	71857	15957	68971	11403
131	65251	07629	37239	33295	05870	01119	92784	26340	18477	65622
132	36815	43625	18637	37509	82444	99005	04921	73701	14707	93997
133	64397	11692	05327	82162	20247	81759	45197	25332	83745	22567
134	04515	25624	95096	67946	48460	85558	15191	18782	16930	33361
135	83761	60873	43253	84145	60833	25983	01291	41349	20368	07126
136	14387	06345	80854	09279	43529	06318	38384	74761	41196	37480
137	51321	92246	80088	77074	88722	56736	66164	49431	66919	31678
138	72472	00008	80890	18002	94813	31900	54155	83436	35352	54131
139	05466	55306	93128	18464	74457	90561	72848	11834	79982	68416
140	39528	72484	82474	25593	48545	35247	18619	13674	18611	19241
141	81616	18711	53342	44276	75122	11724	74627	73707	58319	15997
142	07586	16120	82641	22820	92904	13141	32392	19763	61199	67940
143	90767	04235	13574	17200	69902	63742	78464	22501	18627	90872
144	40188	28193	29593	88627	94972	11598	62095	36787	00441	58997
145	34414	82157	86887	55087	19152	00023	12302	80783	32624	68691
146	63439	75363	44989	16822	36024	00867	76378	41605	65961	73488
147	67049	09070	93399	45547	94458	74284	05041	49807	20288	34060
148	79495	04146	52162	90286	54158	34243	46978	35482	59362	95938
149	91704	30552	04737	21031	75051	93029	47665	64382	99782	93478

Table 2-1. 10,000 Random Digits (Continued)

Line No.	1-5	6-10	11-15	16-20	21-25	26-30	31-35	36-40	41-45	46-50
150	94015	46874	32444	48277	59820	96163	64654	25843	41145	42820
151	74108	88222	88570	74015	25704	91035	02755	14750	48968	38603
152	62880	87873	95160	59221	22304	90314	72877	17334	39283	04149
153	11748	12102	80580	41867	17710	59621	06554	07850	73950	79552
154	17944	05600	60478	03343	25852	58905	57216	39618	49856	99326
155	66067	42792	95043	52680	46780	56487	09971	59481	37006	22186
156	54244	91030	45547	70818	59849	96169	61459	21647	87417	17198
157	30945	57589	31732	57260	47670	07654	46376	25366	94746	49580
158	69170	37403	86995	90307	94304	71803	26825	05511	12459	91314
159	08345	88975	35841	85771	08105	59987	87112	21476	14713	71181
160	27767	43584	85301	88977	29490	69714	73035	41207	74699	09310
161	13025	14338	54066	15243	47724	66733	47431	43905	31048	56699
162	80217	36292	98525	24335	24432	24896	43277	58874	11466	16082
163	10875	62004	90391	61105	57411	06368	53856	30743	08670	84741
164	54127	57326	26629	19087	24472	88779	30540	27886	61732	75454
165	60311	42824	37301	42678	45990	43242	17374	52003	70707	70214
166	49739	71484	92003	98086	76668	73209	59202	11973	02902	33250
167	78626	51594	16453	94614	39014	97066	83012	09832	25571	77628
168	66692	13986	99837	00582	81232	44987	09504	96412	90193	79568
169	44071	28091	07362	97703	76447	42537	98524	97831	65704	09514
170	41468	85149	49554	17994	14924	39650	95294	00556	70481	06905
171	94559	37559	49678	53119	70312	05682	66986	34099	74474	20740
172	41615	70360	64114	58660	90850	64618	80620	51790	11436	38072
173	50273	93113	41794	86861	14781	89683	55411	85667	77535	99892
174	41396	80504	90670	08289	40902	05069	95083	06783	28102	57816
175	25807	24260	71529	78920	72682	07385	90726	57166	98884	08583
176	06170	97965	88302	98041	21443	41808	68984	83620	89747	98882
177	60808	54444	74412	81105	01176	28838	36421	16489	18059	51061
178	80940	44893	10408	36222	80582	71944	92638	40333	67054	16067
179	19516	90120	46759	71643	13177	55292	21036	82808	77501	97427
180	49386	54480	23604	23554	21785	41101	91178	10174	29420	90438
181	06312	88940	15995	69321	47458	64809	98189	81851	29651	84215
182	60942	00307	11897	92674	40405	68032	96717	54244	10701	41393
183	92329	98932	78284	46347	71209	92061	39448	93136	25722	08564
184	77936	63574	31384	51924	85561	29671	58137	17820	22751	36518
185	38101	77756	11657	13897	95889	57067	47648	13885	70669	93406
186	39641	69457	91339	22502	92613	89719	11947	56203	19324	20504
187	84054	40455	99396	63680	67667	60631	69181	96845	38525	11600
188	47468	03577	57649	63266	24700	71594	14004	23153	69249	05747
189	43321	31370	28977	23896	76479	68562	62342	07589	08899	05985
190	64281	61826	18555	64937	13173	33365	78851	16499	87064	13075
191	66847	70495	32350	02985	86716	38746	26313	77463	55387	72681
192	72461	33230	21529	53424	92581	02262	78438	66276	18396	73538
193	21032	91050	13058	16218	12470	56500	15292	76139	59526	52113
194	95362	67011	06651	16136	01016	00857	55018	56374	35824	71708
195	49712	97380	10404	55452	34030	60726	75211	10271	36633	68424
196	58275	61764	97586	54716	50259	46345	87195	46092	26787	60939
197	89514	11788	68224	23417	73959	76145	30342	40277	11049	72049
198	15412	50669	48139	36732	46874	37088	73465	09819	58869	35220
199	12120	86124	51247	44302	60883	52109	21437	36786	49226	77837

FORMATION OF TREATMENT AND CONTROL GROUP.

Would you show how this table could be used to produce a randomized treatment group and a randomized control group?

For simplicity, consider an experiment in which there are just ten subjects in total: A, B, C, D, E, F, G, H, I, and J. Pick a starting point in the table by any method that does not permit you to examine the actual numbers in the table. (Otherwise you could search around to find random configurations that would "rig" the selection.) I decided to start at the beginning of line 77.

Given the starting point, you may proceed to pick numbers in any predetermined sequence. I decided to read horizontally across rows, as in reading a book. Other systematic procedures such as reading down columns would have served just as well.

I thus read off ten consecutive groups of five digits each:

83596 35655 06958 92983 05128
09719 77433 53783 92301 50498.

Then I assigned each successive block of five digits to each of the ten subjects:

A-83596 B-35655 C-06958 D-92983 E-05128 F-09719
G-77433 H-53783 I-92301 J-50498.

Why did you work by blocks of five random digits?

You're one step ahead of the story. I'm going to use the random numbers assigned to the subjects as a method of "thorough shuffling". In particular, I shall sort the subjects from low to high according to the number assigned. By using five digit numbers, I am taking a good precaution against the occurrence of ties in the sorting. Also, it is easy to read off groups of five (as opposed, say, to four or six) because of the typographical organization of this random number table into blocks of five.

By "sort" you mean that the subject with the smallest number will appear first, the subject with the next smallest number, second, and so on?

That's right. The result is:

E-05128 C-06958 F-09719 B-35655 J-50498   Treated group
H-53783 G-77433 A-83596 I-92301 D-92983   Control group

I let the first five constitute the treated group: E, C, F, B, J; the second five, the control group: H, G, A, I, D.

What would you have done if a tie had occurred?

I would have repeated the procedure, using only the tied subjects.

EFFECTIVE USES OF STATISTICS ▢ 21

If there were a larger number of subjects, that procedure would have become a little cumbersome. Could you do the same thing by computer?

Yes. This will give us our first occasion to use IDA, although the command that we shall use is unrelated to IDA's main analytical capabilities and is unusual in a couple of respects.

First, let me explain that a computer does not ordinarily store a table of random numbers such as the one we have seen. Rather, the computer generates random numbers internally by a deterministic scheme or algorithm.

Then how can the numbers be "random"?

If you knew the computing scheme, you *could* predict the numbers without error, which is inconsistent with our earlier discussion. On the other hand, if you just look at the numbers that are produced, their behavior is not distinguishable (or should not be distinguishable) from those of random numbers. The term "pseudo-random" is often used for computer-generated numbers. As a rule, pseudo-random numbers are displayed with a decimal point in front: thus, for example, 05128 would be displayed as .05128.

I've looked at the Computer Notes (Appendix B), but I'm a little hazy about how you get started with IDA.

First, you must log in with a terminal. I'll refer you back to the Notes for the details of this procedure. Then type GET-$IDA (carriage return), then RUN (carriage return). Here's what happens:

```
GET-$IDA
RUN
IDA

VERSION 15AUG.1973

GOOD AFTERNOON.  DO YOU NEED HELP ?NO

> PSAM
```

Notice that after saying that I didn't need help, I got the IDA command readiness symbol ">"; then I issued the command PSAM.

If you hadn't told me about this command, how could I have found out that it was the one I needed?

From the HELP or INFO command, as explained in the Computer Notes. However, you will be learning about commands one at a time, in little doses, as we proceed.

Suppose you weren't around, and I wanted to find out what PSAM did. How could I go about doing this?

To get a brief explanation, use the IDA command EXPLAIN, as shown:

PSEUDO-RANDOM NUMBERS.

IDA COMMAND "PSAM".

22 □ CONVERSATIONAL STATISTICS

```
> EXPLAIN
NAME OF COMMAND TO BE EXPLAINED = ?PSAM
PROBABILITY SAMPLE SELECTION:  SIMPLE RANDOM, STRATIFIED RANDOM
-          ---              OR SYSTEMATIC.
```

The technical terminology in this explanation will be explained in this chapter.

What are those dashes in the bottom left of the explanation?

They indicate the letters used in the abbreviation of the function to a four-letter command name: P, S, A, M.

Once you know the appropriate command name, you can issue the command, and IDA's prompts will take care of you from then on. Here's what happened:

```
> PSAM

WANT LIST OF SUBCOMMAND NAMES ?YES

COMMAND    PROCEDURE
*******    *********
  SYST     SYSTEMATIC SAMPLING
  RAN1     SIMPLE RANDOM SAMPLING BY SORTING
  RAN2     SIMPLE RANDOM SAMPLING BY DRAWING RANDOM NUMBERS
  SRAN     STRATIFIED RANDOM SAMPLING
  NONE     NONE OF THE ABOVE

WANT MORE DETAILED EXPLANATIONS ?YES

WHICH SUBCOMMAND DO YOU WANT EXPLAINED   ?RAN1

SIMPLE RANDOM SAMPLING BY SORTING

     THIS PROCEDURE CAN BE USED TO SORT A POPULATION OF SIZE
N INTO A RANDOM SEQUENCE.  THIS IS OFTEN USEFUL FOR THE
FORMATION OF RANDOM SUBGROUPS IN AN EXPERIMENT.  IT CAN BE
USED ALSO FOR SEQUENTIAL SAMPLING OF SMALL FINITE POPULATIONS.
     THE INDIVIDUALS IN YOUR POPULATION MUST HAVE BEEN NUM-
BERED CONSECUTIVELY FROM 1 TO N FOR IDENTIFICATION.
     WHEN YOU RUN THE PROGRAM YOU ARE ASKED THE SIZE OF
THE POPULATION.  YOU ARE THEN GIVEN THE OPTION OF A PRINT-
OUT OF THE ORIGINAL ORDERING OF THE POPULATION WITH THE
RANDOM NUMBERS ASSIGNED TO EACH ELEMENT THEREOF.  FINALLY,
YOU ARE GIVEN THE PRINTOUT OF THE RANDOM ORDERING OF THE
POPULATION AND THE RANDOM NUMBER OF EACH INDIVIDUAL.  FOR
SPEED OF PRINTOUT, YOU CAN ELECT TO SUPPRESS THE RANDOM
NUMBER, SINCE ONLY THE ORDERING IS NEEDED FOR STATISTICAL
PURPOSES.
     YOU CAN USE THIS PROGRAM FOR SORTS OF UP TO ABOUT 800
ITEMS WITHOUT EXCESSIVE COMPUTER TIME.  FOR N=750 A SORT
TOOK LESS THAN ONE MINUTE OF CONNECT TIME.

WARNING!
USE OF RAN1
WILL PREVENT RETURN TO IDA COMMAND LEVEL
AND ANY DATA ENTERED WILL BE LOST.

MORE EXPLANATIONS ?NO

*WHICH PROCEDURE DO YOU WANT TO USE? RAN1
WARNING!!!
USE OF RAN1
WILL PREVENT RETURN TO IDA COMMAND LEVEL
AND ANY DATA ENTERED WILL BE LOST.
WANT TO CONTINUE ?YES
```

Unlike most IDA commands, PSAM uses a number of subcommands, so you are given the opportunity to list them out.

I see from the brief descriptions above that RAN1 is the subcommand I need, so I have requested explanation.

This warning is unusual in IDA since, for almost all commands, one is returned after execution of command to the IDA command level. A technical requirement of the programming implementing RAN1 made this requirement necessary. No harm is done since RAN1 is ordinarily a separate task.

If I had known how to use RAN1 at the start, I would have gotten into action much more quickly, I trust!

Yes, here's how it would go:

```
> PSAM
WANT LIST OF SUBCOMMAND NAMES ?NO

*WHICH PROCEDURE DO YOU WANT TO USE?  RAN1
WARNING!!!
USE OF RAN1
WILL PREVENT RETURN TO IDA COMMAND LEVEL
AND ANY DATA ENTERED WILL BE LOST.
WANT TO CONTINUE ?YES
```

The actual execution for randomization of a group of 40 is shown in the printout in Tables 2-2 and 2-3.

In this printout, the computer is simply doing for a larger group what we saw earlier in the example of 10 subjects?

Yes. For example, let's look at what happens to the 17th subject in the group of 40. In Table 2-2, we see that he is assigned the random number 0.478371. In Table 2-3, we see that he is 20th in the randomization, and that his random number 0.478371 has been carried along with him in the sorting process.

Table 2-2. Assignment of Printouts for Random Numbers

```
WHAT IS SIZE OF GROUP TO BE SORTED ?40
WANT RANDOM NUMBER PRINTOUT BEFORE SORTING ?YES

GROUP      RANDOM
ORDER      NUMBER

  1        0.444256
  2        0.377221
  3        0.195122
  4        0.081159
  5        0.393722
  6        0.410113
  7        0.061998
  8        0.165411
  9        0.592799
 10        0.368228
 11        0.072373
 12        0.880208
 13        0.271273
 14        0.270861
 15        0.576597
 16        0.503501
 17        0.478371
 18        0.116965
 19        0.531454
 20        0.779598
 21        0.741700
 22        0.134910
 23        0.821200
 24        0.495088
 25        0.472273
 26        0.918956
 27        0.543458
 28        0.285241
 29        0.303134
 30        0.889947
 31        0.947443
 32        0.608367
 33        0.549075
 34        0.517723
 35        0.712046
 36        0.825657
 37        0.974946
 38        0.742752
 39        0.251247
 40        0.323982
```

Table 2-3. Assignment of Printouts for Sorted Random Numbers

```
WANT PRINTOUT OF RANDOM NUMBERS?YES

GROUP      SAMPLE     RANDOM
ORDER      ORDER      NUMBER

  7           1       0.061998
 11           2       0.072373
  4           3       0.081159
 18           4       0.116965
 22           5       0.134910
  8           6       0.165411
  3           7       0.195122
 39           8       0.251247
 14           9       0.270861
 13          10       0.271273
 28          11       0.285241
 29          12       0.303134
 40          13       0.323982
 10          14       0.368228
  2          15       0.377221
  5          16       0.393722
  6          17       0.410113
  1          18       0.444256
 25          19       0.472273
 17          20       0.478371
 24          21       0.495088
 16          22       0.503501
 34          23       0.517723
 19          24       0.531454
 27          25       0.543458
 33          26       0.549075
 15          27       0.576597
  9          28       0.592799
 32          29       0.608367
 35          30       0.712046
 21          31       0.741700
 38          32       0.742752
 20          33       0.779598
 23          34       0.821200
 36          35       0.825657
 12          36       0.880208
 30          37       0.889947
 26          38       0.918956
 31          39       0.947443
 37          40       0.974946

ANOTHER SAMPLE ?NO
TYPE 'GET-$IDA' AND 'RUN' TO USE IDA AGAIN.
DONE
```

Similarly, the first subject of the group had a random number of 0.444256, and he ended up as 18th in the randomization.

Good.

If I hadn't wanted the random numbers but only the final ordering, I suppose that I could have speeded things up?

Yes, as in Table 2-4, where I selected another randomization of 40 subjects, but did not elect to print out the random numbers.

Table 2-4. Final Randomization of Subjects

```
WHAT IS SIZE OF GROUP TO BE SORTED ? 40
WANT RANDOM NUMBER PRINTOUT BEFORE SORTING ? NO

WANT PRINTOUT OF RANDOM NUMBERS? NO

GROUP        SAMPLE       RANDOM
ORDER        ORDER        NUMBER

 10            1
  1            2
  3            3
 21            4
 30            5
 25            6
 36            7
  6            8
 22            9
 39           10
 17           11
 15           12
 12           13
  8           14
 33           15
 24           16
 16           17
  2           18
  4           19
 23           20
  9           21
 27           22
 11           23
 19           24
 34           25
 14           26
 29           27
 37           28
 13           29
 20           30
 38           31
 26           32
  5           33
 40           34
 31           35
 28           36
 32           37
  7           38
 18           39
 35           40

ANOTHER SAMPLE ? NO
```

If, say, you had wanted 4 groups of 10 each rather than 2 groups of 20 each, would you have simply taken the first 10, second 10, and so on, in the randomized listing of subjects?

Yes.

Suppose that I didn't have a random number table or access to a computer. Could I simply put the 40 numbers on little slips of paper, the slips in a hat, then stir thoroughly and draw them out one at a time?

Although thorough stirring is the intuitive idea of randomization, makeshift attempts at stirring are often not sufficiently thorough. Such methods were actually used in the draft lottery of 1940 in order to determine the order in which registrants at local draft boards would be called up to active service. The results were very bad: men with high registration numbers tended to be called up earlier.

Didn't something like that happen again in 1969?

Indeed it did. This time birthdays were to be randomized. The shuffling left much to be desired. Men with birthdays late in the year were much more likely to show up early on the draft priority list.

HAZARDS OF SHUFFLING TO ACHIEVE RANDOMIZATION.

That suggests that people don't learn from history!

Well, the intuitive conception of "random" means "haphazard", but haphazard methods of randomization often work badly. If you want to randomize, march out the random numbers!

Suppose that I simply designate every other subject as a control subject?

Often that procedure, which is an example of systematic sampling, will work quite well, but it can backfire. Suppose, for example, that the subjects are married couples, listed so that every other subject is the husband (or wife). Then you'd get all men in one group and all women in the other.

What happens if you take no special precautions at all in forming the groups?

You risk disaster. One famous illustration was the Lanarkshire milk experiment, which was designed to measure the effect of lunchtime milk supplementation on the growth rate of young school children. Teachers determined which children were to receive the milk. Quite understandably, the teachers tended to give the milk to the children who appeared to need it most. Although this was certainly the humane thing to do, the validity of the experiment was destroyed.

Suppose that there is no control group at all.

That's even worse, since you lose almost completely the ability to tell what would have happened in absence of the experimental treatment. A great deal of "experimentation" is like this, though: something is tried out, without any provision for a control group, and some intuitive judgment is made as to whether or not it "worked", even though the benchmark for comparison is hazy.

Is there ever a possibility for conclusive results without a control group?

Only if the response to the experiment is very different from what would have been expected by any past benchmark. For example, a cure for a previously fatal disease such as acute leukemia could be convincing in the absence of a control group of patients who did not receive the treatment.

I've heard the term "placebo" in connection with experimentation. What does that mean?

A placebo is a treatment that appears genuine but that is in fact inert: for example, a "sugar pill".

Why is it desirable to have a placebo group in an experiment?

There may be a psychological benefit simply because a subject thinks he is getting a valuable treatment, and the psychological benefit leads to measurable improvement in physical well being or performance of some task. In the famous experiments at the Hawthorn plant of Western Electric it was found that worker

IMPORTANCE OF CONTROL GROUPS.

productivity improved with almost any change in the working environment, whether favorable or unfavorable, and it was concluded that the improvement could be attributed to the fact that workers responded favorably to what they interpreted as an interest in them shown by the experimenters. As a result, the term "Hawthorn effect" has become widely used by industrial psychologists.

What is meant by the expression "double-blind"?

Neither the subjects of the experiment nor anyone judging the responses to the treatments--including, in a medical experiment, doctors, laboratory technicians, even statisticians analyzing the data--should know which subjects received which treatment. This information should be made available only at the end of the experiment.

Why is the double-blind feature desirable?

To keep biases, conscious or unconscious, from influencing any measurements or analyses of results that require exercise of judgment. Fortunately, the importance of double-blind experiments has become more widely appreciated. Those on both sides of the controversy over whether or not large amounts of vitamin C decrease the incidence or seriousness of common colds have agreed that the double-blind experiments that have been carried out are the ones deserving of the most serious attention.

The Vitamin and Endurance study in TNOS, assigned at the outset of this chapter, shows that the achievement of the double-blind requirement may not be easy in practice (see TNOS, page 73).

Are there other important ideas besides randomization in the statistical design of experiments?

There are many. To illustrate, consider the device called "blocking", which increases the precision of experiments for a given number of subjects or experimental elements. Instead of randomization of the entire group of subjects, the entire group may be divided judgmentally into smaller, more homogeneous groups called *blocks*. Then randomization is used within each block to allocate subjects to the different treatments.

Example: In the Vitamin and Endurance study, the subjects were a company of soldiers. Each of the four platoons of the company was made into a block, and randomization was carried out within each platoon (see TNOS, page 74.)

Are statistical experiments feasible in business?

They are often feasible in research and development laboratories. To a lesser extent, they are feasible in production and technology.

THE "DOUBLE-BLIND" PRINCIPLE.

EXPERIMENTATION IN BUSINESS.

In other areas of business, such as marketing, a host of practical limitations restrict the role of experimentation. It is possible to evaluate direct mail advertising experimentally, but more difficult to apply experimental methods in measuring the effectiveness of national television.

Is it ever possible to run a statistical experiment without seriously disturbing an ongoing business operation?

One interesting approach along these lines is called Evolutionary Operation or EVOP. A very readable account, with detailed advice for practical implementation, is given in *Evolutionary Operation* by George E. P. Box and Norman R. Draper. [Wiley, 1969.]

In the absence of randomized experimentation, can anything useful be learned about cause and effect by the use of experimental methods?

CAUSAL INFERENCE IN THE ABSENCE OF EXPERIMENTATION.

Yes, but it's enormously more difficult and the conclusions are usually much less clear cut. For example, in spite of a great amount of statistical evidence, economists disagree as to whether the money supply has an independent causal effect on business activity, or whether the money supply simply responds passively to business activity, or whether both directions of causal relationship are in fact at work simultaneously.

The problem of inferring causation in the absence of experimentation is so important that the field of econometrics has been defined informally by one distinguished scholar as the science of drawing causal inferences from non-experimental data. Certainly this is one of the important functions of statistical inference in many fields of application.

Could you suggest some examples that would give me some feeling as to how statisticians go about this?

Several of the articles in SAGTU deal with this problem. Try browsing the following:

1. Donald T. Campbell, "Measuring the Effects of Social Innovation by Means of Time Series", pages 120-127.

2. Lincoln E. Moses and Frederick Mosteller, "Safety of Anesthetics", pages 14-22.

3. B. W. Brown, Jr., "Statistics, Scientific Method, and Smoking", pages 40-51.

4. D. D. Reid, "Does Inheritance Matter in Disease?", pages 77-83.

5. C. A. Whitney, "Statistics, the Sun, and the Stars", pages 385-391.

Will we be placing heavy stress on causal inference in later chapters?

Not unduly, since statistical estimates of quantities and associations can be valuable even when causal interpretations are murky. For example, we may be able to predict sales or employee performance without special causal insight.

But you should always keep in mind the aphorism: statistical associations or relationships do not necessarily imply causation.

When I think of statistics, I usually think of sampling from a population, as in an opinion poll. Could we talk about sampling?

Certainly. In sampling, as in experimental design, many important statistical tools are available.

Let me start by asking what is meant in statistics by the term population?

Any finite collection of elements. Examples: (1) The student body in a college, (2) the dealers of an automobile manufacturer, (3) the fish in a lake, (4) orders received by a mail order company last month, (5) mortar shells in an ammunition dump, and (6) farms in Iowa.

What characteristics of these populations might be of interest?

(1) Fraction of students who are registered voters, (2) total retail sales of station wagons during the second ten days of May, (3) total number of fish in the lake, (4) the fraction of orders that were correctly filled, (5) the fraction of defective shells, and (6) the number of bushels of corn spoiled by southern corn blight.

What is a census?

A complete enumeration of some characteristic of the population. The best known example is perhaps the Decennial Census of Population in the United States.

I should mention that in practice an *actual* complete enumeration of a population is difficult to attain; even the Census Bureau fails to enumerate a small percentage of the United States population.

Is there error in a census?

Typically, yes. Error arises partly because, as just noted, some elements of the population are typically not reached; partly because there is measurement error in the determination of characteristics of the elements that are reached.

All such errors present in an attempted census are grouped under the term *non-sampling error*.

What is a sample?

Any part of a population.

SAMPLES AND POPULATIONS.

Why are we interested in samples?

For what they tell us about the populations from which they come.

But doesn't sampling introduce error that would not be present in a census?

Yes, and as you might expect, this error is called *sampling error*.

Curiously, however, the total error of a sample may be smaller, even substantially smaller, than that of a census.

That sounds like a contradiction in terms, or a defiance of the laws of logic, or what have you.

A good deal of evidence suggests that non-sampling errors, errors that would be present even in an attempted census, are often more substantial than sampling errors. Since a sample often involves a relatively small administrative apparatus as compared with a census, it may be possible to reduce the non-sampling errors. For example, interviewers can be better trained and supervised, and thus may reach a higher fraction of interviewees and may get more accurate answers to questions. One famous example of a sample outperforming a census occurred in the province of Bengal in India, during the late 1930's, when the purpose was to measure the size of the jute harvest. When the true size eventually was ascertained through warehouse records, the census was found to have erred by 20% while the concurrent sample survey came within ½%!

In economic terms, you're saying that there may be diseconomies of scale in an attempted census?

Yes.

Is there any other reason why we might wish to take a sample rather than a census?

Economic logic: even if a census is more accurate, it may not be worth the added cost. The extreme case of destructive testing is the easiest to see: consider the doctor who wants to determine your white blood count!

In illustrating the use of IDA a little while ago, we encountered the term probability sample. What does that mean?

A sample in which random numbers (or equivalent procedure) have been used in such a way that every element of the population is assigned a known, positive probability of being drawn in the sample.

The simplest probability sampling design is called *simple random sampling*. This design assures that each possible sample of n elements in a population of N elements has an equal chance of being chosen.

SAMPLING AND NON-SAMPLING ERRORS.

SIMPLE RANDOM SAMPLING: THOROUGH SHUFFLING.

Isn't that just the thorough shuffling approach illustrated during your discussion of formation of treatment groups in a randomized experiment?

Yes, in the sense that the first n elements of the shuffled population of N constitute a simple random sample of the population.

But this method is not feasible when N is large, because sorting is too expensive.

What do you do then?

It might be called the "one-at-a-time" approach. The idea is this: use random numbers to choose the first element so that each of the N elements in the population has an equal probability of selection. Then choose a second element from the remaining N-1 elements so that each of these has an equal probability of selection. Continue in this way until the desired sample size n is obtained. On each round of selection the remaining elements of the population have an equal chance of being selected.

And mathematically that comes to the same thing as the shuffling approach?

Yes. It also guarantees an equal probability of selection to each possible sample of n.

I find it hard to visualize how this works. Could you give me an example?

Suppose that we wanted to draw a simple random sample of n = 100 from the student body at the University of California, Berkeley, and that the student body comprises 28,000 students, all of whom are listed in the student directory.

1. Number the students from 1 to 28,000. (With ingenuity, this tedious task may be alleviated by a more sophisticated approach that works with the page numbers in the directory and a knowledge of the maximum number of students that can appear on any page.)

2. Work with consecutive five-digit numbers from a table of random numbers, just as before. If the first number exceeds 28000, forget it and go on to the next. If it is less than or equal to 28000, include the corresponding member of the population in the sample. For example, 17385 designates student number 17,385 on the list; 89470 is ignored.

3. Continue until n = 100 *different* students have been chosen. (Thus if 17385 showed up a second time before the sample was complete, it would be ignored.)

Why five-digit random numbers?

Because N = 28000 is a five-digit number.

AN EXAMPLE OF SIMPLE RANDOM SAMPLING.

CONVERSATIONAL STATISTICS

I see: since each random number from 00001 to 28000 is equally probable, each student has the same probability of inclusion on the first draw.

That's right. And each of the remaining 27999 will have an equal chance on the second draw, and so on.

It looks to me like nearly three-fourths of the five digit numbers will exceed 28000 and will thus be "wasted". Are there any shortcuts to reduce this waste?

Yes, but forget about them. Random numbers are cheap, and the shortcuts are treacherous for novices.

Can you implement this method of sample selection on IDA?

Yes, by the command PSAM and the subcommand RAN2. Here's the explanation given by IDA:

```
WHICH SUBCOMMAND DO YOU WANT EXPLAINED   ?RAN2

SIMPLE RANDOM SAMPLING BY DRAWING OF RANDOM NUMBERS UNTIL
DESIRED UNDUPLICATED SAMPLE SIZE IS REACHED.

    THIS PROCEDURE SELECTS A RANDOM SAMPLE OF N1 FROM A FINITE
POPULATION OF N, SELECTION WITHOUT REPLACEMENT.  THE USER IS
ASKED FOR N AND N1, AND THE SAMPLE ELEMENTS ARE PRINTED OUT IN
ORDER OF THEIR LISTING IN THE POPULATION. IT SHOULD BE POSSIBLE
TO SELECT SAMPLES OF SIZE 1000 OR SO WITHOUT USING EXCESSIVE
COMPUTER TIME, UNLESS BOTH N1 AND N1/N ARE LARGE.

WARNING!
USE OF RAN2
WILL PREVENT RETURN TO IDA COMMAND LEVEL
AND ANY DATA ENTERED WILL BE LOST.

MORE EXPLANATIONS ?N
```

And here's an implementation:

```
> PSAM

WANT LIST OF SUBCOMMAND NAMES ?NO

*WHICH PROCEDURE DO YOU WANT TO USE?   RAN2
WARNING!!!
USE OF RAN2
WILL PREVENT RETURN TO IDA COMMAND LEVEL
AND ANY DATA ENTERED WILL BE LOST.
WANT TO CONTINUE ?YES

*SIZE OF POPULATION TO BE SAMPLED   ?28000

DESIRED SAMPLE SIZE ?100

THE FOLLOWING MEMBERS OF THE POPULATION OF  28000
COMPRISE THE SAMPLE OF   100  :

32       732      771     1200     1386     1392     1674     2724     2779
2892     3708     3724     4306     4446     5383     5673     5884
6179     6319     6636     6697     7171     7415     7419     7698
8163     8342     8661     8860     9269     9340     9859     9888
10630    10722    10799    10898    10977    11118    11397    11399
11752    11931    13298    13412    13467    13617    13873    15012
15022    15083    15304    15472    16759    16931    17029    17379
17656    17715    18568    18734    18945    19100    19662    19776
20096    20750    20915    21054    21328    21456    21461    22108
22121    22212    22327    22485    22526    22549    22639    22723
22806    22942    23190    23454    23751    24425    24504    24561
24786    25099    25184    25270    25389    25418    25944    26159
27123    27462    27758

ANOTHER SAMPLE ?NO

TYPE 'GET-$IDA' AND 'RUN' TO ENTER IDA AGAIN.
```

As I look at the sample numbers, it seems to me that they tend to bunch up or cluster. Am I imagining things?

No, the clustering that you have observed is to be expected from probability theory. Random numbers display many anti-intuitive properties, and so do random samples. People are often astounded, for example, by the *non*-representativeness of *very small* random samples. One practical consequence of this latter fact about random sampling is that the argument for random sampling becomes weak if for some reason the sample size n must be very small, say n =5. But we won't have time to explore the fascinating issues connected with the anti-intuitive behavior of simple random samples.

But before we leave the subject, I want to be sure that I understand one point that you've just made: Is it true that a random sample is not necessarily representative?

That's true. But the probability is high that a medium or large random sample will be representative.

Let me warn you, though, that "representative" is a weasel word. One statistician says that the word leaves him very itchy. The point is that if you are sampling to learn something about a population that you didn't know in the first place, *no* sampling method can *guarantee* representativeness. If there are exactly 100 students at Berkeley who are Jewish Republicans who voted for McGovern in 1972, there's no ironclad guarantee that we wouldn't get all of them in our sample of 100: the probability of this happening is not zero.

But the probability is minuscule?

Yes, but don't lose sight of the point. At best, the person who claims that his sample is "representative" is indulging in a bit of advertising puffery. It's my impression, moreover, that this word is more likely to be used when the sampling methods are sloppy.

When you speak of a "large" sample, do you mean that n/N is large or that n is large.

That n is large, almost regardless of n/N.

You said that the argument for simple random sampling gets weak for small n. What's the alternative?

One possibility is some other form of probability sampling, along lines that I will suggest later.

How about just picking the sample in any way that you judge to be likely to produce (excuse the expression) a representative result?

If n is very small, you may be forced to it. If I had to get by with a sample of one county of the roughly 3100 counties in the United States, I'd prefer to pick it judgmentally, without any reliance on probability sampling.

At the same time, I must stress that human judgment, unaided by concepts of statistical sampling, is often grossly unreliable in sample selection. Example:

CLUSTERING OF RANDOM NUMBERS AND RANDOM SAMPLES.

RANDOMNESS AND REPRESENTATIVENESS.

A former student has told us of a sampling blunder in his company which stemmed from using a convenient but nonrandom sampling method. The company had found from past experience that about 90 percent of certain castings it was buying were defective. The defects usually showed up only after some machining had been done. One supplier claimed that he had developed a new method which would virtually eliminate defective castings.

When the first new lot was received it was decided to take a sample of the lot and have the sample items X-rayed before any machining was done. A sample of 20 castings was taken from the top of the box containing the entire lot and the X-ray inspection did show a great improvement in quality (actually no flaws were detected). On the basis of this the lot was accepted.

The lot was machined and 75 percent of the castings had to be scrapped. Subsequent inquiry showed that by an error of the supplier, the box was filled mostly with castings from the old method, with the new ones only on top.

Had the 20 castings been chosen randomly from the entire box, not just the top layer, there would have been only one chance in a trillion of finding no defective castings if the lot contained 75 percent defectives.

Let's go back to the Berkeley example. How about skipping systematically through the Berkeley listing taking every 280th name in order to get the sample of 100?

SYSTEMATIC SAMPLING.

That method is called *systematic sampling*. Usually a random start in the first interval--in the example here, in the first 280 students--is chosen. Systematic sampling is implemented on IDA as the subcommand SYST of PSAM, as illustrated:

```
*WHICH PROCEDURE DO YOU WANT TO USE?  SYST

*SIZE OF POPULATION TO BE SAMPLED  ?28000

DESIRED MINIMUM SAMPLE SIZE  ?100

THE SAMPLING INTERVAL IS  280
SO  100  OBSERVATIONS WILL BE DRAWN
FROM THE FIRST  100  TIMES  280  ELEMENTS OF THE POPULATION.

THE FOLLOWING ARE THE SAMPLE ELEMENTS:

  36     316    596    876    1156   1436   1716   1996   2276
  2556   2836   3116   3396   3676   3956   4236   4516
  4796   5076   5356   5636   5916   6196   6476   6756
  7036   7316   7596   7876   8156   8436   8716   8996
  9276   9556   9836   10116  10396  10676  10956  11236
  11516  11796  12076  12356  12636  12916  13196  13476
  13756  14036  14316  14596  14876  15156  15436  15716
  15996  16276  16556  16836  17116  17396  17676  17956
  18236  18516  18796  19076  19356  19636  19916  20196
  20476  20756  21036  21316  21596  21876  22156  22436
  22716  22996  23276  23556  23836  24116  24396  24676
  24956  25236  25516  25796  26076  26356  26636  26916
  27196  27476  27756
                      ACTUAL SAMPLE SIZE IS  100

WANT ANOTHER SAMPLE  ?NO

>
```

Note that you go back to the IDA command level, as indicated by the readiness symbol ">". This is typical of IDA, although the subcommands RAN1 and RAN2 did not do this.

Well, that really spreads out the sample. No clustering here. Is that good?

It depends on the population. In many applications, the uniform spreading-out tends to give precision equal to or better than that of simple random sampling with the same sample size.

But you have to be careful, for the same reason I mentioned when you made a similar suggestion for formation of treatment groups in randomized experimentation (in the discussion of the hazards of shuffling): there may be an unfortunate periodicity in the population listing.

For example, suppose that you wished to estimate the total lineage of classified advertising carried by the Chicago *Tribune* during 1972 by sampling every seventh issue for that year.

The Sunday edition carries much more classified advertising than other days, and I'd expect some kind of systematic pattern for the other days as well.

So systematic sampling could be disastrous. The examples of this kind, however, seem to arise mainly with populations listed in a time sequence. In other applications, the simplicity of systematic sampling can be a strong advantage, especially when sample selection must be done in the field by untrained people.

Is there any sampling scheme that achieves the advantages of systematic sampling, yet offers protection against periodicities?

Consider the following idea: sample one issue of the *Tribune* at random each week. This is simple, tends to spread out the sample through the year, yet protects against systematic patterns within the week.

This method can be applied with IDA by the subcommand SRAN of PSAM, as illustrated below for the Berkeley example:

```
*WHICH PROCEDURE DO YOU WANT TO USE? SRAN

*SIZE OF POPULATION TO BE SAMPLED  ?28000

DESIRED MINIMUM SAMPLE SIZE  ?100

THE SAMPLING INTERVAL IS  280
SO  100  OBSERVATIONS WILL BE DRAWN
FROM THE FIRST  100  TIMES  280  ELEMENTS OF THE POPULATION.

THE FOLLOWING ARE THE SAMPLE ELEMENTS:
  35      500     610     971    1175    1520    1895    2008    2265
 2735    2960    3318    3557    3789    4129    4458    4734
 4774    5294    5401    5826    6044    6362    6649    6965
 7169    7474    7642    7952    8326    8445    8854    9164
 9250    9620   10045   10271   10440   10793   11172   11367
11490   11793   12102   12456   12602   13052   13394   13623
13891   14169   14307   14649   14908   15217   15638   15839
15968   16298   16792   16849   17287   17560   17798   18108
18349   18543   19021   19065   19383   19646   20040   20409
20663   20917   21093   21399   21773   22087   22389   22520
22724   22980   23293   23788   24027   24290   24523   24697
25074   25338   25570   25944   26064   26354   26624   26882
27225   27719   27795
                      ACTUAL SAMPLE SIZE IS  100
```

What's the significance of the subcommand SRAN?

It is taken from *strat*ified *ran*dom sampling, of which this is a special case. This is another method of probability sampling.

The population is first divided up into subpopulations or strata. In the example above, the strata comprise consecutive groups of 280 students in the listing. When feasible, strata are formed in such a way that the elements within each stratum are relatively homogeneous with respect to the characteristic of interest.

PITFALLS OF SYSTEMATIC SAMPLING.

STRATIFIED RANDOM SAMPLING.

Then a simple random sample is taken from *each* stratum and used to estimate the characteristic of interest for the entire stratum. In the Berkeley example, only one observation per stratum was randomly drawn; in other applications of stratified sampling, more than one might be drawn.

Finally, the estimates of each stratum are combined to make an estimate for the entire population.

For example, suppose that a company wished to ascertain attitudes of its employees towards a proposed four-day work week. The monthly payroll provides a listing of the employees. The strata might be departments of the company.

Do you always take the same fraction of sample elements in each stratum?

Very commonly this is done. Among other things it simplifies the subsequent statistical analysis.

There is one common type of business application in which non-proportional sampling is desirable. Suppose that a company wished to estimate monthly retail sales of its product by taking a sample of its dealers and getting a special report from them.

Intuitively, I'd guess that you'd want to give high representation to dealers with high sales in the past.

Formal statistical reasoning will back up your intuition. One could stratify by past sales, and sample higher fractions of dealers in the strata with high past sales. A rough rule-of-thumb is to allocate the sample across strata in proportion to the total dollar sales (as opposed to number of dealers) in each stratum. This might even require bringing in the largest dealers with certainty.

Then you'd have to make some allowance for that in your analysis.

Yes. If your largest 10 dealers, say, came in with certainty, then they would naturally represent only themselves--which might mean, of course, a high fraction of total company sales. On the other hand, if you sampled only 1/25 at random of the smallest 5000 dealers, you'd multiply the total sales of the 200 sample dealers from this stratum by 25: each dealer represents himself and 24 others, so to speak.

That gives me some feeling for stratified random sampling. Are there any other types of probability sampling that can be used safely by beginners?

I might mention *cluster sampling*, and in particular a special kind of cluster sampling called *area sampling*. This approach is often useful when the sampling elements of interest are widely spread out geographically and you don't have a good list or *frame* of these elements.

Suppose, for example, that you want to study a suburb consisting mainly of single family homes. The natural sampling units would be families. Your questions are such that you feel that personal

AREA SAMPLING.

interviews in the home are essential. But there is no village directory, and, let us suppose, no other usable list or frame of families.

From the village map, however, you can easily make a frame of larger sampling units, blocks or similar small areas. Then you can draw blocks at random, sample families at random within the blocks so drawn. In this example, the families living in a block constitute a cluster, whence the name cluster sampling.

Why is cluster sampling a probability sampling approach?

Random selection is applied at the stage of selection of sample clusters (blocks), and then again in the choice of families within the sample clusters. It is possible to figure out the probability of selection for any family.

It sounds a little tricky for a beginner.

That's why I used the example of a suburb consisting mainly of single family homes. It would be desirable if the number of homes per block didn't vary greatly. Then a reasonable approach is to take the same number of families within each sample block, say two families per block. All you have to do is to know how to use random numbers, which you do at two stages, first in choosing blocks and then in choosing families within blocks.

Might I want first to stratify the blocks, say by the part of town in which they were located?

That might be desirable. That design is called *stratified cluster sampling*.

But a national probability sample would require an expert?

Yes. For an example, read the article by Conrad Taeuber, "Information for the Nation from a Sample Survey", pages 285-296 of SAGTU. This article describes the large and intricate probability sample used to obtain monthly estimates of unemployment and other population characteristics of the entire United States. Randomization, stratification, clustering, and other statistical devices are interwoven intricately in the design of this sample.

Are non-probability sampling methods used?

Widely.

Are they statistically sound?

They can't be given the same theoretical support and guarantees justified by probability sampling. Some of them are terrible: gross offenses against common sense, as when the sales manager asks his wife and her social club what they think of the new cold cereal his company is thinking of putting on the market.

On the other hand, many applications arise in which it simply doesn't appear economic to execute strict probability sampling. For example, suppose that you were given the task of designing a sample of current users of computer terminals.

NON-PROBABILITY SAMPLING.

38 CONVERSATIONAL STATISTICS

Non-probability sampling can, however, borrow many of the ideas of probability sampling. I have considerable sympathy with such mixed designs, and there is evidence that they sometimes work pretty well.

What is quota sampling?

That's one of the rather well-defined species of non-probability sampling. It superficially resembles a stratified random sample but selection within the strata is made by interviewer judgment rather than by random selection. Quota samples have been used extensively in political polling, where they have worked pretty well on the whole, at least at the national level. Read George Gallup, "Opinion Polling in a Democracy", pages 146-152 of SAGTU.

What is meant by the problem of "non-response" in statistical studies?

Some elements designated by the sample design may not be reached. For example, in a survey of voters, some designated voters may not be found at home while others may refuse to answer. The first category of non-response is called "not-at-home" and the second is called "refusal". Non-response may arise even in sampling studies of inanimate objects; for example, some buried telephone cable may be missed in a study designed to estimate the condition of telephone cable.

Why is non-response a problem?

Often the non-respondents differ substantially from the respondents.

Would you classify non-response as a source of non-sampling error?

Yes, it is a source of error that is present equally in attempted censuses and in samples.

What can be done about non-response?

The most important thing is to plan the study in such a way as to keep non-response small.

For example?

It is desirable to provide an adequate incentive for response. Sometimes this is simply a clear explanation of the purpose of the study. Sometimes monetary or other material incentives may be used.

In personal interviews, people not found at home on the first attempt should be sought later on a callback; sometimes several callbacks may be provided.

Why not just substitute the next-door neighbor?

The sample would then tend to underrepresent the people who are less frequently at home; for example, working wives.

NON-RESPONSE.

The analogue of the callback in a mailed questionnaire is the follow-up questionnaire. Roberts' rule for mail questionnaires: throw the first questionnaire in the waste basket; reply to the follow-up if they take the trouble to send one.

But couldn't you get a bigger sample size if you didn't fool around with callbacks or follow-ups?

Yes, but then you'd be playing a meaningless numbers game. A larger nominal sample size does not assure greater accuracy; often just the reverse is true if the smaller sample is more carefully designed and implemented. From an economic point of view, you want to maximize the accuracy attainable within the research budget, not to maximize the number of completed interviews.

That reminds me of something I've been meaning to ask you: How big should a sample be?

That's a commonly-asked question. Unfortunately I can't answer it. I'd like to say something like "17", "100", or "1000", but I can't.

Why not? Can't a statistician tell about the accuracy of a proposed sample?

In principle, he can do that. That is, given the sampling method and the sample size, and background information about the population, the statistician can provide an estimate of the precision with which a proposed sample will mirror the population. (Even so, however, he can't treat non-sampling errors in a very satisfactory way in many circumstances.) You will see some sampling error formulas in Chapter 7.

But the statistician, as a statistician, usually can't say how much precision is worth paying for. That is a judgment that must be made by the user of sampling.

What then is the role of the statistician in helping a user to determine sample size?

He can point out that the decision is essentially an economic choice in which the user must balance at the margin the value of additional sample information against the cost of obtaining it. And, as a part of this choice, he can say something, based on sampling error formulas, about the incremental improvement in precision that comes with any incremental increase in the sample size. But he can't say how much the incremental improvement in precision is worth to the user.

By analogy, to ask a statistician how big a sample should be, in the abstract, is like asking a bakery how big a cake should be baked for a wedding. It depends on the wedding, the kind of cake contemplated, the value of having second helpings available at the reception as opposed to the cost of not having enough first helpings to go around, and so on.

SAMPLE SIZE.

Hasn't statistical decision theory made important contributions to the determination of sample size?

For certain types of applications, the economics of sample size have been well worked out. Even here, however, the user has to provide some of the key inputs to the analysis.

In many applications, however, the problem of sample size is not as perplexing as you might think. This is because an arbitrary judgment has been made very early in the planning of the study as to how large the research budget shall be, and there isn't usually much play with that constraint. For example, a congressional committee may authorize an appropriation of $100,000 for a particular study, and the statistician must design as well as he can within that limit. Even when a specific budget is not mentioned, the statistician can often sense quite accurately what the user is willing to spend.

To clinch a point touched on earlier: a statistician thinks of sample size in absolute units (n) rather than the percentage sample size (100n/N)?

Yes.

Can you suggest a good source describing statistical applications in business and other fields of interest to us?

The book SAGTU has many examples besides the ones I've mentioned already. Here are some of the most useful:

1. W. Edwards Deming, "Making Things Right", pages 229-236.

2. Charles W. Dunnett, "Drug Screening: The Never-Ending Search for New and Better Drugs", pages 23-33.

3. D. G. Chapman, "The Plight of the Whales", pages 84-91.

4. S. James Press, "Police Manpower Versus Crime", pages 112-119.

5. Frank A. Haight, "Do Speed Limits Reduce Traffic Accidents?", pages 130-136.

6. Frederick Mosteller and David L. Wallace, "Deciding Authorship", pages 164-175.

7. Elisabeth Street and Mavis G. Carroll, "Preliminary Evaluation of a New Food Product", pages 220-228.

8. John Neter, "How Accountants Save Money by Sampling", pages 203-209.

9. Philip J. McCarthy, "The Consumer Price Index", pages 266-275.

10. Morris H. Hansen, "How to Count Better: Using Statistics to Improve the Census", pages 276-284.

11. Geoffrey H. Moore and Julius Shiskin, "Early Warning Signals for the Economy", pages 310-320.

MORE EXAMPLES.

CHAPTER THREE: OBSERVATION AND MEASUREMENT

READING: TNOS, Chapters 7 and 8

Are problems of measurement and observation a part of the subject matter of statistics?

If we carefully delineate the boundaries between specialties, these problems are at best on the boundaries of statistics with other disciplines. Nothing in the specialized training of a Ph.D. in mathematical statistics will help him to write a sensible questionnaire or make a chemical determination in the laboratory.

However, statistical principles of the design of experiments and investigations may point the way toward the improvement of the process of data gathering. A trained statistician is likely to think of the desirability of making duplicate or triplicate measurements, of setting up the data gathering so that error of measurement can be assessed statistically, and so on.

But a statistician--or one learning to emulate a statistician--need pay little or no attention to how measurements and observations are actually made? That's someone else's concern?

I couldn't disagree more. If you are seriously concerned with a statistical investigation, whether as producer or consumer, you must learn as much as you can about the process by which the data are collected. In many statistical investigations, this is the weakest link. When it has been possible to estimate the extent of non-sampling errors in statistical studies, it has usually been found that they are more substantial than the sampling errors. Ask questions of technical people, read about the methods used, in short, do anything that you can to learn about the measurements.

Suppose, however, that I have a study done by someone else. It doesn't go into details about the methods of data collection, and the author isn't around to be queried. What can I do?

You can be suspicious. If the study is important to you, and you have time to write the author, write him and ask for documentation.

If the author gives his data or reasonably detailed tabulations or summary computations, you may be able to do a little detective work. We'll see an example of that in the next chapter.

How does one write a good questionnaire?

Start with the right attitude: few tasks are harder for the novice. Just think back on the questionnaires you have filled out or been asked, including application forms and the like. How often were you confused about what was being asked or skeptical about its relevance?

ACCURACY OF DATA.

42 □ CONVERSATIONAL STATISTICS

I agree with you about the prevalence of poor questionnaires, but I don't see how that applies to me. After all, I've been asking questions since I was about two years old.

And often getting inaccurate, ambiguous answers. The art of designing a questionnaire consists of phrasing questions that are exactly understood, and that will elicit answers relevant to sensible objectives, not just to vague curiosity.

All right, then. Do you have any tips on how to write a questionnaire?

1. Pretest your first draft--that is, try out your questions on a small convenience sample and find out the difficulties of respondents in answering them.

2. Revise the questionnaire on the basis of the pretest. Pretest again.

3. Repeat Step 2 as often as necessary. As a computer expert once said about computer programming by beginners, this can go on for quite a while.

QUESTIONNAIRES.

CHAPTER FOUR: ANALYSIS OF SEQUENCE

How does one get started with a statistical analysis?

Plot the data so that you can see what is going on.

That surprises me. I thought that you would begin with computations.

Computations without plotting of data are like flying an airplane in bad weather with very limited instrumentation. Even when you might wish to deviate from my suggestion of plotting first, you will be well advised to do a good deal of plotting during the course of the analysis.

How about an example?

Let's start with a set of readings of something called "hardness" of each coil in a sample of 100 steel coils produced a few years ago at a steel company in the Chicago area. The measurements were made by a procedure known as a Rockwell hardness test. I have saved the data in a file called $ROCKY in the system library of the HP-2000F.

Before we continue the discussion of this example, could you give me some idea of how one sets up a file on the computer?

That's covered in some detail in the Computer Notes. I strongly urge that you review this carefully and try out a simple example of your own.

To show you what the data look like, I'll bring them into IDA by use of the command ENTER and then use the command PRTS to print them out in sequence. The sequence of the observations is the sequence of their actual occurrence.

```
GET-$IDA
RUN
IDA

30AUG73 VERSION

GOOD AFTERNOON.  DO YOU NEED HELP ?N

> ENTER

* MODE OF INPUT : FROM 'FILE' OR 'TERMINAL' ?FILE

NAME OF INPUT FILE IS ?$ROCKY

* ARE THE FIRST TWO ELEMENTS OF YOUR DATA FILE
    VALUES FOR N AND K (SIZE OF DATA MATRIX) TO FOLLOW ?YES

DATA MATRIX HAS 100 OBSERVATIONS ON  1 VARIABLES
COMPUTING ...
```

PLOTTING THE DATA.

Table 4-1. Rockwell Hardness Data

```
YOU HAVE THE OPTION TO GIVE EACH VARIABLE A NAME
OF 1 TO 6 CHARACTERS.  IF THIS OPTION IS NOT EXER-
CISED, THE VARIABLES WILL AUTOMATICALLY BE NAMED
1,2, ETC., ACCORDING TO THE COLUMNS IN WHICH
THEY ARE STORED IN THE DATA MATRIX

WANT TO SUPPLY NAMES ?YES
VAR.  1 = ?HARDNS

> PRTS
DO VARIABLES TO BE PRINTED OCCUPY
CONSECUTIVE COLUMNS OF DATA MATRIX ?YES
PRINT DATA SUBMATRIX BETWEEN ROWS I1 TO I2 AND COLUMNS
J1 TO J2, INCLUSIVE.  GIVE FOUR VALUES, SEPARATED BY
COMMAS, FOR I1, I2, J1, J2 : ?1,100,1,1
```

** 1 **	58.00000	** 52 **	57.00000
** 2 **	49.00000	** 53 **	65.00000
** 3 **	58.00000	** 54 **	60.00000
** 4 **	57.00000	** 55 **	55.00000
** 5 **	50.00000	** 56 **	64.00000
** 6 **	60.00000	** 57 **	65.00000
** 7 **	64.00000	** 58 **	59.00000
** 8 **	65.00000	** 59 **	62.00000
** 9 **	64.00000	** 60 **	65.00000
** 10 **	59.00000	** 61 **	64.00000
** 11 **	65.00000	** 62 **	54.00000
** 12 **	65.00000	** 63 **	56.00000
** 13 **	45.00000	** 64 **	58.00000
** 14 **	54.00000	** 65 **	40.00000
** 15 **	52.00000	** 66 **	85.00000
** 16 **	59.00000	** 67 **	53.00000
** 17 **	65.00000	** 68 **	61.00000
** 18 **	57.00000	** 69 **	56.00000
** 19 **	63.00000	** 70 **	65.00000
** 20 **	54.00000	** 71 **	58.00000
** 21 **	65.00000	** 72 **	58.00000
** 22 **	60.00000	** 73 **	55.00000
** 23 **	61.00000	** 74 **	52.00000
** 24 **	47.00000	** 75 **	65.00000
** 25 **	60.00000	** 76 **	60.00000
** 26 **	52.00000	** 77 **	65.00000
** 27 **	63.00000	** 78 **	63.00000
** 28 **	61.00000	** 79 **	64.00000
** 29 **	54.00000	** 80 **	63.00000
** 30 **	63.00000	** 81 **	60.00000
** 31 **	62.00000	** 82 **	61.00000
** 32 **	56.00000	** 83 **	61.00000
** 33 **	56.00000	** 84 **	65.00000
** 34 **	65.00000	** 85 **	56.00000
** 35 **	56.00000	** 86 **	62.00000
** 36 **	64.00000	** 87 **	65.00000
** 37 **	65.00000	** 88 **	54.00000
** 38 **	55.00000	** 89 **	64.00000
** 39 **	59.00000	** 90 **	63.00000
** 40 **	65.00000	** 91 **	57.00000
** 41 **	64.00000	** 92 **	64.00000
** 42 **	49.00000	** 93 **	62.00000
** 43 **	65.00000	** 94 **	58.00000
** 44 **	50.00000	** 95 **	60.00000
** 45 **	65.00000	** 96 **	52.00000
** 46 **	61.00000	** 97 **	53.00000
** 47 **	64.00000	** 98 **	62.00000
** 48 **	61.00000	** 99 **	56.00000
** 49 **	59.00000	** 100 **	65.00000
** 50 **	63.00000		
** 51 **	58.00000		

This output, of course, came out of the terminal in two columns printed at the left-hand side of the page; from time to time, computer output will be slightly rearranged to save space on the printed page.

I can see that the hardness readings are all two-digit numbers and that most are in the 50's and 60's, but otherwise, it looks like pretty much of a jumble, doesn't it?

Precisely. That's why it's helpful to plot the data.

Before you get to that, I'd like to have a better idea of what we are trying to accomplish? What can we hope to learn by plotting the data--or, for that matter, by a numerical analysis later?

I find the following orientation helpful. The data arose in a particular time sequence, and are so indicated by their ordering in the file. For this reason, incidentally, it would be correct to refer to this sample of 100 numbers as a *time series*. We are trying to see what can be learned from this time series of 100 observations as to what will happen in the future: we want to learn what kind of predictions can be justified on the basis of the evidence reflected in our data.

Do you mean that we will try to predict exactly what is going to happen?

Ordinarily, errorless prediction is far too much to hope for. We must be content with some kind of estimates or informed guesses, together with some indication of the reliability of these estimates or guesses.

Even so, wouldn't you want to consider more than just the statistical data?

By all means. But it is helpful to start by seeing what the data seem to be telling us, without drawing on other information.

One approach to statistics--the Bayesian--tries to give an explicit rationale for reconciling data with other kinds of information in arriving at statistical predictions.

How do we start?

A natural first step is to see whether the *sequence* in which observations have occurred gives any guidance as to what our predictions ought to be.

I don't follow you.

I'm glad you spoke out. The idea is simple but elusive. Let's illustrate by considering the 99th and 100th readings: you see from Table 4-1 that these were 56 and 65 respectively.

Suppose that instead the sequence had been 65 and then 56, and that you were trying to make an estimate of the next observation, the 101st. Would you make the same estimate now as you would have made if the actual sequence of 56, then 65 had been observed?

Am I supposed to be able to answer that?

Well, no, not yet. But at the conclusion of your analysis, you will have to answer it--even if by default.

If you say that your prediction would be different if the last two observations had been interchanged, then you have concluded that the sequence does matter in making predictions.

If you say that your prediction would be the same if the two observations had been interchanged and that, further, nothing about your prediction would have been changed if the sequence had been rearranged in any way, then you have concluded that sequence does not matter.

Now that I think about it a bit, I would guess that I'd always want my estimate to reflect sequence to some extent. I'd make a lower estimate for the sequence 65, 56 than for the sequence 56, 65.

Then you would be making the judgment that sequence does matter--and that further it always matters. Why do you feel that way?

Well, in my experience, the future seems to resemble the immediate past more closely than the more remote past.

That is often true, to be sure. Are you confident that your generalization applies to the hardness readings of steel coils in this particular mill?

STATISTICAL PREDICTION.

46 □ CONVERSATIONAL STATISTICS

Well, I'll admit that I don't know much about steel production.

I don't either, so let's look at the data together.

Since we've been talking about the significance of sequence, a natural first data plot is one called a *sequence plot*. A sequence plot is a graph of the observations in time sequence on ordinary graph paper. The observations themselves are scaled vertically; the sequence is scaled horizontally. Each observation appears as a dot on the graph in a rectilinear coordinate system. Figure 4-1, which I drew by hand, shows the Rockwell hardness data displayed in this way.

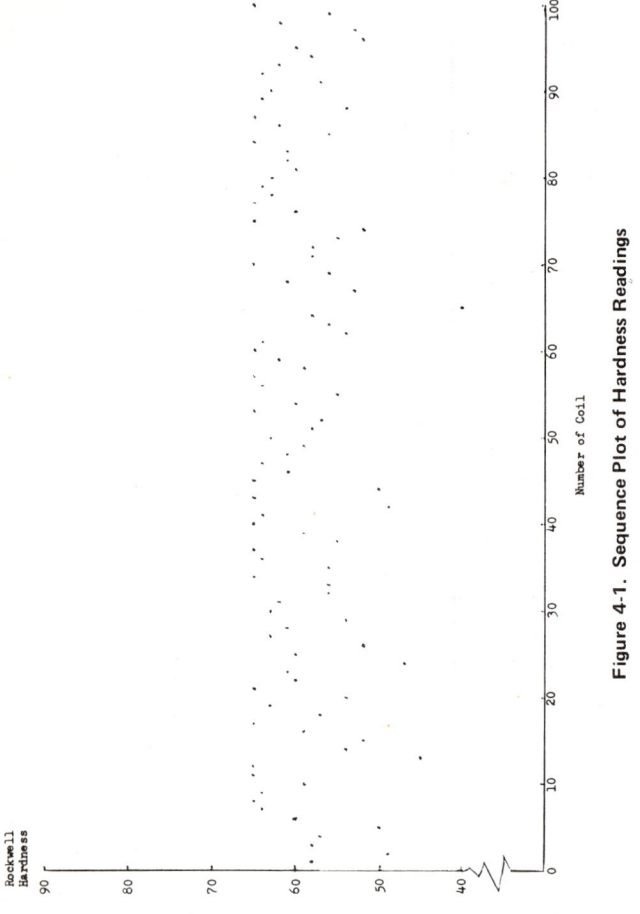

Figure 4-1. Sequence Plot of Hardness Readings

How am I supposed to look at this sequence plot?

The idea is to see if you can find *patterns in the sequence* of the observations that will bear out your conjecture that you'd want to consider the sequence observed in the data as one factor to be given weight in your prediction.

One good way to start is by using what I call the unfolding technique (not a standard technical term). Cover the plot with a card or sheet of paper, draw the paper slowly to the right, revealing one observation at a time.

Just as if I were plotting the data myself?

That's the idea. Now, as you do this, keep studying the observations just revealed, asking yourself if you can discern patterns in the sequence of these points that might give a clue as to what will happen when further points are revealed. Try to invent systems for doing this; let your imagination run freely. For example:

1. Do you see a tendency for points close together to be more similar than points further removed?

SEQUENCE PLOT.

EXAMINING A SEQUENCE PLOT.

2. Do you see any systematic tendency for the points to drift mainly upward or mainly downward? Alternatively, although the word is easily misunderstood in a more technical context, do you see a trend?

3. Do you see any systematic wave-like movements that seem to display some degree of regularity? (The word "cycle", another word easily misunderstood, may help you here to see what I am driving at.)

4. Do you see any abrupt shifts of level?

5. Does the vertical scatter of the points appear to be different at different segments of the sequence?

But when I first look at Figure 4-1, the thing that catches my attention is the apparent upper barrier at about 65--there are lots of points at or near this level and one far above it? Is that high reading a mistake?

You've made a good start with the visual analysis. You've applied what the statistician Berkson has called the *Interocular Traumatic Test*--that is, the data have hit you between the eyes. But let's defer discussion of this point for a while; we'll return to it in Chapter 5.

For although your remark was a good one, it was completely off the logical track we were following. Note that what struck you had nothing to do with the sequence as such. When we explore a data plot later in which sequence is lost completely, you will see the barrier at least as clearly.

Now go back to the unfolding technique. Tell me your reaction after the first ten observations?

It looks to me that there's an upward drift. If I were predicting the 11th observation, I'd say that it ought to be 60 or higher.

All right, now unfold 10 more observations and give me your reaction based on the first 20 observations.

My prediction of the 11th observation looks pretty good, but after that the updrift idea pretty much falls apart. But now I think that I can detect a pattern that looks cyclical.

How do you feel at the end of 30 observations?

I'll cling to my conjecture about a cycle, but I'm getting pretty uneasy.

And after 40?

I give up. I guess my imagination isn't vivid enough. Even after I look at the five items on your checklist above, I am pretty well beaten.

Then you concede that one might find a situation in which one would conclude that sequence doesn't matter insofar as prediction is concerned?

I guess that I'll have to. But looking ahead, I do notice that the smallest reading--40--occurs immediately before the largest reading--85.

A good observation. If the example were a current one so that we could conduct a further investigation, that would be a good lead for a little statistical detective work. Interocularly speaking, however, this seems to be the main evidence of anything systematic about the sequence.

BERKSON'S INTEROCULAR TRAUMATIC TEST.

48 □ CONVERSATIONAL STATISTICS

RANDOMNESS.

You're saying, then, that there appears to be little or no pattern in the sequence of the observations?

That's the tentative conclusion at this stage.

If we stick with this, we will be saying that the data behave as if they had arisen *at random*. The essence of the judgment of randomness of the process that gave rise to the data is this: we are unwilling to make any attempt to exploit the sequence of the observations when we make predictions of the future.

But how can you make a judgment of randomness when there is such a striking barrier at 65?

Grip yourself calmly and remember that we have seen that this has nothing to do with sequence. I speak forcefully because I find that some people stay fuzzy on this point long after we have gone on to more advanced material!

I thought that randomness referred to what statisticians call the normal distribution. From what I know about the normal distribution, the hardness data don't resemble it at all.

We'll study the normal distribution, and you're right: the hardness data don't conform to it at all.

But there is no simple connection between the terms *normal distribution* and *random*. Don't get hung up by the false association you have made!

What you've been saying about examination of the sequence is interesting, but I haven't had experience in looking at data. Can you suggest any safeguards or checks on my traumatic impressions?

Although I like to stress visual examination of data, I certainly feel that checks are desirable. There are three main kinds of checks:

1. How well does the visual interpretation agree with what we would have thought on the basis of other information?

2. How well does the interpretation agree with certain conformity measures that can be computed numerically?

3. How well is the interpretation borne out when we apply to new data a prediction scheme based upon the interpretation?

From what you have said, I suppose that the third is the most important.

It is, but it also is unavailable at the time we are analyzing the data. Predictive checks, when possible, are the ultimate checks on the judgments that go into statistical analyses; they point the way both to statistical wisdom and to humility.

Is there anything concrete to be said about the first check, consistency with other information?

Let's illustrate by the current example. The judgment was made that the data behave as if they had arisen at random. A little while ago your intuition was telling you that this was an implausible judgment under any circumstances because you felt that the present was more likely to resemble the recent past than the more distant past.

And the judgment of randomness contradicts my feeling, does it not?

Yes. If you make that judgment about the data, you are saying that the early observations are to be considered on an equal footing with the more recent ones.

And when you put it that way, I feel uneasy about our assessment that the data behave as if they had arisen at random.

I find your reaction understandable. Also, I find it interesting to hear your intuitive resistance to the idea of randomness, because ordinarily people studying statistics swallow the notion without serious reflection, and they often get the erroneous notion that there's almost a natural law that data should behave randomly.

But let me add this to your information. The judgment or assessment of randomness that we have made here is equivalent to an assessment of what is called by quality control experts a "state of statistical control". They describe a process in statistical control as a process with no "assignable causes" of variation.

And do you ordinarily find that industrial processes are in a state of statistical control?

My impression is that it takes a pretty taut management to achieve this result. But if the process is carefully watched, mainly by use of sequence plots called control charts, and if grossly non-random behavior is exploited for the discovery of assignable causes of variation, the assignable causes can gradually be removed. Experience seems to suggest that a state of statistical control can be frequently achieved in this way.

In the course of moving toward a state of statistical control, is it possible to improve the typical performance of the process as well? I mean, can the process yield be enhanced or its quality level improved?

Often this is possible.

When a process reaches a state of statistical control, does it stay that way?

There's no automatic guarantee that it will. Taut ships have a way of working loose. Quality control experts advocate continuing surveillance of the process by means of control charts. In this way, there is a kind of fire alarm system to alert management to possible problems.

RANDOMNESS AND STATISTICAL CONTROL.

50 ☐ CONVERSATIONAL STATISTICS

If a process is in a state of statistical control, is it possible to make further improvements in typical yield or quality?

It may be, but the direction of search for such improvements will no longer be indicated by the control chart alone.

Was the Rockwell hardness process under study by quality control experts?

Yes, that's where the data came from.

In summary, then, the assessment of randomness does not conflict seriously with information other than that of the data?

A reasonable summary.

All right, then, would you tell me about the "conformity measures" that you referred to under your second type of check mentioned above?

One such measure is called a *runs count*. I'll tell you how to do it first, then work in the rationale as we discuss the interpretation.

1. Go back to Figure 4-1. Draw a light horizontal line somewhere near the "middle" of the dots. For example, let the line have height between 59 and 60.

2. Proceed from left to right, and record a + for each observation above the line and a - for each observation below it.

3. Separate consecutive strings of +'s and -'s by vertical lines. Thus:

----|++++|-| ++|----|+|-|+|-| +++|-|+|-|++|-|+ +|--|+|-|++|--|+ +|-|+|-|++++|-|+|

--|++|-|++|-|++ +|----|+|-|+|-|+| ----|++++++ ++++|-|++|-|++| -|++|-|+|--|+|-|+|

4. Each vertical line marks the end of a *run*--that is, of a consecutive string either of +'s or -'s.

5. Count the runs. There are 54 in this example.

How do I know if 54 is good or bad?

First, a point of terminology. By making a tentative assessment that the data behave as if they had arisen at random, we have commenced the process of specification of a *statistical model*. I won't try to give a formal definition of the term statistical model; it will become meaningful by example as we proceed. Suffice it to say that any reasoned statistical analysis of a set of data must be based on a specification of a model.

Do you mean that our analysis would take one direction if we specified randomness, and another direction if we did not?

Precisely.

RUNS COUNT.

Now to your question of whether 54 is good or bad. If we specify randomness as a part of our model, it is possible to compare the observed number of runs, 54, with the number of runs "expected", given that the specification of randomness is correct and that there are 56 +'s and 44 -'s in the 100 observations. There's a simple formula for that. Let n be the sample size and r the number of +'s (*not* the number of runs). Then the expected number of runs is given by

$$\frac{2r(n-r)}{n} + 1 = \frac{2(56)(44)}{100} + 1$$

$$= 50.28 .$$

I'll take your word for the arithmetic, but what do you mean when you say that the resulting number, 50.28, is the expected number, given the specification of randomness?

Start with 56 +'s and 44 -'s. Imagine all possible sequences in which these 56 +'s and 44 -'s might have occurred. Under the specification of randomness, each of these possible sequences is equally probable, as we learned in Chapter 2.

Do you mean that a sequence of 56 +'s followed by 44 -'s, which would give only two runs, has the same probability of occurrence as the sequence we actually observed, which has 54 runs?

Given the correctness of the random specification, yes.

But many more sequences yield 54 runs than yield two runs. In fact, you can see immediately that there are just two sequences that yield two runs.

56 +'s and 44 -'s, or the other way around?

Yes.

A number of runs close to the expected number of 50.28, such as 50, has a greater probability than, say, 55 or 45, and these latter have greater probability than 40 or 60, and so on.

All right, I see that. But you still haven't told me how to interpret the expected number, 50.28.

Each possible sequence has a definite number of runs. If you average the number of runs over all possible sequences, the result is the expected number of runs.

How do you know that?

It can be proven to be true by application of probability theory.

I'm relieved that you didn't have to do it by direct enumeration! But now I want to know how to compare the observed runs count of 54 with the expected number of 50.28.

First, note that the actual number of runs is larger than the expected number. This means that the average length of run in the sample is a little shorter than the average length of run in a sample for which the number of runs was closer to the expected, say a sample in which the number of runs was 50.

EXPECTED NUMBER OF RUNS GIVEN THE SPECIFICATION OF RANDOMNESS.

OBSERVED VERSUS EXPECTED NUMBER OF RUNS.

If there were 50 runs in a sample of 100 observations, there would be, on the average, 100/50 = 2 observations per run. Is that what you're talking about?

That's the idea. And there were 100/54 = 1.85 observations per run in our actual sample. Thus we can say the following about our sample:

1. There is somewhat more alternation of observations above and below the horizontal line between 59 and 60 than expected under the specification of randomness.

2. The persistence of observations on a given side of the line is less than expected.

3. There is less clustering of similar outcomes than expected.

But how do I judge the deviation between the observed and expected number of runs? Does 54 - 50.28 = 3.72 support or oppose our specification of randomness?

Patience. We must take one step at a time. The next is a formula, derived from probability theory, for a measure of the expected dispersion of runs among the different samples that could be turned up given the specification of randomness. This measure of dispersion will help us to judge how seriously we should take any given departure of the actual number of runs from the expected number. The measure is called the standard deviation of the number of runs (we'll be talking about the meaning of "standard deviation" in a little bit), and it is computed as follows in this application:

$$\sqrt{\frac{2r(n-r)(2r(n-r)-n)}{n^2(n-1)}} = \sqrt{\frac{(2 \times 56 \times 44)(2 \times 56 \times 44 - 100)}{100^2 \times 99}}$$

$$= 4.90.$$

We will compare the actual deviation of 54 - 50.28 = 3.72 against the standard deviation of 4.90. In particular, we look at the ratio 3.72/4.90 = 0.76.

And this ratio 0.76 tells us what?

I'll start by giving you, with some apology, my personal rule-of-thumb. If the ratio is no more than 1 in absolute value (that is, not larger than +1 and not smaller than -1), the runs measure gives little evidence against the specification of randomness.

So in the example of the hardness measures, the runs measure confirms our interocular traumatic examination of the sequence plot?

Yes.

Suppose the ratio had been greater than one in absolute value?

The larger the absolute value of the ratio, the greater the reservation about the specification of randomness. When we get to 2 or 3, the reservation would be very strong.

EVALUATING RUNS COUNTS.

What do you mean by "strong reservation"?

I mean that I would be inclined to jettison the specification of randomness for the observations, and search for an alternative statistical model.

Do you mean that statistical treatment is possible when you judge that the data did not arise randomly?

Often a very satisfactory statistical treatment is possible. I'll give you an example later in this chapter.

Is this runs measure an example of a statistical test of significance?

It is, although I would prefer to call it a measure of conformity of data to a tentative model specification. It serves as a guide as to whether we should rest content with the present specification or search for a more satisfactory one.

And why do larger absolute values of the conformity measure incline you more strongly to look for alternative specifications?

First, given the tentative specification--in this example, the model of randomness--the most probable numbers of runs are those close to the expected number. For these, the conformity measure will be small in absolute value, because the conformity measure is proportional to the discrepancy between the observed and the expected number of runs.

Second, given most plausible alternative specifications, the most probable number of runs is higher (less clustering) or lower (more clustering) than the expected number given the specification of randomness. For these, the conformity measure will be larger in absolute value.

The larger the conformity measure, then, the more probable is the sample result given plausible alternative models by comparison to its probability given the specification of randomness.

Roughly, then, small values of the conformity measure are reasonably compatible with the specification of randomness, but large values are more compatible with other plausible specifications?

That's the idea. And you must say "absolute values", because a -3 would have the same importance, roughly, as a +3.

We haven't mentioned the interpretation of negative values. These would occur when the observed number of runs is less than expected?

Yes. Then the persistence of observations on a given side of the line would be more than expected given the random specification; there would be more clustering of similar outcomes, and less alternation.

I hope that IDA will do the computational work for us in analysis of runs.

The command RUNS is what you need:

CONFORMITY MEASURES.

```
> RUNS
* GIVE NAME OR COLUMN NUMBER OF VARIABLE :
HARDNS
      OBSERVED NUMBER OF RUNS= 54
      EXPECTED NUMBER OF RUNS= 50.28
   STANDARD DEVIATION OF RUNS= 4.90232
       (OBS.-EXP.)/(STD.DEV.)= .758823
```

How does IDA draw the line to define the +'s and the -'s?

At the arithmetic mean of all the observations.

What is the arithmetic mean?

Add up all the observations, then divide by the sample size n. Sometimes the mean is denoted by the symbol $\bar{x}$, whence the formula:

$$\bar{x} = \frac{\Sigma x}{n},$$

which says, "Add up the individual observations (the x's) and then divide by the sample size".

Will IDA compute the mean if I want it?

Use the command MEAN, as illustrated:

```
> MEAN
VARIABLE    MEAN        STD. DEV.
 HARDNS   5.95800E+01   5.82433E+00
```

I don't understand the E+01 that appears after 5.95800.

This is called floating point notation. E+01 means that you must move the decimal point one place to the right to get the result: thus 5.95800E+01 = 59.5800. The mean of the 100 hardness readings is 59.58.

And since E+00 appears in the result for the standard deviation, we don't have to move the decimal at all?

Yes. The answer is 5.82433.

And, say, an E-02 would tell us to move the decimal point two places to the left?

You have the hang of it.

Before we get away from the runs analysis, how did you decide in IDA to draw the line at the mean of the observations?

The method is valid wherever you draw the line, and it will be highly informative so long as you draw the line somewhere in the "middle" of the range of the observations. Since the exact location is not critical, IDA uses the mean, which is calculated automatically when data are entered.

ARITHMETIC MEAN.

Notice that you get also the standard deviation; this time the standard deviation is computed for the 100 Rockwell hardness readings. We will explain this soon.

ANALYSIS OF SEQUENCE □ 55

Are we going to be able to do all our computations on the computer?

Nearly all. However, you would be wise to acquire two supplemental capabilities:

1. You should be able to make quick mental approximations. For example, if r/n is not too far from 0.5, then n/2 + 1 is a good approximation to the expected number of runs. And if n is not too small, the standard deviation of the number of runs is approximately the expected number divided by the square root of the sample size. In our example, these approximations give 51 and 5, which are not too far from 50.28 and 4.90, the correct values.

2. You should have access to a good modern electronic desk calculator, and know how to use it. This will serve you for simple calculations for which it's not worth the trouble or expense to enter the data in a computer, or even to log in on a computer. The HP-35, HP-45, and HP-65 are electronic calculators compact enough to be carried in a shirt pocket.

Is IDA of any help for miscellaneous calculations of the sort one might do on a desk calculator?

The command CALC serves this purpose to some degree.

Can IDA be used to plot the data? For example, can it make a sequence plot like Figure 4-1?

Very nicely. Assuming that you have entered the data as I illustrated earlier, you need only issue the command PLTS. The result is shown as Figure 4-2.

If you turn Figure 4-2 on its side, you will see that it looks very much like Figure 4-1. The dashed horizontal line is to help guide the eye, particularly to see whether there is any drift in the level of observations.

What are the units of the observations? The scale is labeled -5, -4, -3, etc.

These are called standardized units. You get the sequence plot for any set of data without the need to scale explicitly for the units involved. Also, as you get used to them, you will be able to compare patterns seen in different sets of data, even when the units are entirely different.

How does the standardization work?

1. First, each observation is expressed as a deviation from its mean.

    In the Rockwell example the mean $\bar{x}$ = 59.58. The first observation was x = 58. Hence the first deviation is

    $$x - \bar{x} = 58 - 59.58$$

    $$= -1.58.$$

COMPUTATIONS WITHOUT A COMPUTER.

56 □ CONVERSATIONAL STATISTICS

```
> PLTS
* GIVE VARIABLE NAME OR COLUMN NUMBER FOR THE
  VARIABLE TO BE PLOTTED : HARDNS
WANT TO PLOT ALL ROWS OF ACTIVE DATA MATRIX ?YES

SEQUENCE PLOT OF STANDARDIZED VALUES
```

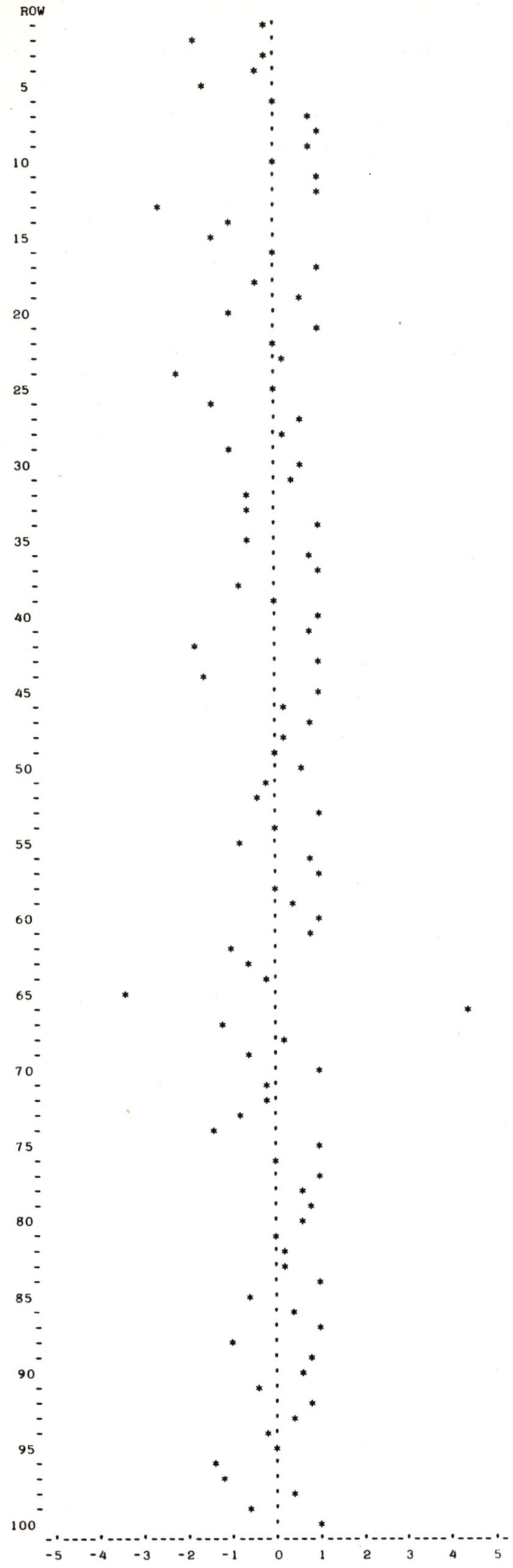

[sequence plot of 100 standardized values ranging approximately from -5 to 5]

```
MEAN =   59.58
STD. DEV. =  5.82433
SAMPLE SIZE =   100
```

Figure 4-2. Example of IDA Command PLTS

2. The second step involves a measure of dispersion or spread of the 100 measurements known as the standard deviation, which will be abbreviated s. We will say more about the standard deviation later; now we will just state its formula and show how it is used.

$$s = \sqrt{\frac{\Sigma (x - \bar{x})^2}{n - 1}}$$

$$= 5.82433 \, .$$

In words, sum the squared deviations from the mean; divide this sum by n-1; and finally take the square root.

3. Divide the deviation (Step 1) by the dispersion measure (Step 2) to obtain the standardized observation:

$$\frac{-1.58}{5.82433} = -0.2713 \, .$$

In words, the standardized observation is the distance of the actual observation from the mean, measured in units of the dispersion measure, the standard deviation. Thus, the first reading of 58 is 0.27 below the mean in units of the standard deviation.

4. The computer output medium is grainy--it moves by steps of 1/10 inch horizontally and 1/6 inch vertically, as the paper is viewed by the user. To allow for graininess, we round off the observation in standardized units to the nearest even tenth. Thus -0.2713 becomes -0.2. This is the first interval to the left of zero on the standardized scale, as you can verify from the actual plot in Figure 4-2.

5. If you want to translate from the standardized scale back to the original units, proceed as in the following example. Take the sixth interval to the right of 0 on the standardized scale; in standardized units, this is +1.2. Now calculate

59.58 + 1.2×5.82433 = 66.6 .

An actual observation of 66.6, and any close enough to round to 1.2 in standardized units, would be plotted at 1.2.

If I follow you correctly, then, the zero point on the scale at the bottom of Figure 4-2 corresponds to the mean of the observations plotted?

And the mean is printed out below the scale so that you can see what it is.

Similarly, each unit step on the scale is equal to one standard deviation of the observations. So +1 means the mean plus one standard deviation, or $\bar{x} + s$.

And the standard deviation s is also printed beneath the scale. Always get your bearings by noting both $\bar{x}$ and s.

STANDARD DEVIATION.

STANDARDIZED SCALING OF DATA.

All right, now let's get back to our original objective. It's been a while, but we started out with the idea of finding out how the data from 100 coils could help us in predicting, statistically of course, the hardness of future coils. Can we do something now?

The specification of randomness, which we've pretty well vindicated, is only part of the specification of the statistical model. So we'll be able to meet your questions more fully later on, after our study has proceeded further. To give you an idea, however, we can now say that the mean 59.58 of the first 100 observations would be a sensible estimate of the next observation: we say that 59.58 would be a *point estimate*.

And because of our assessment of random behavior, we would not pay any special attention to the most recent observations?

That's right. The 65 that occurred on the 100th observation is simply one of 100 numbers that determine our point estimate.

Incidentally, since the Rockwell hardness readings seem to be given as whole numbers, wouldn't you want to round off the 59.58 to 60?

Under most circumstances that would be sensible.

One question has been nagging at me throughout this discussion. What happens if we decide that the specification of randomness is not tenable?

We try to find a specification that is tenable. There's no simple way to formulate a strategy for doing that under all circumstances, but that's the idea.

Could you give an example at this stage to illustrate a successful search for an alternative specification?

We'll take an example that elicits a strategy that is often useful. We shall study the monthly closings of the Dow-Jones Industrial Index from August, 1968, through July, 1972, a 48-month period.

In the process, we shall introduce some more features of IDA. The data are in a file called $DJ. After entering the data as before, using ENTER, I decided to print them out with control over the format of the printed numbers. This is done by FPRS, which is simply a formatting variation of PRTS. The result is shown in Table 4-2. For those unfamiliar with formatting conventions, I've appended a brief discussion.

Figure 4-3 shows the sequence plot.

I don't even have to unfold Figure 4-3. The interocular approach tells me immediately that the behavior is decidedly non-random. It looks cyclical, in fact.

Of course, you're right in saying that the behavior is obviously non-random. But be careful not to be carried away by the word "cyclical". We'll see very soon that "pseudo-cyclical" is more accurate.

DOW-JONES INDEX EXAMPLE.

What does the runs count show?

It supports you:

```
> RUNS
* GIVE NAME OR COLUMN NUMBER OF VARIABLE :
DJ

    OBSERVED NUMBER OF RUNS= 7
    EXPECTED NUMBER OF RUNS= 24.625
STANDARD DEVIATION OF RUNS= 3.37234
    (OBS.-EXP.)/(STD.DEV.)=-5.22634


> FPRS
DO VARIABLES TO BE PRINTED OCCUPY
CONSECUTIVE COLUMNS OF DATA MATRIX ?YES
PRINT DATA SUBMATRIX BETWEEN ROWS I1 TO I2 AND COLUMNS
J1 TO J2, INCLUSIVE.  GIVE FOUR VALUES, SEPARATED BY
COMMAS, FOR I1, I2, J1, J2 : ?1,48,1,1
GIVE FORMAT FOR PRINTING AN OBSERVATION
SUCH AS      '#, DD.4D,2X'
FORMAT MUST BE PRECEDED BY '#,' :
?#,6D.2D

    **  1 **    896.01
    **  2 **    935.79
    **  3 **    952.39
    **  4 **    985.08
    **  5 **    943.75
    **  6 **    946.05
    **  7 **    905.21
    **  8 **    935.48
    **  9 **    950.18
    ** 10 **    937.56
    ** 11 **    873.25
    ** 12 **    815.47
    ** 13 **    836.72
    ** 14 **    813.09
    ** 15 **    855.99
    ** 16 **    812.30
    ** 17 **    800.36
    ** 18 **    744.06
    ** 19 **    779.59
    ** 20 **    785.57
    ** 21 **    736.07
    ** 22 **    700.44
    ** 23 **    683.50
    ** 24 **    734.12
    ** 25 **    764.58
    ** 26 **    760.68
    ** 27 **    755.61
    ** 28 **    794.09
    ** 29 **    838.90
    ** 30 **    868.50
    ** 31 **    878.80
    ** 32 **    904.37
    ** 33 **    941.75
    ** 34 **    907.80
    ** 35 **    891.14
    ** 36 **    858.40
    ** 37 **    898.07
    ** 38 **    887.19
    ** 39 **    839.00
    ** 40 **    831.75
    ** 41 **    890.20
    ** 42 **    902.17
    ** 43 **    928.13
    ** 44 **    940.70
    ** 45 **    954.17
    ** 46 **    960.72
    ** 47 **    929.03
    ** 48 **    924.74
```

Table 4-2. Dow-Jones Monthly Closing Data

The suggestion DD.4D,2X would print out the first two digits to the left of the decimal point, four digits to the right, and then two spaces. My choice, 6D.2D, prints up to six digits to the left of the decimal, and two after it; the latter corresponds to the reporting of the data. Since the DJ index was never over 1000 during this period, only three digits appear to the left of the decimal. Be sure to provide plenty of room to the left of the decimal to provide for the maximum number of digits and also for a "-" in case some of your numbers are negative.

See the HP-2000F Manual for more information on the specification of formats.

On the other hand, if you find this puzzling, don't brood over it. The command PRTS will do nearly as good a job for you, anyway.

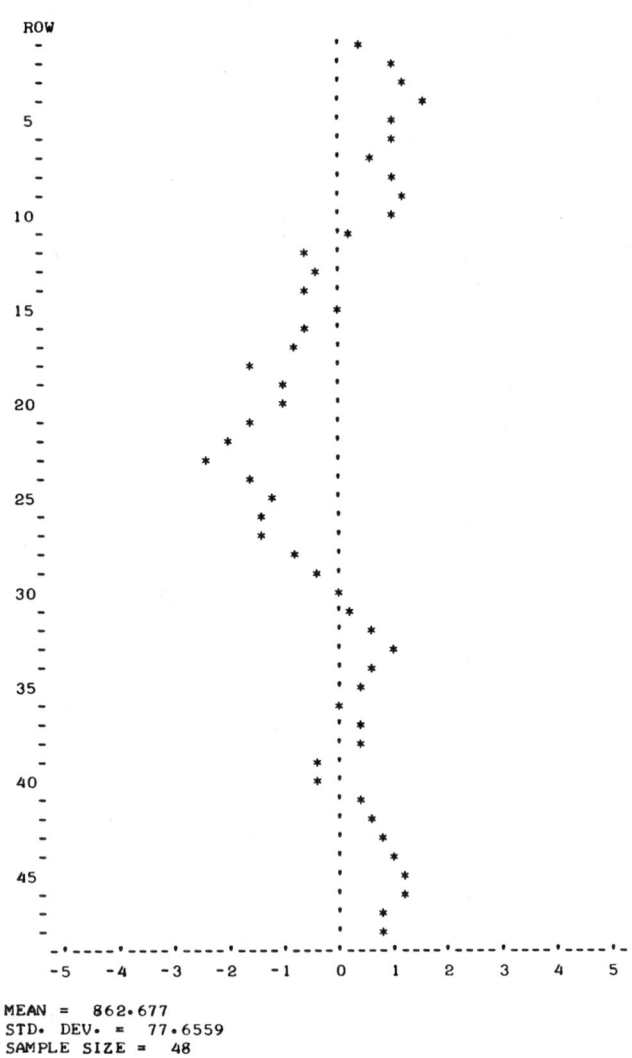

Figure 4-3. Sequence Plot of Standardized Dow-Jones Values

Where do we go from here?

Figure 4-3 shows *levels* of the Dow-Jones Industrial Index. I suggest that we look at month-to-month *changes* of the index instead of the monthly levels. We'll call these DIFDJ.

Do you mean that a change is equal to this month's level minus last month's level?

That's right. The change is the difference between two successive levels.

But what does that do for us? If the levels are behaving non-randomly, the changes would be non-random.

Wrong: it is possible, though of course not necessary, that the changes of non-randomly behaving levels would themselves behave randomly.

TRANSFORMATION FROM LEVELS TO CHANGES, DJ TO DIFDJ.

That sounds crazy to me.

You will get used to the idea. Rather than convince you by abstract argument, however, I'll ask you to follow me in a further development of the example. And please keep your mind open, even if you are skeptical!

All right. But how will you compute the changes?

By use of the command DIFF, as shown below. As you look at this, refer to the discussion in the Computer Notes of the working data matrix in IDA.

```
> DIFF
  *WHICH COLUMN WILL THE TRANSFORMED
  VARIABLE BE PLACED?  GIVE NUMBER : 2
  * COLUMN TO BE TRANSFORMED :1
  GIVE A POSITIVE INTEGER FOR GAP :
  ?1
  UPDATING MEAN, STD, ...
  GIVE NAME OF NEW VARIABLE
  VAR.  2 ?DIFDJ
```

Since only one variable, DJ, was originally entered, it was placed in column 1 of the data matrix; column 2 is free. This means that we are taking the change or difference with respect to the level of one period back. If we had wanted change with respect to a year ago, we would have specified a gap of 12 (since there are 12 months in a year).

What does the data matrix look like now?

I used the command FPRS to print it out. The results are shown in Table 4-3.

What happened to the first row?

The change from month 1 to month 2 is 935.79 -896.01 = +39.78, and this is displayed on row 2 of the data matrix. Since month 0 is not given, no change from month 0 to month 1 is defined. Under these circumstances, IDA simply masks out row 1 for all variables; then all variables in the matrix are defined for exactly the same time period.

Table 4-3. Revised Data Matrix for Dow-Jones Data

```
> FPRS
DO VARIABLES TO BE PRINTED OCCUPY
CONSECUTIVE COLUMNS OF DATA MATRIX ?YES
PRINT DATA SUBMATRIX BETWEEN ROWS I1 TO I2 AND COLUMNS
J1 TO J2, INCLUSIVE.  GIVE FOUR VALUES, SEPARATED BY
COMMAS, FOR I1, I2, J1, J2 : ?1,48,1,2
GIVE FORMAT FOR PRINTING AN OBSERVATION
SUCH AS      '#, DD.4D,2X'
FORMAT MUST BE PRECEDED BY '#,' :
?#,6D.2D
    **  2 **    935.79      39.78
    **  3 **    952.39      16.60
    **  4 **    985.08      32.69
    **  5 **    943.75     -41.33
    **  6 **    946.05       2.30
    **  7 **    905.21     -40.84
    **  8 **    935.48      30.27
    **  9 **    950.18      14.70
    ** 10 **    937.56     -12.62
    ** 11 **    873.25     -64.31
    ** 12 **    815.47     -57.78
    ** 13 **    836.72      21.25
    ** 14 **    813.09     -23.63
    ** 15 **    855.99      42.90
    ** 16 **    812.30     -43.69
    ** 17 **    800.36     -11.94
    ** 18 **    744.06     -56.30
    ** 19 **    779.59      35.53
    ** 20 **    785.57       5.98
    ** 21 **    736.07     -49.50
    ** 22 **    700.44     -35.63
    ** 23 **    683.50     -16.94
    ** 24 **    734.12      50.62
    ** 25 **    764.58      30.46
    ** 26 **    760.68      -3.90
    ** 27 **    755.61       5.07
    ** 28 **    794.09      38.48
    ** 29 **    838.90      44.81
    ** 30 **    868.50      29.60
    ** 31 **    878.80      10.30
    ** 32 **    904.37      25.57
    ** 33 **    941.75      37.38
    ** 34 **    907.80     -33.95
    ** 35 **    891.14     -16.66
    ** 36 **    858.40     -32.74
    ** 37 **    898.07      39.67
    ** 38 **    887.19     -10.88
    ** 39 **    839.00     -48.19
    ** 40 **    831.75      -7.25
    ** 41 **    890.20      58.45
    ** 42 **    902.17      11.97
    ** 43 **    928.13      25.96
    ** 44 **    940.70      12.57
    ** 45 **    954.17      13.47
    ** 46 **    960.72       6.55
    ** 47 **    929.03     -31.69
    ** 48 **    924.74      -4.29
```

Check me to be sure I'm following: the change 16.60 in row 3, column 2, is obtained by subtracting 935.79 from 952.39?

That's it. If you're clear, then, on how the changes DIFDJ are defined, let's look at the sequence plot in Figure 4-4.

I'll have to admit that my interocular examination fails to turn up anything that suggests gross departures from randomness.

Nor does mine.

What does the RUNS command show?

Slightly fewer observed than expected runs; persistence of positive or negative changes is slightly more than the random expectation.

```
> RUNS
* GIVE NAME OR COLUMN NUMBER OF VARIABLE :
DIFDJ

    OBSERVED NUMBER OF RUNS= 20
    EXPECTED NUMBER OF RUNS= 24.4043
 STANDARD DEVIATION OF RUNS= 3.37624
       (OBS.-EXP.)/(STD.DEV.)=-1.30448
```

I notice that the ratio of the deviation between observed and expected and the standard deviation is -1.3. That's barely into your zone for suspicion according to the rule-of-thumb you gave. How do you react to it?

It gives a hint that we might pursue the search for a model somewhat further. In combination with other information that I have, it tells me that the search would be rewarded. This other information is the *Fisher Effect*, which explains why one can observe a little positive persistence in the index even if the *individual stock* changes are behaving randomly.

For our present purposes, however, we can say that the specification of random behavior of DIFDJ is pretty satisfactory. Let's discuss it on that basis.

DIFDJ CONFORMS REASONABLY TO SPECIFICATION OF RANDOMNESS.

I'll have to concede that the changes behave pretty much as we would expect from the specification of randomness, but I feel that you must have tricked me in some subtle way. Haven't we lost something by looking only at changes?

Nothing at all. Look at it this way: if you can forecast next month's change, you can forecast next month's level, and you know this month's level for sure when you are forecasting next month's change.

But can't we somehow exploit the systematic movements we see in Figure 4-3?

Not unless they can tell us something useful in our attempt to predict changes. Figure 4-4 makes that appear doubtful.

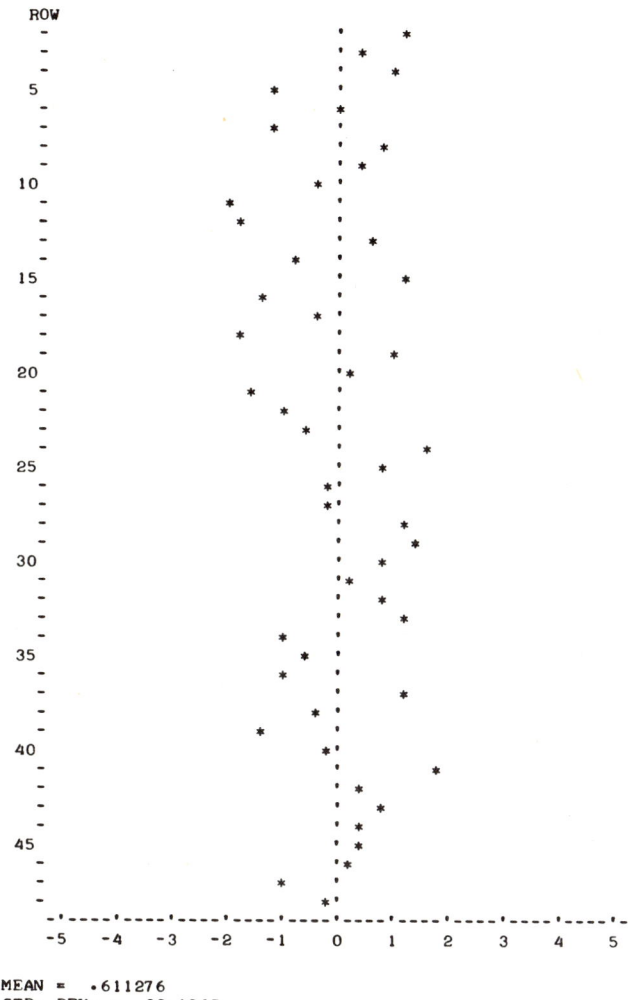

Figure 4-4. Sequence Plot of Standardized Values of 'DIFDJ'

Does the use of DIFF deal with many of the non-random time series we meet in practice?

It often works for securities prices and prices on other organized markets, and sometimes for other kinds of data. When you see wavy plots like Figure 4-3, it's usually worth trying DIFF.

Sometimes, as we shall see, DIFF is only a first step in a series of steps designed to place the data in a form that the specification of randomness will hold.

There are also occasions in which differencing is not appropriate at all. We shall see examples.

Do all sets of data have a natural time sequence?

In some applications, the data arise essentially simultaneously. For example, current monthly sales for each Chicken Unlimited franchise in the Chicago area. Such data are called *cross sectional*. The procedures of this chapter are not ordinarily applicable to cross sectional data.

However, it is often useful to study the *geographical* sequence for cross sectional data, although this is somewhat complicated by the fact that we must then work with two dimensions rather than one. We can, however, plot the data by hand on a map. Or we may single out a geographical dimension of particular interest, such as the distance of each franchise from the downtown shopping district of Chicago. If the data are listed in this sequence, then PLTS and RUNS can be used to study the relationship of the data to distance just as we studied time relationships in this lecture.

CROSS SECTIONAL DATA.

READING: TNOS, Chapters 5 and 6. This reading gives a slightly different perspective, and includes some points not here considered, but you can read it with advantage at this time.

Excuse me. I've been brooding, as you might put it, about the stock market example. It rather offends my intuition to think that changes occur randomly. Is there really evidence that this is true?

With some qualifications, yes, there is a lot of evidence.

What do you mean by qualifications?

Such things as these: there must be some allowance for dividends, often we work with percentage changes rather than dollar changes, and we usually are thinking of individual stocks rather than stock indexes.

And is this evidence the basis for what I've heard called the random walk hypothesis?

Yes. The summation or cumulation of random observations is called technically a random walk.

Do most people interested in the market who have heard of the random walk hypothesis believe that it is well-supported?

No, and for the same reasons that you are resisting the idea.

Do you believe it?

I have seen a lot of evidence consistent with the random walk hypothesis and very little that is inconsistent. I retain a little residual skepticism, but I have great respect for the hypothesis.

THE RANDOM WALK MODEL.

ANALYSIS OF SEQUENCE 65

Is it ever possible to predict a random walk statistically, in the sense of exploiting the sequence of changes for prediction?

Not from the past history of the variable that you are trying to predict. It may be possible to find some other variable, called a *leading indicator*, that could be exploited by appropriate statistical analysis. For example, it is conceivable that past changes in the money supply have a statistical connection with current changes in the stock market.

Now please don't think that I'm being skeptical of what you're telling me, but I would be pleased if you could prescribe something for my ailing intuition. Then I might be more able to believe you.

The best medicine is known as simulation. That is, try things out in the statistical equivalent of a laboratory and see if things work out as the theory suggests.

IDA has a simulation capability as a part of the command RAND. I'll show you a simulation, with appropriate annotations, of the stock market situation. Don't get into it unless you are clear on all the fundamentals brought up during this chapter. In fact, I should warn you that even if you are clear, there's a danger in reading ahead:

WARNING--simulations can be habit-forming!

SIMULATION OF DOW-JONES INDEX.

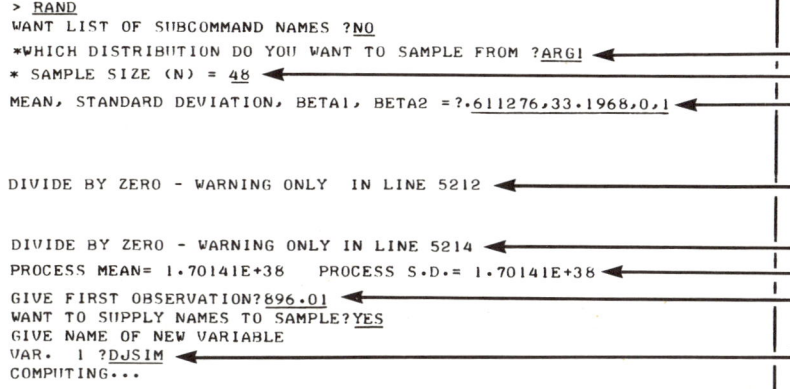

```
> RAND
WANT LIST OF SUBCOMMAND NAMES ?NO
*WHICH DISTRIBUTION DO YOU WANT TO SAMPLE FROM ?ARG1  ←     Subcommand we need.
* SAMPLE SIZE (N) = 48  ←                                   48 months, as in actual example.
MEAN, STANDARD DEVIATION, BETA1, BETA2 =?.611276,33.1968,0,1  ←  The first two numbers are the x̄ and s of Figure 4-4; 0
                                                                 and 1 are needed to make ARG1 do what we want
                                                                 now.
DIVIDE BY ZERO - WARNING ONLY  IN LINE 5212  ←              Ignore this message--it's irrelevant to our simulation.

DIVIDE BY ZERO - WARNING ONLY IN LINE 5214  ←               Ignore, for same reason.
PROCESS MEAN= 1.70141E+38    PROCESS S.D.= 1.70141E+38  ←   Ignore.
GIVE FIRST OBSERVATION?896.01  ←                            Actual DJ, August, 1968.
WANT TO SUPPLY NAMES TO SAMPLE?YES
GIVE NAME OF NEW VARIABLE
VAR.  1 ?DJSIM  ←                                           Will be automatically saved in IDA data matrix.
COMPUTING...
```

All right, I'll follow along. But please tell me what you just did.

I requested a simulation of random changes from a population with mean and standard deviation equal to the sample mean and standard deviation of the actual Dow-Jones changes. Then the computer converted these changes to levels by successive additions, starting with the actual starting level of 896.01. The resulting variable was called DJSIM, the "SIM" suggesting "simulated".

In other words, we have some artificial data for which we know the differences are drawn randomly, and we can do the same kind of thing we did with the actual data.

Yes. Figure 4-5 gives PLTS, together with the results of FPRS.

I'll have to admit that the sequence plot does remind me of the stock market, pseudo-cycles and all.

Though the simulation does not, of course, mimic Figure 4-3 in fine detail, the statistical resemblance is unmistakable.

```
> PLTS
* GIVE VARIABLE NAME OR COLUMN NUMBER FOR THE
  VARIABLE TO BE PLOTTED : DJSIM
WANT TO PLOT ALL ROWS OF ACTIVE DATA MATRIX ?YES

SEQUENCE PLOT OF STANDARDIZED VALUES

 ROW
    -                   *        '              **   1 **    896.01
    -                    *       '              **   2 **    914.50
    -                      *     '              **   3 **    969.90
    -                       *'                  **   4 **    981.76
    5                      *'                   **   5 **    961.26
    -                    *  '                   **   6 **    920.41
    -                  *     '                  **   7 **    894.33
    -                       *'                  **   8 **    967.86
    -                       *'                  **   9 **    963.66
   10                       ' *                 **  10 **   1009.15
    -                       *                   **  11 **    961.69
    -                       '  *                **  12 **   1024.67
    -                       '  *                **  13 **   1016.46
    -                       '   *               **  14 **   1057.88
   15                       '   *               **  15 **   1048.43
    -                       '    *              **  16 **   1069.44
    -                       '   *               **  17 **   1045.56
    -                       '    *              **  18 **   1061.98
    -                       '   *               **  19 **   1051.34
   20                       ' *                 **  20 **   1004.70
    -                       *                   **  21 **    990.41
    -                     *  '                  **  22 **    941.75
    -                    *'                     **  23 **    936.66
    -                       *                   **  24 **    990.63
   25                       '*                  **  25 **    999.50
    -                       ' *                 **  26 **   1018.46
    -                       '   *               **  27 **   1052.15
    -                       ' *                 **  28 **   1013.76
    -                       ' *                 **  29 **   1025.58
   30                       ' *                 **  30 **    967.74
    -                      *                    **  31 **    948.81
    -                     * '                   **  32 **    940.23
    -                     * '                   **  33 **    932.93
    -                    *  '                   **  34 **    911.70
   35                  *     '                  **  35 **    891.69
    -                    *   '                  **  36 **    953.72
    -                       '*                  **  37 **   1022.67
    -                       '  *                **  38 **   1034.85
    -                       '  *                **  39 **   1039.19
   40                       '  *                **  40 **   1038.76
    -                       ' *                 **  41 **   1016.79
    -                       '*                  **  42 **   1006.83
    -                       *'                  **  43 **    987.21
    -                       ' *                 **  44 **   1016.13
   45                       '  *                **  45 **   1040.02
    -                       '   *               **  46 **   1081.11
    -                       ' *                 **  47 **   1018.11
    -                       ' *                 **  48 **   1009.58
         -'----'----'----'----'----'----'----'----'----'-
         -5   -4   -3   -2   -1    0    1    2    3    4    5

MEAN        =  993.708
STD. DEV.   =   49.5761
SAMPLE SIZE =   48
```

Figure 4-5. Sequence Plot of Standardized Values of 'DJSIM'

And we can check up on the computer by using DIFF to get back the random changes that IDA was supposed to have generated in the first place?

See below and Figure 4-6.

```
> DIFF

* WHICH COLUMN WILL THE TRANSFORMED
VARIABLE BE PLACED?  GIVE NUMBER : 2
* COLUMN TO BE TRANSFORMED : 1
GIVE A POSITIVE INTEGER FOR GAP :
?1
UPDATING MEAN, STD, ...
GIVE NAME OF NEW VARIABLE
VAR.  2 ?DDJSIM

> RUNS
* GIVE NAME OR COLUMN NUMBER OF VARIABLE :
DDJSIM

EXPECTED NO. OF RUNS      =       24.4043
STANDARD DEVIATION (R)    =        3.37624
OBSERVED NO. OF RUNS      =       22
STANDARD NORMAL APPROX.   =        -.712111
```

Well, I don't know whether I'm converted or not, but I'll have to admit that the simulation provides food for thought.

Simulations have a way of doing that.

```
> PLTS
* GIVE VARIABLE NAME OR COLUMN NUMBER FOR THE
  VARIABLE TO BE PLOTTED : DDJSIM
WANT TO PLOT ALL ROWS OF ACTIVE DATA MATRIX ?YES
```

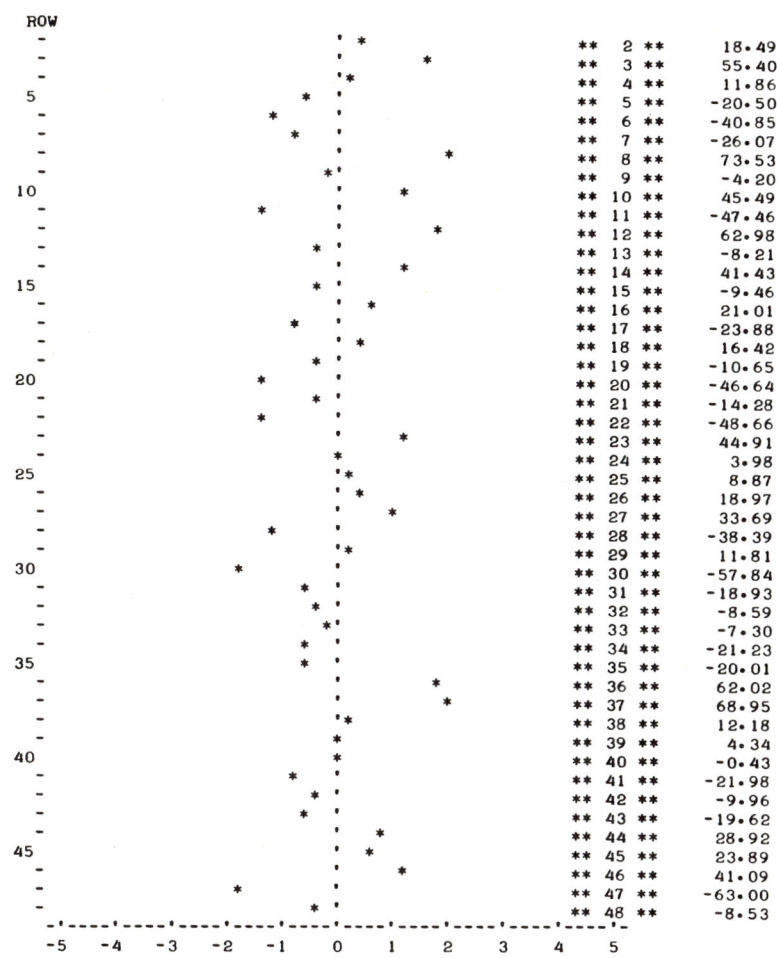

```
MEAN    =   2.41635
STD. DEV. = 34.4351
SAMPLE SIZE =   47
```

Figure 4-6. Sequence Plot of Standardized Values of 'DDJSIM'

CHAPTER FIVE: DISTRIBUTIONS

If we decide that nothing serious will be lost by ignoring the time sequence of the observations, what is the next step in a statistical analysis?

It is often helpful to collapse the sequence plot into a *histogram*, which shows the general pattern or distribution of the data.

I'm confused. I thought that you said that randomly arising observations were patternless.

Patternless *in sequence only*. The Rockwell hardness data, recall, displayed a pattern with respect to an apparent barrier at 65, even though there was little evidence of pattern in the sequence.

Can you give an example of what you mean by "collapsing a sequence plot"?

I'll use the example of changes or differences of the Dow-Jones Industrial Index that we examined in the last part of Chapter 4. I'll actually be using the command PLTS again, as I did for Figure 4-4, but we'll get something additional this time.

Before showing you the example, however, I'll pause to show you a feature of IDA that you will want to know about. Once you become familiar with an IDA command, such as PLTS, you will probably become impatient with the length of the prompts given to you by IDA. A simple command CHGP permits you to change the level of prompt; see the General Comments in the Computer Notes. For this example, I decided to change the prompt level to level 3, very brief prompts. I did this first, as shown, and then I repeated the ENTER and DIFF commands at prompt level 3, so that you can compare with these same commands in Chapter 4:

```
> CHGP
* LEVEL = 3

> ENTER
* MODE = FILE
NAME IS ?$DJ
* N,K GIVEN IN FILE ?YES
DATA MATRIX HAS  48 OBSERVATIONS ON   1 VARIABLES
COMPUTING ...
WANT TO SUPPLY NAMES ?YES
VAR.  1 = ?DJ

> DIFF
* COL# TO PLACE VAR. : 2
* COLUMN TO BE TRANSFORMED :1
GAP = ?1
UPDATING MEAN, STD, ...
GIVE NAME OF NEW VARIABLE
VAR.  2 ?DIFDJ
```

HISTOGRAMS

You had to give exactly the same information as before?

That's right. The prompts are shorter; that's the only difference.

And then PLTS?

Yes, as shown in Figure 5-1.

Figure 5-1 is the same as Figure 4-4 except that an additional data plot is given at the end of the sequence plot. This plot shows the histogram that results when the data are collapsed.

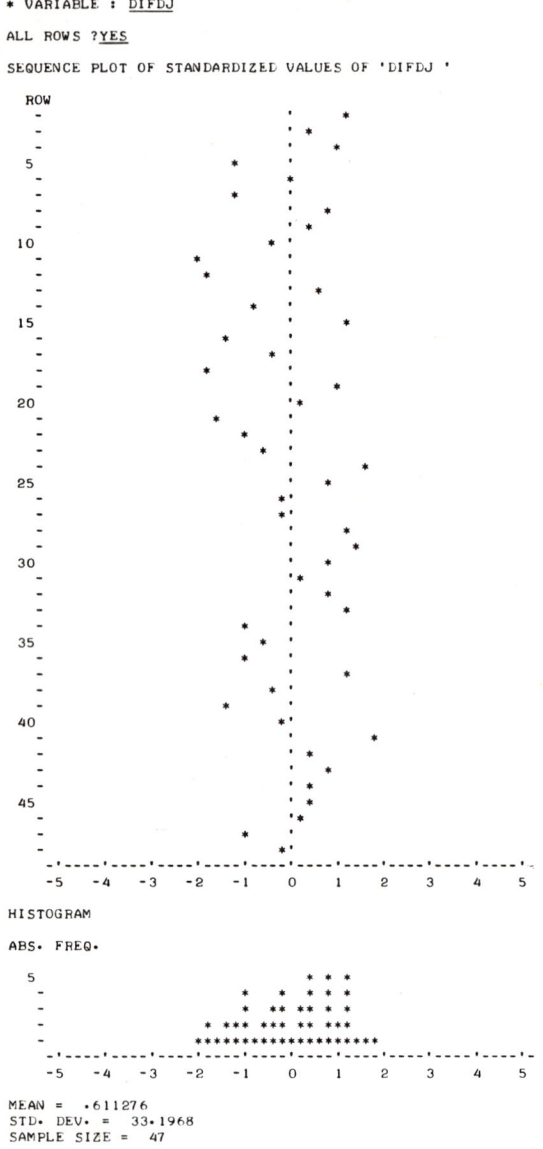

Figure 5-1. Sequence Plot Plus Histogram of Standardized Values of 'DIFDJ'

The histogram was actually produced before, but you didn't show it?

That's right. I wanted you to focus singlemindedly on the sequence plot before even mentioning the histogram.

Now I'm ready to hear an explanation of "collapsing"?

Simply imagine that the asterisks of the sequence plot are brought straight down and stacked up. For example, in the interval that rounds to -1, you can count four asterisks on the sequence plot; there are four asterisks stacked at -1 on the histogram. The absolute frequency is scaled on the vertical axis of the histogram.

What does the histogram tell me in this example?

It shows that all the changes are between -2 and +2, after rounding, in standardized units. There is a tendency for greater frequencies of observations toward the middle of this interval than toward the ends of it, but there is a good deal of raggedness in the picture.

Would you review the meaning of standardized units?

From each observation x, subtract the mean $\bar{x} = 0.611276$; then divide each deviation $x - \bar{x}$ by the standard deviation $s = 33.1968$. Thus, a standardized observation is expressed as a deviation from the mean in units of the standard deviation.

Then an observation of -1 in standardized units is one standard deviation below the mean?

That's right. And in the histogram, all observations from -0.9 to -1.1 are counted as -1. The interval (-1.1, -0.9) is said to be a *class interval*. The width of the class interval is 0.2 standard deviations.

Why did you set the class interval to be 0.2 standard deviations?

For a wide range of applications, this choice gives about the right amount of detail in the picture.

And this histogram is provided automatically with PLTS by IDA?

Yes.

Suppose that I want to use the computer to make a histogram with a set of classes that I specify, rather than using the all-purpose choice that comes with PLTS.

Use a command called HIST. Since this is a new command for you, I'll change back to prompt level 1 so that you can see the detailed prompts. (see Figure 5-2).

```
> CHGP
* LEVEL = 1

> HIST
* GIVE VARIABLE NAME OR COLUMN NUMBER FOR THE
  VARIABLE TO BE PLOTTED : DIFDJ

MIN. OBS. =-64.3101    MAX. OBS. = 58.45
MEAN = .611276    STD. DEV. = 33.1968
SAMPLE SIZE= 47

* GIVE MIDPOINT OF A CENTRAL INTERVAL OF HISTOGRAM: 0
* GIVE WIDTH OF EACH CLASS INTERVAL OF HISTOGRAM: 5

HISTOGRAM

ABS. FREQ.
         -
         -                    *   *
         -             *    * *   *  ***
         -          *  **  *** *** *****
         -       ********* *************** *
          -'----'----'----'----'----'----'----'----'----'----'-
              -1E+02     -5E+01    00E+00    50E+00    10E+01
                                  DIFDJ
MEAN = .611276
STD. DEV. = 33.1968
SAMPLE SIZE = 47
```

Figure 5-2. Histogram for Dow-Jones Difference Data

THE HISTOGRAM SHOWS THE PATTERN OF THE DATA WHEN SEQUENCE IS LOST.

CLASS INTERVAL IS CHOSEN TO SHOW THE RIGHT AMOUNT OF DETAIL.

Here I was given the smallest observation (MIN. OBS.), the largest observation (MAX. OBS.), the mean, standard deviation, and sample size to aid my choice of class intervals. I had to specify MIDPOINT, which is the central base, so to speak, and WIDTH OF INTERVAL.

1. MIDPOINT fixes the center of the scale. In this example, I chose 0 because it is near the mean, not too far from halfway between the smallest and largest observation, and a convenient number to think of when one is working with changes.

2. WIDTH OF INTERVAL fixes the scale. Although there may sometimes be good reasons for much broader intervals, it is usually well to take an interval width that is a relatively small fraction of a standard deviation. Subject to that requirement, it is well to pick a "round number". In this example, I chose 5.

This histogram looks somewhat different to me than the one in Figure 5-1.

Each of the sets of intervals gives a somewhat different impression; the one in Figure 5-2 gives a little more detail. At the same time, there's a definite resemblance.

Which is the better choice?

I have no strong preference, although I would give a slight edge to the one in Figure 5-2 because it does give more detail.

The main danger, however, is in the opposite direction: choice of classes with such broad width that they cover up important detail.

Please explain the numbers under the horizontal axis of the histogram in Figure 5-2.

These are in the original units rather than standardized units. When you chose your own intervals, the scale is in terms of the units of the problem.

Note that these numbers are printed in the floating-point notation explained in Chapter 4. Thus -1E + 02 means -100; 00E + 00 means 0; 10E + 01 means 100.

What do the histograms for the Rockwell hardness data look like?

The automatic choice, the one that comes with PLTS, is shown in Figure 5-3. It shows a striking pattern: a buildup of frequency to the sharp spike at +1, and then a wide gap from +1 to the deviate plotted at 4.4.

How well does this automatic choice of intervals represent what is going on?

Pretty well. Even so, it would be slightly better to make your own choice. The reason is that each hardness reading is a whole number or integer, and it is well to adjust the intervals to correspond with this graininess of the data. Figure 5-4 does this, using HIST (short prompts) with a MIDPOINT of 60 and WIDTH OF INTERVAL of 1.

```
HISTOGRAM

ABS. FREQ.

         -                              *
         -                              *
         -                              *
         -                              *
      15 -                              *
         -                              *
         -                              *
         -                      *       *
         -                      *       *
      10 -                      *      **
         -                      *      **
         -                      *      **
         -                  *  ***  ***
         -                  *  ***  ***
       5 -                * *  *******
         -                * ***********
         -              * *************
         -              ***************
         -  *   * * *  ****************                       *
          -'----'----'----'----'----'----'----'----'----'----'-
            -5   -4   -3   -2   -1    0    1    2    3    4    5

MEAN     =  59.58
STD. DEV. =  5.82433
SAMPLE SIZE =   100
```

Figure 5-3. Automatic Histogram Provided by PLTS for Rockwell Hardness Data

```
> HIST
* VARIABLE : HARDNS


MIN. OBS. = 40    MAX. OBS. = 85
MEAN = 59.58      STD. DEV. = 5.82433
SAMPLE SIZE= 100

* MIDPOINT: 60
* WIDTH OF INTERVAL: 1

HISTOGRAM

ABS. FREQ.

         -                         *
         -                         *
         -                         *
         -                         *
      15 -                         *
         -                         *
         -                         *
         -                         *
         -                        **
      10 -                        **
         -                        **
         -                        **
         -              *  * **  ***
         -              *  * **  ***
         -              * * ********
       5 -            * * **********
         -            * ************
         -            **************
         -            **************
         -  *      * * ** ***************                *
          -'----'----'----'----'----'----'----'----'----'-
              40E+00     50E+00     60E+00     70E+00     80E+00
                                   HARDNS

MEAN     =  59.58
STD. DEV. =  5.82433
SAMPLE SIZE =   100
```

Figure 5-4. Histogram for the Rockwell Hardness Data Adjusted for Graininess

The two histograms show pretty much the same thing.

Yes, but the bottom histogram is truer to the actual situation in one important respect the top histogram shows a spurious secondary spike at 0, in standardized units.

74 □ CONVERSATIONAL STATISTICS

Why do you say "spurious"?

You should check closely to satisfy yourself that this spike results from the consolidation of five 59's and seven 60's in a single class designated by "0" in the standardized units.

Could not both of these histograms be criticized for showing too much detail?

I prefer to give relatively fine detail in a histogram and then to let the reader mentally smooth the histogram, which is usually not hard to do. Look for yourself: you can see the general pattern perfectly well in either histogram. Overly broad class intervals, on the other hand, cover up detail that may be important, as we shall see later in this chapter.

What are histograms good for?

There are two important uses:

1. Histograms may provide clues as to further investigation of the data or of the underlying conditions that gave rise to the data.

2. Histograms are useful for what they may tell us about probability distributions, a new concept that we'll have to explain.

Let's start with the first.

We'll stick with the example. Suppose that you, as a production executive in the steel company, had received a chart containing either of these histograms, but preferably the second, as a part of a report from the quality control department. I would hope that your interocular analysis would instantly pick up the buildup of frequencies culminating in the sharp spike at 65, and also the gap on the right before the lone outlier at 85 shows up.

So what?

Here's where your imagination should go to work with some quick speculation as to how such a bizarre pattern might have arisen. What possible explanations occur to you?

Maybe the 85 is a mistake--for example, a transposition of 58.

That's possible. Even so, however, the spike at 65 and the truncation to the right is curious.

Maybe coils had been screened for hardness before they had reached this stage of production, and all coils with hardness readings above 65 had been removed.

That's clever, but would you then expect such a sharp buildup at 64 and 65?

TOO MUCH DETAIL IS PREFERABLE TO TOO LITTLE.

STATISTICAL DETECTIVE WORK.

It seems intuitively plausible that frequencies should change more smoothly than that, so I concede your point. In fact, as I think of it, it would also take a very efficient screening process to achieve such a sharp cutoff.

Yes, they would have had to have used the Rockwell hardness test earlier, and we would also have to assume that the test is perfectly reproducible. Neither seems reasonable.

Could the Rockwell hardness scale have a maximum at 65--no matter what the true hardness, measured hardness cannot exceed 65?

A good suggestion that reminds us how important it is to know as much as possible about how data are collected. Otherwise we may rush prematurely to fancy numerical analysis.

In this application, my understanding is this: On one widely used Rockwell test, the Rockwell A, 65 is indeed the maximum possible reading. On Rockwell B, however, 85 is possible.

This test was Rockwell B. A small but important detail.

Maybe there is a upper tolerance limit at 65, and the quality control department is fudging readings above that number by reporting them as 65 or 64.

How does the 85 reading fit in with this interpretation?

Oh, that was too high to fudge--a statistical felony as opposed to a misdemeanor!

By thinking of these possible explanations for the unusual pattern displayed by this histogram, and making some conjectures about their plausibility, you have illustrated how one should put his head to work in looking at statistical evidence. Don't just stare--think!

But what good is this armchair reasoning when looking at data?

In a live situation, you are protected against naive conclusions that often follow passive acceptance of the data at face value. Equally important, you can get out of your armchair to make a direct investigation as to what went wrong.

Do you know what was wrong in the actual application?

The correct explanation of the spike was that the data were being fudged. It was felt that coils only slightly over the specification of 65 would perform satisfactorily for customers.

Would it have been better to modify the tolerance limit of 65?

If it could be demonstrated that a coil slightly over 65 would perform satisfactorily, this modification would deserve serious consideration.

How about the 85? Was it a mistake?

It was genuine. It arose because of a scheduling error at the mill that caused a billet of mixed steel to be rolled into a coil.

ARMCHAIR ANALYSIS PAYS OFF.

DEALING WITH OUTLIERS.

It was just as well, then, that the 85 was not discarded as a measurement error.

Yes, because it was evidence of something going wrong with the manufacturing process.

Should one ever discard extreme observations like the 85?

The problem of dealing with outliers is usually tough, and I don't know of an easy generalization. Certainly, one should check carefully into the circumstances that gave rise to the outlier to see, if possible, whether the problem was solely one of measurement error. Often, however, it will turn out that the outlier was genuine, but that it reflected an unusual condition, such as the scheduling error. In such an instance, one must make a judgment whether he wants to study what goes on when such unusual conditions don't occur, or whether he wants to let the effects of such unusual conditions be reflected in the data.

This might depend on whether one thinks the unusual condition will ever be repeated?

Yes. If we decided to regard the scheduling error as an assignable cause that has been removed, we could discard the 85.

I would suppose, however, that one often fails to identify an assignable cause for an outlier. What can be done then?

One useful trick is this. Conduct the subsequent analysis with the outlier and then again without it. Then you can see whether or not the final conclusions are sensitive to the treatment of the outlier.

That sounds like a lot of extra work.

Not as much as you might think. IDA has a command called DELO that permits you to delete a single observation. You may wish to try it.

Suppose that you decide later that you want to get back the observation?

Use the IDA command RECO, which permits you to recoup.

You said also that histograms tell us something about probability distributions. Would you explain that?

We will think of probability in two different senses.

First, we draw on our intuition about the behavior of relative frequencies in random samples of astronomical size. For example, imagine that the process of manufacture of steel coils could continue for a few eons under the same basic conditions that prevailed during the sample of 100 that we have been studying. By "random sample", I mean that we would be willing to judge that the enormous sequence of observations thus generated arose under random conditions--the specification of randomness again.

Now imagine making a histogram of this enormous sample. There are some technical problems of how to scale the histogram so that it would appear to be about the same size as the histograms of Figures 5-3 and 5-4, so that we could compare the patterns. I'll defer these technical points until the start of Chapter 6.

What do you think this histogram would look like?

Something like the ones we saw for the sample of 100 but smoother or less jagged.

That's my intuitive feeling also. Probability theory backs up this intuition, fortunately.

Think of the histogram of the enormous random sample as a close approximation to the *probability distribution* of Rockwell hardness measures. The *relative frequencies* of individual outcomes, such as a reading of 59, would be approximations to the *probabilities* of those outcomes.

Then a probability, according to this interpretation, is a kind of long-run relative frequency?

That's the idea. I must warn you, however, that there's some logical fuzziness about this interpretation, since we can never observe the long-run, which means an infinite number of observations. Despite this problem, however, this intuitive approach is often useful, so long as you don't take it too literally.

Now when we explore further into the specification of the statistical model that serves as the basis for analysis, we shall see that this specification often involves some judgment about the *shape* of the probability distribution underlying the sample. One common specification, to be studied in detail in Chapter 6, is that the underlying probability distribution has the shape of what is known as a normal distribution.

The histogram of a sample, then, tells you something about the appropriate specification of the shape of the probability distribution, just as the sequence plot tells you something about the wisdom of the specification of randomness?

That's the idea. The histogram of samples of the relatively small sizes that we typically work with will be ragged, but they contain useful information. We shall see more of this later.

You said that we will think of probability in two senses. Long-run relative frequency is one. What is the other?

Personal or subjective probability is the other. I'll define this in the context of an example.

Suppose that you could buy a contract that would pay you $10 if the National Football Conference wins the next Super Bowl game. After careful reflection, you say $5.10 is the amount you would pay for the contract.

PROBABILITY AS LONG-RUN RELATIVE FREQUENCY

SUBJECTIVE PROBABILITY

You have then made an assessment of personal probability. For you, the probability of the NFC winning is

$5.10/$10 = 0.51 .

Your probability of an event is then the ratio of the amount you would be willing to pay for a prize contingent on the occurrence of the event to the amount of the prize itself.

Would this work if the contingent prize were $100,000 rather than $10?

Not for most of us. The amount of the prize should be small compared with your total assets, yet large enough that you would take seriously the task of assessing how much you would be willing to pay.

Then there's some logical fuzziness in the definition of subjective probability, just as in the definition of long-run frequency probability?

It's possible to remove the fuzziness about subjective probability by an alternative definition to the one I gave, but the one I gave is intuitively satisfactory and it's very simple to grasp.

I see another kind of problem with subjective probability, however. Suppose that I were vague about the amount I would offer for the prize.

The vagueness you refer to is no different than the kind of vagueness you feel in any real life situation when you have to specify a price--say, an offering price for 100 shares of stock in a closely-held company. In statistical analysis, as in other real-life activities, you cannot avoid such decisions. In particular, if you want to draw probabilistic inferences from a statistical analysis, you must make some probability assessments to put into it.

But can't subjective probability be used in a ridiculous way? Suppose that I don't know anything at all about football but make the snap judgment that the NFC (whatever that is) can hardly lose--probability 0.999.

I'd take you just as seriously as if you had just said that you're so thirsty that you'd give a million dollars for a cold beer right now.

All right, then. How would you go about assessing probabilities for the hardness-reading of a future coil?

One way would be to make a formal statistical analysis of the data. The answer to your question would be a simple by-product of that analysis, as we shall see later.

But I might want to make a direct judgmental assessment, without going to the time and expense of a formal analysis. In this event, I would use the relative frequencies in our sample of 100 as a starting point in my assessment. For example, in the 100 observations, there were 19 readings of 65. I might tentatively assess the probability of a 65 on the next observation as 0.19.

VAGUENESS IS UNAVOIDABLE IN PROBABILITY ASSESSMENT

And for a reading of 60, you would assess the probability at 7/100 = 0.07?

Tentatively, yes.

Your tentative assessments could be read directly from the histogram if the vertical axis were scaled in terms of relative frequency rather than absolute frequency?

Yes. Sometimes histograms are shown in terms of relative frequency rather than absolute frequency.

But you say that these assessments are only tentative, only a starting point. Would you explain that?

First, let me remind you that it is a sensible starting point only if the sample observations are specified to have arisen at random.

The reason I emphasized the tentativeness of the assessment is that there is almost always additional information besides that reflected in the sample, and careful probability assessments should take this information into account. Only for very large samples, such as those sometimes arising in actuarial applications, can we simply neglect this additional information.

For example, consider part of our sample information:

Hardness Value	Relative Frequency
50	0.02
51	0.00
52	0.04
53	0.02

Would you be willing to assess the probability of a 51 at 0.00 because no 51's had turned up in the sample of 100?

I think that I would somehow smooth out the probability assessments so that there was no dip--or at least not a dip clear down to zero--at 51.

Good. Now smoothing is an example of what I meant when I said that you had information other than that contained in the sample frequencies themselves. You judge it reasonable that probability assessments at nearby values should not change too drastically or too irregularly.

On the other hand, I wouldn't want to smooth off the peak at 65?

No, at least not after our discussion.

Are there other types of non-sample information that we would bring to bear in making probability assessments?

You might have other experience with the process. For example, even without precise records, you might feel that the process usually tends to produce lower hardness readings than those shown in the sample of 100. You might want to nudge the histogram a little to the left, so to speak, because of this information.

SMOOTHING OF RELATIVE FREQUENCIES IN PROBABILITY ASSESSMENT.

Or you might have information about a basic change in the process. For example, you may know that the quality control department has just received orders to cut out the fudging of observations above 65. With this information, you would want to take away some of the tentative probability at 65, 64, and possibly 63. and redistribute it at 66, 67, 68, etc.

All of this involves judgment, and that makes me uneasy. Isn't there some mathematical way to avoid these judgments?

I'll have to chide you a little for being so uneasy about judgment since, after all, you are studying management!

Mathematics never displaces judgment. Judgmental inputs are needed for all quantitative procedures used by management. In statistics, for example, the judgments about the appropriate statistical model with which to analyze the data are especially crucial.

On the other hand, we sometimes may be able to apply judgment directly and intuitively, as in the judgmental smoothing of sample frequencies.

What is involved in this judgmental smoothing?

Essentially one just draws a freehand curve to smooth out the ragged edges of the histogram, using judgment as to which peaks and valleys are ragged, and to what extent. We shall not actually pursue the technique in detail since for us the probability assessments will be by-products of formal statistical analyses. But you may be interested in reading, "Assessment of Probabilities by Smoothing Historical Frequencies", which is Chapter 6 of *Probability and Statistics for Business Decisions,* by Robert Schlaifer [McGraw-Hill, 1959].

If you were to smooth historical frequencies, however, there is an alternative mode of presentation that is easier than the histogram to deal with. Since this alternative is useful for other statistical purposes as well, this might be a good time to take it up.

What is it?

It is called a *cumulative distribution.* The terminology arises as follows. A histogram is just one device for showing the frequency pattern for a set of data. The pattern itself has a general name: *distribution* or *frequency distribution.*

A histogram displays the actual frequencies in each individual class. An alternative display exhibits the *cumulative* frequencies up to and including the present interval. This display is the *cumulative distribution.* Statisticians often call it by the more elaborate name *cumulative distribution function,* which is abbreviated c.d.f.

How does one get the c.d.f. with IDA?

The HIST command offers it as an option. After the histogram is printed out, you are asked, WANT CDF ALSO? Both the histogram and the c.d.f. for the Dow-Jones changes are shown in Figure 5-5.

JUDGMENT IS UNAVOIDABLE

CUMULATIVE DISTRIBUTION: THE C.D.F.

DISTRIBUTIONS □ 81

Are the horizontal scales aligned for the histogram and the c.d.f. in Figure 5-5?

Yes. That makes it easy to see the relationship between the two modes of presentation of the same basic information.

For example, you can see that the c.d.f. is zero until it jumps to 1 in the interval centered on -65. If you now look up at the histogram, you will see that the smallest observation occurred in this interval.

And in the next interval, centered on -60, I see from the histogram that the second smallest observation occurred here. Then when I look at the same interval on the horizontal scale of the c.d.f., I see that the cumulative frequency is two. Am I interpreting things correctly?

Good.

You'd better not compliment me too quickly, because I'm suddenly confused. I just noticed that there are two kinds of dots in the c.d.f. What does that mean?

Simply a typographical device to get a finer scale: I have used ' and . , which print high and low on each line.

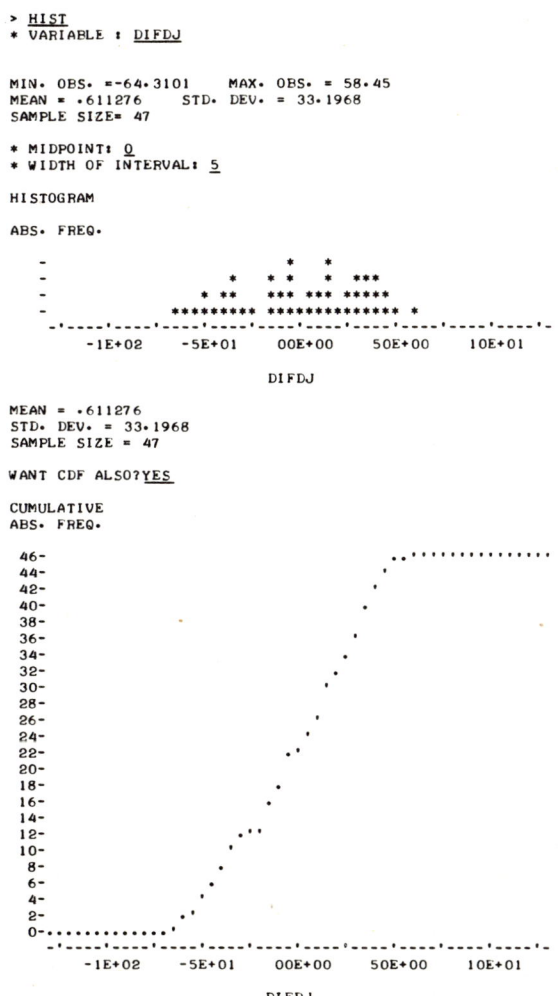

Figure 5-5. Histogram and C.D.F. for Dow-Jones Changes

82 ☐ CONVERSATIONAL STATISTICS

All right, then. As I now understand it, each jump of the c.d.f. corresponds to the same class just above it on the histogram, and the height of the jump equals the number of observations in that class of of the histogram.

That's right. And when we pass the last non-zero frequency class of the histogram, the c.d.f. stays at 47, the sample size, from there on to the right.

You said that hand-smoothing is easier with a c.d.f. than with a histogram. Why?

If you smooth a histogram directly, the smoothed frequencies will not usually add to the sample size, and an additional computation is needed so that they will. If you smooth a c.d.f., on the other hand, the smoothed curve will rise exactly to the sample size and no higher. When you then take the difference of the heights of the smoothed curve to get frequencies (or probability assessments) for individual intervals, these values will add up correctly without adjustment.

Further, I find it easier to exercise my judgment in drawing the curve when I am using the c.d.f. than when I am using the histogram. Look again at Figure 5-5 to see if you agree.

What does the c.d.f. look like for the Rockwell hardness data?

It is shown in Figure 5-6. You can clinch your understanding of the relationship between histogram and c.d.f. by study of that chart. Note especially the enormous jump of the c.d.f. in going from 64 to 65, and then the little blip at 85.

In all the distributions so far, you have shown absolute frequencies. If relative frequencies are directly related to probability assessments, why don't you show relative frequencies instead?

To see the general pattern of the data, it doesn't matter, since the two forms of representation have the same shape either for histograms or c.d.f.'s. The only difference is the labeling of the vertical axis: for absolute frequencies the maximum height is n, the sample size; for relative frequencies, it is 1.00 or 100 percent, depending on whether you choose decimals or percents to represent the relative frequencies.

The absolute frequencies are a little more convenient for display on the computer terminal.

How about the reporting of statistical findings?

Relative frequencies are often preferable for reporting, and they are often shown in tables rather than graphs. The command FREQ permits this presentation.

The example shown in Table 5-1 displays the changes of the Dow-Jones index. This table gives the same information as Figure 5-5, except that relative frequencies (in percentage form) are displayed. As is usual for the first display of a new command, the first level prompts are used.

HAND-SMOOTHING A C.D.F.

```
> HIST
* VARIABLE : HARDNS

MIN. OBS. = 40    MAX. OBS. = 85
MEAN = 59.58      STD. DEV. = 5.82433
SAMPLE SIZE= 100

* MIDPOINT: 60
* WIDTH OF INTERVAL: 1

HISTOGRAM

ABS. FREQ.
         -                         *
         -                         *
         -                         *
         -                         *
      15 -                         *
         -                         *
         -                         *
         -                         *
         -                         *
      10 -                        **
         -                        **
         -                        **
         -              *  * ** ***
         -              *  * *** ***
       5 -             *  * ********
         -             * * *********
         -              ** ***********
         -     *    * * ** *************                            *
           -'----'----'----'----'----'----'----'----'----'-
                 40E+00     50E+00     60E+00     70E+00     80E+00
                                    HARDNS

MEAN = 59.58
STD. DEV. = 5.82433
SAMPLE SIZE = 100

WANT CDF ALSO?YES

CUMULATIVE
ABS. FREQ.

    100-
     98-                           ..................   ..
     96-
     94-
     92-
     90-
     88-
     86-
     84-
     82-
     80-                          .
     78-
     76-
     74-
     72-
     70-                         .
     68-
     66-
     64-
     62-                        .
     60-
     58-                       .
     56-
     54-
     52-
     50-                      .
     48-
     46-
     44-                     .
     42-
     40-
     38-                    .
     36-
     34-
     32-
     30-                   .
     28-                  .
     26-
     24-
     22-
     20-                 .
     18-                .
     16-
     14-
     12-               .
     10-              .
      8-             ..
      6-            ..
      4-          ...
      2-       ....
      0-.......
        -'----'----'----'----'----'----'----'----'----'-
              40E+00     50E+00     60E+00     70E+00     80E+00
                                 HARDNS
```

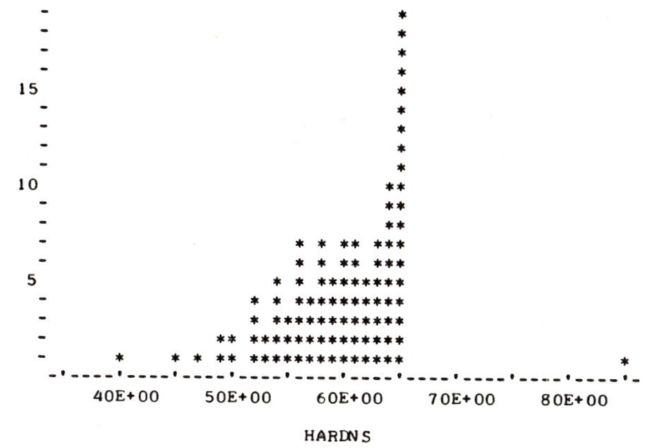

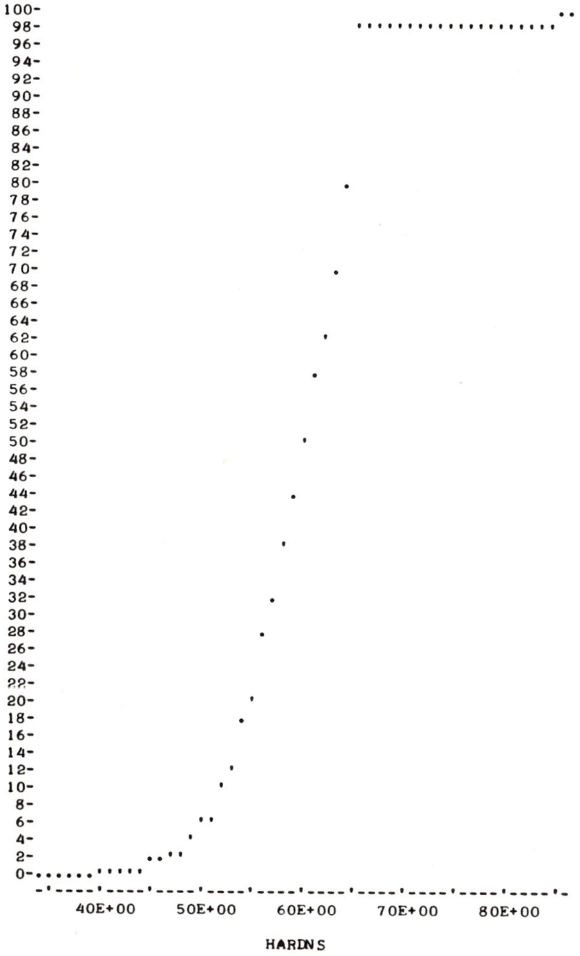

Figure 5-6. Histogram and C.D.F. for Rockwell Hardness Data

DISTRIBUTIONS □ 83

84 □ CONVERSATIONAL STATISTICS

I find it harder to see the pattern of the data in this table than in the histogram and c.d.f. of Figure 5-5.

Charts are usually better than tables for indicating the pattern of the data. In addition, however, the use of narrow class intervals, which was fine for the histogram, gives a cluttered impression in the table. Tables based on two wider intervals are shown in Tables 5-2 and 5-3. Both are preferable to Table 5-1 for indicating the pattern of the data, but Table 5-1 gives more detail that may be of interest to some readers of the report.

Table 5-1. Frequency Distribution of 'DIFDJ'
(Class Interval Width = 5)

```
> FREQ
* GIVE VARIABLE NAME OR COLUMN NUMBER FOR THE
  VARIABLE TO BE TABULATED : DIFDJ

MIN. OBS. =-64.3101    MAX. OBS. = 58.45
MEAN = ..611276        STD. DEV. = 33.1968
SAMPLE SIZE= 47

* GIVE MIDPOINT OF A CENTRAL INTERVAL OF TABLE: 0
* GIVE WIDTH OF EACH CLASS INTERVAL OF TABLE: 5

FREQUENCY DISTRIBUTION OF 'DIFDJ '
```

CLASS MIDPOINT	PERCENTAGE	CUMULATIVE PERCENTAGE
-6.5E+01	2.1	2.1
-6.0E+01	2.1	4.3
-5.5E+01	2.1	6.4
-5.0E+01	4.3	10.6
-4.5E+01	2.1	12.8
-4.0E+01	4.3	17.0
-3.5E+01	6.4	23.4
-3.0E+01	2.1	25.5
-2.5E+01	2.1	27.7
-2.0E+01	0.0	27.7
-1.5E+01	6.4	34.0
-1.0E+01	4.3	38.3
-5.0E+00	8.5	46.8
00.0E+00	2.1	48.9
50.0E-01	4.3	53.2
10.0E+00	4.3	57.4
15.0E+00	8.5	66.0
20.0E+00	2.1	68.1
25.0E+00	4.3	72.3
30.0E+00	6.4	78.7
35.0E+00	6.4	85.1
40.0E+00	6.4	91.5
45.0E+00	4.3	95.7
50.0E+00	2.1	97.9
55.0E+00	0.0	97.9
60.0E+00	2.1	100.0
TOTAL	100.0	
(NUMBER)	(47)	

Table 5-2. Frequency Distribution of 'DIFDJ'
(Class Interval Width = 10)

```
> FREQ
* VARIABLE : DIFDJ

MIN. OBS. =-64.3101    MAX. OBS. = 58.45
MEAN = .611276         STD. DEV. = 33.1968
SAMPLE SIZE= 47

* MIDPOINT: 0
* WIDTH OF INTERVAL: 10

FREQUENCY DISTRIBUTION OF 'DIFDJ '
```

CLASS MIDPOINT	PERCENTAGE	CUMULATIVE PERCENTAGE
-6.0E+01	6.4	6.4
-5.0E+01	4.3	10.6
-4.0E+01	8.5	19.1
-3.0E+01	6.4	25.5
-2.0E+01	6.4	31.9
-1.0E+01	10.6	42.6
00.0E+00	6.4	48.9
10.0E+00	14.9	63.8
20.0E+00	4.3	68.1
30.0E+00	12.8	80.9
40.0E+00	14.9	95.7
50.0E+00	2.1	97.9
60.0E+00	2.1	100.0
TOTAL	100.0	
(NUMBER)	(47)	

Table 5-3. Frequency Distribution of 'DIFDJ'
(Class Interval Width = 20)

```
> FREQ
* VARIABLE : DIFDJ

MIN. OBS. =-64.3101     MAX. OBS. = 58.45
MEAN = .611276    STD. DEV. = 33.1968
SAMPLE SIZE= 47

* MIDPOINT: 0
* WIDTH OF INTERVAL: 20

FREQUENCY DISTRIBUTION OF 'DIFDJ '

    CLASS                       CUMULATIVE
   MIDPOINT      PERCENTAGE     PERCENTAGE

   -6.0E+01          6.4            6.4
   -4.0E+01         19.1           25.5
   -2.0E+01         12.8           38.3
   00.0E+00         14.9           53.2
    20.0E+00        21.3           74.5

    40.0E+00        21.3           95.7
    60.0E+00         4.3          100.0

       TOTAL       100.0
     (NUMBER)       ( 47)
```

RULES FOR CHOOSING CLASS INTERVALS

Can you give rules for the selection of intervals for charts and tables?

For histograms and c.d.f.'s, the interval width should be a small fraction of the standard deviation, such as 1/5; furthermore, it should be based on "round numbers", and it should take account of any graininess in the data. Also, the midpoint of the axis should be chosen so as to center the plot.

For tables, rules-of-thumb are harder to give. The intervals should be relatively broad, but broad intervals can effect too much smoothing and thus can obscure essential details. Here, as in much of statistics, you must rely on your common sense rather than commit yourself to one stereotyped procedure.

To give yourself an idea of the need to use your common sense, examine Tables 5-4 and 5-5 made by FREQ from the Rockwell hardness data. Both intervals have width five; only the centering is different. Yet the first table preserves the important features that we have noticed during our study of the histogram, while the second hides them!

The problem with the second set of intervals is that the 19 observations at the reading of 65 are all assigned to the interval 65-69?

That's right.

Table 5-4. Frequency Distribution of 'HARDNS' (Midpoint = 60)

```
> FREQ
* VARIABLE : HARDNS

MIN. OBS. = 40   MAX. OBS. = 85
MEAN = 59.58      STD. DEV. = 5.82433
SAMPLE SIZE= 100

* MIDPOINT: 60
* WIDTH OF INTERVAL: 5

FREQUENCY DISTRIBUTION OF 'HARDNS'

    CLASS                         CUMULATIVE
   MIDPOINT       PERCENTAGE      PERCENTAGE

   40.0E+00           1.0             1.0
   45.0E+00           2.0             3.0
   50.0E+00           8.0            11.0
   55.0E+00          21.0            32.0
   60.0E+00          31.0            63.0

   65.0E+00          36.0            99.0
   70.0E+00           0.0            99.0
   75.0E+00           0.0            99.0
   80.0E+00           0.0            99.0
   85.0E+00           1.0           100.0

    TOTAL           100.0
   (NUMBER)         (100)
```

Table 5-5. Frequency Distribution of 'HARDNS' (Midpoint = 62)

```
> FREQ
* VARIABLE : HARDNS

MIN. OBS. = 40   MAX. OBS. = 85
MEAN = 59.58      STD. DEV. = 5.82433
SAMPLE SIZE= 100

* MIDPOINT: 62
* WIDTH OF INTERVAL: 5

FREQUENCY DISTRIBUTION OF 'HARDNS'

    CLASS                         CUMULATIVE
   MIDPOINT       PERCENTAGE      PERCENTAGE

   42.0E+00           1.0             1.0
   47.0E+00           4.0             5.0
   52.0E+00          13.0            18.0
   57.0E+00          26.0            44.0
   62.0E+00          36.0            80.0

   67.0E+00          19.0            99.0
   72.0E+00           0.0            99.0
   77.0E+00           0.0            99.0
   82.0E+00           0.0            99.0
   87.0E+00           1.0           100.0

    TOTAL           100.0
   (NUMBER)         (100)
```

CHAPTER SIX: NORMAL DISTRIBUTION

Some time ago I mentioned the term "normal distribution" and you said that we would talk about it later. Is now a good time?

Yes. The normal distribution represents a theoretical pattern to which data *sometime* correspond more or less closely. Imagine a random sample of truly astronomical size. The observations are measurements that can be made to any degree of fineness: there is no graininess in the data.

Now consider what the histogram would look like if the process generating the data conformed to the statistical model expressed by the normal distribution. With a sample of astronomical size, we choose class intervals of microscopic width, a process you can think of as follows: Start out with a set of class intervals that would be appropriate for the display of the histogram of a sample of moderate size, say 100. Then quadruple the sample size, halve the radius of the asterisks (*), halve the width of the class intervals, and halve the length of a unit of frequency on the vertical axis. Repeat until you have the desired sample size, such as 104,857,600; of course, any enormous number would do.

What would the histogram look like for this enormous sample?

We'll have to be a little careful about plotting conventions. Instead of using asterisks, as we have done on the computer plots, we shall use little black rectangles of height equal to one frequency unit and width equal to the width of the class interval.

Then, if the normal model is applicable, the histogram will almost certainly closely resemble the one shown below:

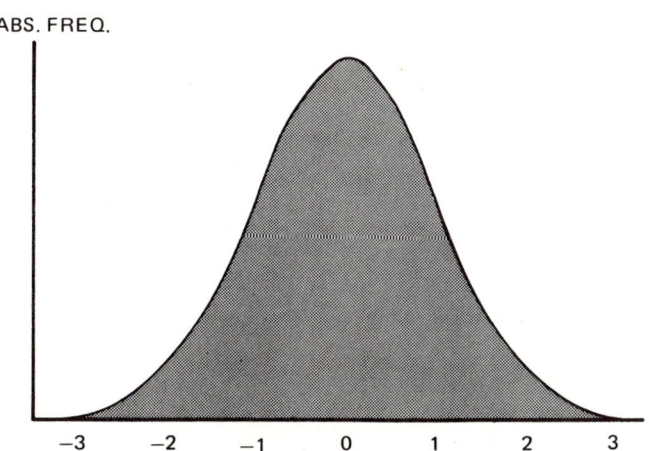

Figure 6-1. Histogram that Closely Follows Normal Distribution

NORMAL DISTRIBUTION AS A SPECIAL SHAPE OF HISTOGRAM

Then the histogram is smooth, solid, and shaped like a bell?

Smooth and solid, yes; but to my eye it's shaped more like a cocked hat than a well-designed bell!

How can you be so literal about the shape of this histogram? Didn't you use artistic license to draw it?

The upper boundary of the histogram has a precise mathematical formula. We don't need to discuss it, but it exists. Figure 6-1 is drawn in conformity with that formula.

If our eyes could actually resolve the tiny rectangles you mentioned, and if we could then count them, the total number of rectangles would equal the astronomical sample size that you mentioned?

That's right. The total area of the histogram would be the sample size n if each tiny rectangle counted as one unit.

Suppose that you had plotted relative (or percentage) frequency instead of absolute frequency on the vertical axis. What then would have been the area?

For relative frequency, 1.00; for percentage frequency, 100%.

How much of this total area lies between -1 and +1 along the horizontal axis?

About two-thirds; more precisely, 0.6827.

How do you know that?

It can be calculated mathematically from the formula mentioned above.

Would it be correct to say that the probability is about two-thirds that an observation will be no more distant from the mean than one standard deviation? I'm assuming the normal model, of course.

It would be correct. Although I introduced the idea in the context of a sample, the sample was astronomical in size. Under those circumstances, recall, relative frequencies are reasonable assessments of probabilities for individual observations.

Suppose that I wanted to make probability calculations for a normal distribution. How could I do it?

As you might guess, there's a command in IDA, called GAUS.

Where did that name come from?

The normal distribution is sometimes called the Gaussian distribution, after the great German mathematician Gauss.

FOR A NORMAL DISTRIBUTION THE PROBABILITY IS ABOUT 2/3 THAT AN OBSERVATION WILL BE WITHIN ONE STANDARD DEVIATION OF THE MEAN

Would you show me how to use GAUS to make the calculation that we just discussed?

Here it is:

```
> GAUS
*WHAT IS THE MEAN OF THE NORMAL VARIABLE X? 0
WHAT IS THE STANDARD DEVIATION OF X? 1

*WHAT IS THE LOWER LIMIT L FOR THE CALCULATION? -1
WHAT IS THE UPPER LIMIT U? 1

PR(L<X<U) =   0.68269
```

How did you decide to give 0 and 1 for the mean and standard deviation?

We are working in standardized units here, so the mean is 0 and the standard deviation is 1, as we have seen.

And the expression "PR(L<X<U)" simply means the probability of an observation between the limits just specified, L = -1 and U = +1?

Yes.

How would you use GAUS if you weren't working in standardized units?

Let's take an example. Suppose that heights of young American men can be reasonably well approximated by a normal distribution with mean of 70" (5'10") and standard deviation of 3". What is the probability that a randomly selected man's height will differ by no more than one standard deviation from the mean?

You'd give 70 and 3 for the first two prompts in GAUS. But what would be L and U?

L would be one standard deviation below the mean, or 70 - 3 = 67. U would be one standard deviation above the mean, or 70 + 3 = 73. Hence:

```
> GAUS
*WHAT IS THE MEAN OF THE NORMAL VARIABLE X? 70
WHAT IS THE STANDARD DEVIATION OF X? 3

*WHAT IS THE LOWER LIMIT L FOR THE CALCULATION? 67
WHAT IS THE UPPER LIMIT U? 73

PR(L<X<U) =   0.68269
```

What's the probability of an observation no farther than two standard deviations from the mean in a normal distribution?

Here's the application of GAUS to answer that question, and also the question for plus or minus three standard deviations. Prompt level 3 is used here.

USE "GAUS" TO CALCULATE NORMAL PROBABILITIES

BENCHMARK PROBABILITIES FOR THE NORMAL DISTRIBUTION

```
> CHGP
* LEVEL = 3

> GAUS

*MEAN= ?0
STD. DEV.= ?1

*LOWER LIMIT L= ?-2
UPPER LIMIT U =?2

PR(L<X<U) =   0.95450


> GAUS

*MEAN= ?0
STD. DEV.= ?1

*LOWER LIMIT L= ?-3
UPPER LIMIT U =?3

PR(L<X<U) =   0.99730
```

Is there any easy way to visualize these computations?

The chart below summarizes the results:

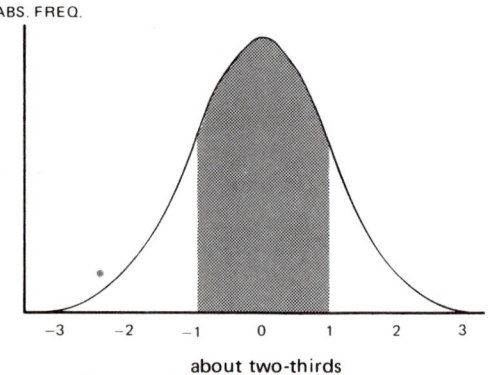

about two-thirds

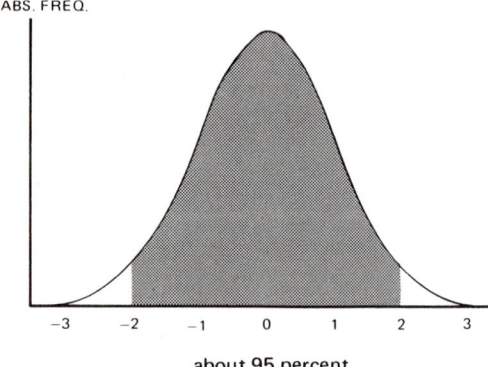

about 95 percent

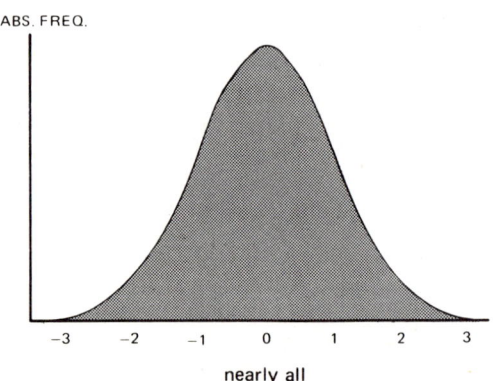

nearly all

Why did you write "about two-thirds", "about 95 percent", and "nearly all"?

These key facts, which are easily remembered, serve to orient you. They are worth memorizing; in fact, they are hard to forget!

But I can use GAUS to answer any specific question about normal probabilities?

Yes, and you may wish to experiment. Here is an example that involves one minor new wrinkle: given our assumptions about the mean and standard deviation, what is the probability of a man taller than 78-1/2" (that is, 6 feet 6-1/2 inches)? Here L = 78.5, but what is U? Actually, any value of U more than 4.5 standard deviations above the mean will serve; I chose 100" in the computation below so that I wouldn't have to make a close mental calculation.

```
> GAUS
*MEAN= ?70
STD. DEV.= ?3

*LOWER LIMIT L= ?78.5
UPPER LIMIT U =?100

PR(L<X<U) =   0.00230
```

Do you mean that there is no probability to the right of 4.5 standard deviations above the mean?

No, and there's an interesting subtlety here: there *is* positive probability to the right of 4.5, or indeed to the right of *any* number, but it's too small to affect the calculation to the number of decimal places shown.

A minor precaution: if *both* L and U are more than 4 standard deviations from the mean, there may be a small numerical error in the fifth decimal place using GAUS. That may seem puzzling at first, but it's a problem that arises because of rounding error in the HP-2000F.

Could I use GAUS to get the heights of the histogram in Figure 6-1?

Expressed in relative frequency units, you could. You set the mean = 0 and the standard deviation = 1. Then, for the horizontal or X value desired, set L = U = X. The heights thus calculated are called *densities*. However, we shall have no need to calculate densities in this book.

Is the normal distribution the "normal" or usual distribution that we encounter in data?

The statistician John W. Tukey has said, "There never was and never will be a normal distribution". This can be interpreted in two ways:

NORMAL DISTRIBUTION IS NOT NECESSARILY "NORMAL"

1. The specification of a *strictly* normal distribution (or of any other theoretical data pattern) can be thought of as entailing not merely an astronomically large (but finite) sample size but an infinite sample size. This we cannot observe.

2. If we looked closely enough at the histogram of data for which the sample size was enormous, we would under the best of circumstances find irregularities difficult to reconcile with the model of a strictly normal distribution. Professor Tukey actually had this second interpretation in mind.

Then what good is the model of a normal distribution?

Under the favorable circumstances in which it is a good approximation to what is going on in the process giving rise to the data, the normal model (like any other statistical model) can facilitate inferences that will be adequate for practical purposes.

All right, when would the circumstances be favorable?

Suppose that each observation we make is the resultant of a substantial number of random factors that are not observed. These factors act independently of each other in the sense that your probability assessment of the distribution of any one would be unaffected by knowledge of the outcomes of the others. Moreover, the factors are additive in their contributions to the observation actually made, and no one factor makes a contribution that is large by comparison with the sum of the contributions of all the others.

If these conditions hold, it would be reasonable to suppose that the histogram of a very large sample could be closely approximated by a normal distribution with the same mean and standard deviation. We would then say that we believe that the normal distribution would be a good statistical model for the data actually to be analyzed.

Is there any mathematical theory to back up what you said?

By stating the conditions leading to approximate normality of data, I have simply tried to translate a mathematical theorem, called the *central limit theorem,* into intuitive language.

Can you give examples in which the normal distribution might be a good model?

I'm glad that you said "might" because each application has to be considered on its merits and because, as we shall see, *a priori* reasoning about the plausibility of normality is only part of the story. But here are examples in which the normal model may be serviceable.

1. Errors of measurement in careful scientific studies. The classical example is that of errors of measurement in astronomy. Each of a number of independent factors may contribute independently a relatively small part of the overall error on any observation.

CENTRAL LIMIT THEOREM

2. Variations of dimensions, weights, yields, etc., of carefully controlled industrial processes in which "assignable causes" of variation have been removed.

3. Variations of biological measurements, such as heights, of a homogeneous group, such as adult American males.

4. We shall be studying a body of techniques discussed under the name "regression". By application of these techniques, we may be able to remove systematic sources of variation that can be related to variables actually measured; what is left over may behave consistently with the normal model.

When the normal model appears to be satisfactory, is it possible to identify the underlying factors that contribute to the approximately normal outcome?

Sometimes not at all. Sometimes with a degree of success. For example, if height is largely determined by hereditary factors, genetic theory may contribute some insight into the way things work. The underlying factors may be identified with genes.

In practical statistical applications, how do we go about deciding whether the normal distribution is a good model?

We should apply any background knowledge or theory that may illuminate the appropriateness of the normal model: is it reasonable to think of the outcome as the summation of a substantial number of independently-acting underlying factors, none of which predominates?

Also, just as in assessing the appropriateness of the specification of random sequence, we want to look at the data to be analyzed.

How do we go about that?

For an example, I have taken artificially-generated data for which the normal model is very plausible a *priori*. I had the computer generate 100 numbers that should behave very much like random drawings from a normal distribution with mean 100 and standard deviation 15. The data are in a file $IQ that is listed in Table 6-1. (Since I.Q.'s of adults are supposed to be approximately normally distributed, I've used the name IQ, but remember that these data are simulated, not real.)

Could I use IDA to create a simulation like this?

Use the command RAND, subcommand NORM.

Do the data in Table 6-1 behave randomly in their sequence?

They behave well in light of the criteria of Chapter 4. It would be good practice for you to check this for yourself.

THE SPECIFICATION OF NORMALITY IN APPLICATIONS

94 ☐ **CONVERSATIONAL STATISTICS**

Table 6-1. Simulated I.Q. Data

```
> FPRS
CONSECUTIVE COL. ?Y
I1,I2,J1,J2 = ?1,100,1,1
FMT = ?0,5D.4D
```

** 1 **	107.1860		** 51 **	98.8670
** 2 **	126.1480		** 52 **	96.3980
** 3 **	89.8249		** 53 **	82.2757
** 4 **	98.6254		** 54 **	109.7160
** 5 **	78.9818		** 55 **	101.6520
** 6 **	126.6310		** 56 **	98.8266
** 7 **	122.7680		** 57 **	87.1504
** 8 **	61.8811		** 58 **	105.1560
** 9 **	108.2520		** 59 **	108.2790
** 10 **	97.5731		** 60 **	86.4665
** 11 **	85.1665		** 61 **	114.4500
** 12 **	112.4630		** 62 **	133.4340
** 13 **	96.7346		** 63 **	82.5008
** 14 **	117.4290		** 64 **	106.3940
** 15 **	100.4810		** 65 **	97.6918
** 16 **	94.9139		** 66 **	98.2634
** 17 **	100.6330		** 67 **	107.9740
** 18 **	114.0610		** 68 **	113.1800
** 19 **	108.7330		** 69 **	91.3707
** 20 **	125.9960		** 70 **	102.3890
** 21 **	114.0490		** 71 **	137.1040
** 22 **	110.7690		** 72 **	72.1542
** 23 **	83.7672		** 73 **	94.0793
** 24 **	103.2380		** 74 **	91.5595
** 25 **	98.6888		** 75 **	114.6160
** 26 **	108.7340		** 76 **	84.8731
** 27 **	95.8594		** 77 **	112.8220
** 28 **	104.0900		** 78 **	109.3180
** 29 **	112.4580		** 79 **	103.6090
** 30 **	103.0530		** 80 **	85.7529
** 31 **	97.1651		** 81 **	81.0519
** 32 **	121.6580		** 82 **	76.8783
** 33 **	84.6681		** 83 **	108.5430
** 34 **	106.8670		** 84 **	94.0413
** 35 **	63.8016		** 85 **	75.0380
** 36 **	111.1510		** 86 **	74.0235
** 37 **	77.4866		** 87 **	82.1519
** 38 **	145.2860		** 88 **	97.5725
** 39 **	104.7830		** 89 **	126.2430
** 40 **	98.8794		** 90 **	103.3320
** 41 **	99.8494		** 91 **	112.6780
** 42 **	130.5520		** 92 **	114.9320
** 43 **	121.5160		** 93 **	110.3110
** 44 **	81.0886		** 94 **	128.3360
** 45 **	114.6260		** 95 **	98.1287
** 46 **	110.5620		** 96 **	113.7490
** 47 **	108.6570		** 97 **	99.9784
** 48 **	115.1750		** 98 **	112.9050
** 49 **	119.4710		** 99 **	95.9556
** 50 **	104.3500		**100 **	129.6880

Would a histogram give some insight into the appropriateness of the normal model?

Try the interocular test for the execution of HIST shown in Figure 6-2.

```
> HIST
* VARIABLE : IQ

MIN. OBS. = 61.8811    MAX. OBS. = 145.286
MEAN = 102.906    STD. DEV. = 15.9522
SAMPLE SIZE= 100

* MIDPOINT: 100
* WIDTH OF INTERVAL: 3

HISTOGRAM
ABS. FREQ.

  10                         *  *
   -                        ** **
   -                        *** ***
   -                        *** ***
   -                  *     *** ***
   5                 **   ******   *
   -                 **   ******   *
   -             *  ** *********   *
   -             ** ** *********  **
   -         **  **************** *
     -'----'----'----'----'----'----'----'----'-
         40E+00    70E+00    10E+01    13E+01    16E+01
                                IQ

MEAN = 102.906
STD. DEV. = 15.9522
SAMPLE SIZE = 100
```

Figure 6-2. Histogram of Simulated I.Q. Data

First, let me ask a question. I see that the sample mean is 102.9 and the sample standard deviation is 15.95. I thought you specified 100 and 15.

I specified 100 and 15 for the mean and standard deviation of the underlying normal distribution. The mean and standard deviation of a particular sample of 100 are subject to a good deal of sampling variation.

How about your interocular impression of the histogram?

It's hard to tell. There are lots of peaks and valleys. But if I imagine a curve smoothed over the tops of the peaks, I suppose that it might look something like Figure 6-1?

The unaided eye is not reliable in one important respect: it does tend to connect the peaks of the histogram, as you did. To see how this tendency should be compensated, study the histogram below. It is the same as before, but I have superimposed a normal histogram or "curve". This fitted curve represents a distribution with the same mean and standard deviation as the sample data.

COMPARING A SAMPLE HISTOGRAM WITH A FITTED NORMAL "CURVE".

```
> HIST
* VARIABLE : IQ

MIN. OBS. = 61.8811    MAX. OBS. = 145.286
MEAN = 102.906    STD. DEV. = 15.9522
SAMPLE SIZE= 100

* MIDPOINT: 100
* WIDTH OF INTERVAL: 3

HISTOGRAM

ABS. FREQ.

 10                  *  *
  -                 ** **
  -                *** ***
  -                *** ***
  -           *   ******* *
  5          **  *********
  -          **  ********* *
  -        * **  *********** *
  -        *** ***************
  -     ** ************************   *
     '----'----'----'----'----'----'----'
      40E+00  70E+00  10E+01  13E+01  16E+01
                        IQ

MEAN = 102.906
STD. DEV. = 15.9522
SAMPLE SIZE = 100
```

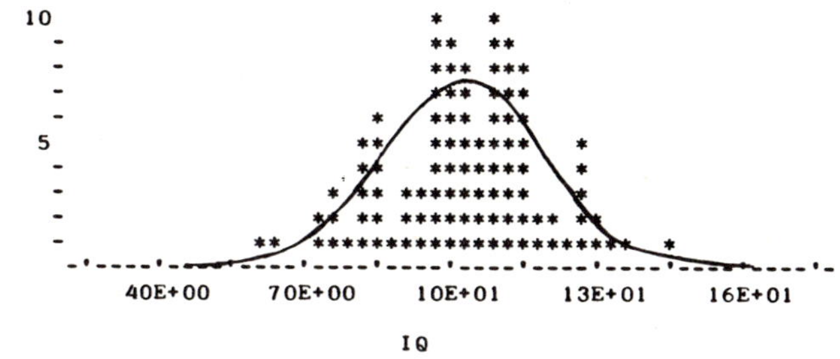

Figure 6-3. Histogram of Simulated I.Q. Data with Superimposed "Normal Curve"

I'm surprised to see how low the curve lies with respect to the sample histogram.

As a matter of fact, I was surprised myself, because I haven't very often fitted histograms by smooth curves using mathematical computations. If you imagine the asterisks in each column as forming a rectangular figure, as is done in hand drawings of histograms, then the protrusions of the rectangles above the curve would have the same total area as the total space between the curve and the rectangles at those intervals for which the height of the rectangles is less than that of the curve.

I'm also surprised at the raggedness of the peaks and valleys of the histogram by comparison with the curve.

For a sample size of 100, sampling variation is so substantial that the normal approximation is very crude. This is one application for which the interocular test has limited usefulness.

There is a much better way of making a visual comparison, however. It is based on a special representation of the c.d.f. Before introducing it, however, I want you to look at the ordinary plot of a c.d.f. with a smooth curve superimposed. The smooth curve represents a normal distribution with the same mean and standard deviation as that of our sample; see Figure 6-4.

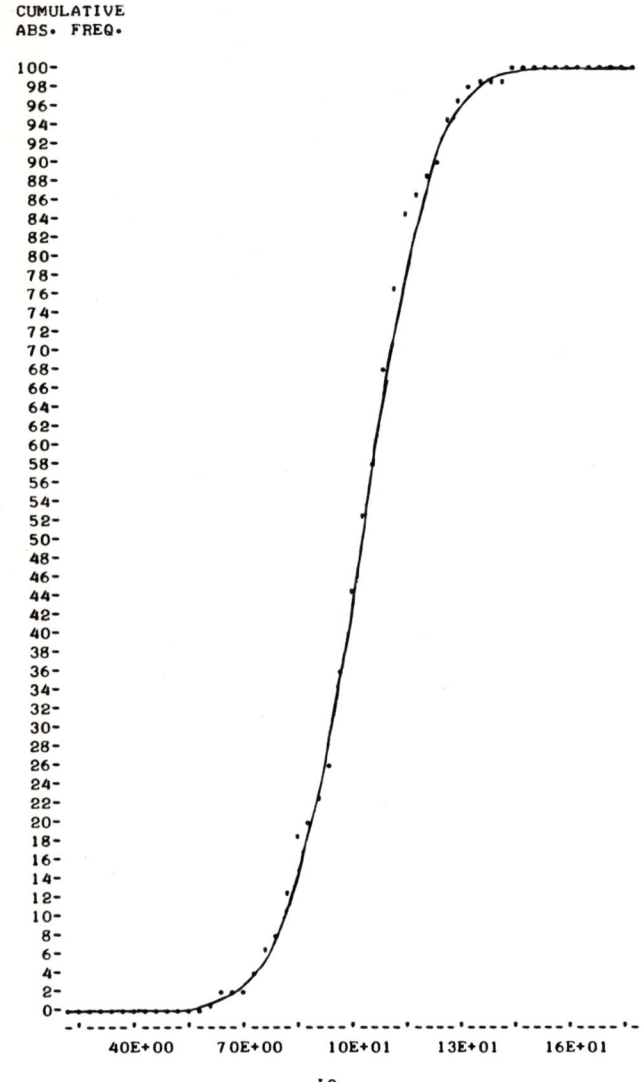

Figure 6-4. Fitted Normal C.D.F. of I.Q. Data

The fitted normal c.d.f. in Figure 6-4 looks like an elongated S-shaped curve.

That's the way it's usually described.

Note that the sample points tend to follow the fitted normal c.d.f., with gentle undulations above and below the curve.

I get more visual reassurance about the fit from Figure 6-4 than I did from Figure 6-3.

The graphical representation shown in Figure 6-4 does facilitate the interocular analysis more than the one based on the histogram. But there's still a better way.

Start from Figure 6-4. Suppose that we alter the vertical cumulative frequency scale in such a way that a strictly normal c.d.f. would plot as a straight line. Then we can plot the sample c.d.f. on this graph. A special kind of graph paper called *arithmetic probability paper* facilitates this procedure when it must be done by hand.

The same plot can be made by computer with little loss of accuracy and great economy of time. This plot is called a *normal probability plot*.

The IDA command to produce this plot is called NORM. An execution for the I.Q. data is shown as Figure 6-5. Figure 6-5 also has a typewritten legend at the extreme left that I shall explain shortly.

The first-level prompt for NORM (which will produce Figure 6-5) is shown below:

THE NORMAL PROBABILITY PLOT

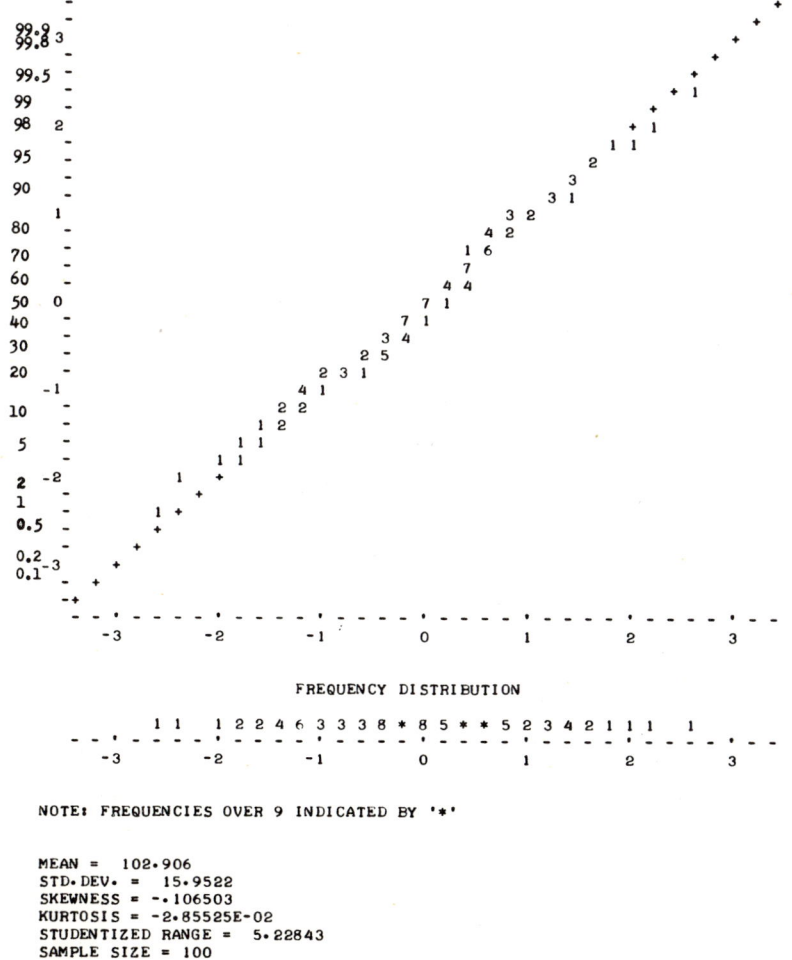

Figure 6-5. Normal Probability Plot of I.Q. Data

What are all the little numbers on the graph itself?

These indicate the number of points that, in a very careful hand plot, would fall in a rectangular area 1/6 inch high and 1/5 inch wide about the number itself on the actual teletype output.

So a "1" means a single dot in a small area, while a digit greater than 1 indicates multiple points?

That's right. If there are more than 9 points, an asterisk is printed.

How about the numbers typed to the left of the vertical axis?

These are cumulative relative frequencies or probabilities, expressed in percentage form. These are actually printed on the arithmetic probability paper used for hand plots, but it isn't feasible to print them with terminals of the kind we are using. So I typed them in the first time to show you what they look like.

But what are the numbers "3", "2", etc. plotted by the terminal down the vertical axis?

For a strictly normal distribution, this is an indication of the standardized scaling that would correspond with the cumulative probability scale typed to the left.

For example, "50" appears to the left of "0". This corresponds to the fact that for a normal distribution of a variable in standardized units, 50 percent of the probability is to the left of 0 (recall that in standardized units, the mean is always at 0).

Similarly, between 99.8 and 99.9 percent of the probability is to the left of +3?

Yes, for a strictly normal distribution.

And I.Q. is plotted in standardized units along the horizontal axis?

That's right. I remind you that the normal probability plot is simply a variant of the ordinary cumulative distribution, as shown in Figure 6-4. The only essential difference is that the vertical axis is so scaled that the theoretical normal cumulative curve plots as a straight line rather than as an S-shaped curve.

Where do I find the straight line on the normal probability plot?

It is outlined by the "+" signs printed on the graph. You can easily connect these by eye.

It looks to me that the data follow the straight line pretty well.

I agree.

How about the frequency distribution shown beneath the probability plot?

NORMAL PROBABILITY PLOT IS A SPECIAL C.D.F.

That is another way to convey the information displayed by a histogram. Note that each number in the frequency distribution is the sum of the numbers in the same column of the normal probability plot. This gives you another way to see what is going on in the data, and to grasp more securely the connection between the frequency distribution and the cumulative distribution.

What is the meaning of the terms "skewness", "kurtosis", and "Studentised range" that I see in the summary information printed below the graph?

These are numerical measures that give some idea of the conformity of the data to the normal model. They are designed as supplements to the interocular test, not as substitutes for it.

1. The *skewness* coefficient is sensitive to departures from symmetry of the data about the mean. Figure 6-1, for example, is symmetric about 0 in the sense that the left half of the figure is the mirror image of the right half.

2. The *kurtosis* coefficient is sensitive to a tendency of data to be more or less "fat-tailed" than would be expected from the normal model; we shall discuss this concept in a little while. Alternatively, the kurtosis coefficient measures the systematic tendency of the data to follow an S-shape or a reverse S-shape on the *normal probability plot*.

3. The *Studentised* range is the ratio of the sample range (largest observation minus smallest observation) to the sample standard deviation, s. Its purpose is the same as the kurtosis coefficient.

But how do I interpret the numerical values of these measures?

If the specification of the normal model is correct, the expected value of the skewness coefficient is 0. If, by contrast, the underlying histogram is relatively blunt on the left and has a long tail pointing to the right, the expected value of the skewness coefficient is positive. This situation is called *positive* skewness, or skewness to the right.

What would this look like in the normal probability plot?

The following schematic sketch conveys the idea. I describe the plot as "bow-shaped":

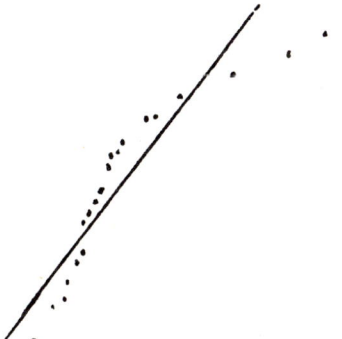

SUPPLEMENTARY MEASURES OF CONFORMITY.

The corresponding frequency distribution at the bottom of the normal probability plot would then show the greater density of points at the left and a thinning-out on the right?

That's correct. And the skewness coefficient would ordinarily have a positive value.

How about negative skewness or skewness to the left?

Just reflect the above sketch about the line so that the bulge of the bow is pointing southeast rather than northwest. The skewness coefficient would ordinarily be negative.

Then the farther the skewness coefficient departs from 0, in either direction, the more substantial the evidence against the normal model?

Yes.

How big a departure from 0 would you tolerate before feeling that an alternative model should be sought?

I can only offer, with apology, another rule-of-thumb. For sample sizes in the range that we shall usually be working with, say 20-100, my personal rule is ± 0.5.

However, I regard a numerical measure such as a coefficient of skewness as secondary to the visual analysis of the normal probability plot. As D. Kerridge put it, "After all, nobody really cares whether a distribution is 'significantly' different from normal; all you want to know is whether you can get away with using it as normal. I, personally, can judge that more easily from a diagram than from anything else... ."

Can you make parallel remarks about the interpretation of the kurtosis coefficient?

If the specification of the normal model is correct, the expected value of the kurtosis coefficient is 0. If, by contrast, for a given standard deviation, the tails of the distribution are relatively fat by comparison with the normal, then the expected value of the kurtosis coefficient is positive.

As before, I want to know what this would look like in the normal probability plot.

In terms of the normal probability plot, a fat-tailed distribution tends to yield an S-shape as suggested by schematic sketch below.

EVALUATING SKEWNESS AND KURTOSIS MEASURES

NORMAL DISTRIBUTION □ 101

And the corresponding frequency distribution at the bottom of the normal probability plot would show a straggling of points at both the right and the left extremes?

Yes. And if I may anticipate your next question, negative kurtosis implies a blunt-tailed, flat-topped distribution, which would have a reverse S-shape on the normal probability plot. To see what it would look like, just reflect the above sketch about the straight line.

And how about your rule-of-thumb for numerical values of the kurtosis coefficient?

With apologies paralleling those for the skewness coefficient, I suggest ±1 as warning limits.

How about the Studentised range?

That measure is necessarily positive. It tends to be "large" for fat-tailed distributions and "small" for blunt-tailed distributions. It is not easy to interpret for beginners; it is offered by IDA mainly for more advanced users. If you press me, however, I'll say that for sample sizes in the range 20-100, I'd begin to suspect an underlying fat-tailed distribution if this measure gets as high as 6 or 7.

In summary, would you say that the I.Q. example is one in which the data conform pretty well to the normal model?

The skewness coefficient is -0.11, the kurtosis coefficient is -0.03, and, more importantly, the normal probability plot looks good. So I'd say yes.

Could you say just a little more about the interocular examination of the normal probability plot?

The main thing to look for is *systematic* and *substantial* variation of the sample points from the straight line. The bow-shaped and S-shaped patterns are especially suspicious.

The main limitation of the plot as a diagnostic check on model adequacy arises when the sample size is very small. Then even if the normal model is seriously inadequate, it's rare that the normal probability plot will give much of a warning.

What do we do about the specification of a normal model if the sample size is very small?

You'll still want to look at the normal probability plot, but you must of necessity rely more heavily on prior knowledge, theory, and experience in related applications.

Could we look at normal probability plots of earlier sets of data to get a better feel for interocular interpretation? For example, what do the Rockwell hardness data look like?

This is shown in Figure 6-6. Here the departure from the straight line is systematic and substantial. Part of this departure comes from the truncation at 65, previously discussed, which shows up as a sharp upward jump.

NON-NORMALITY IS HARD TO SPOT FOR SMALL SAMPLES

102 □ CONVERSATIONAL STATISTICS

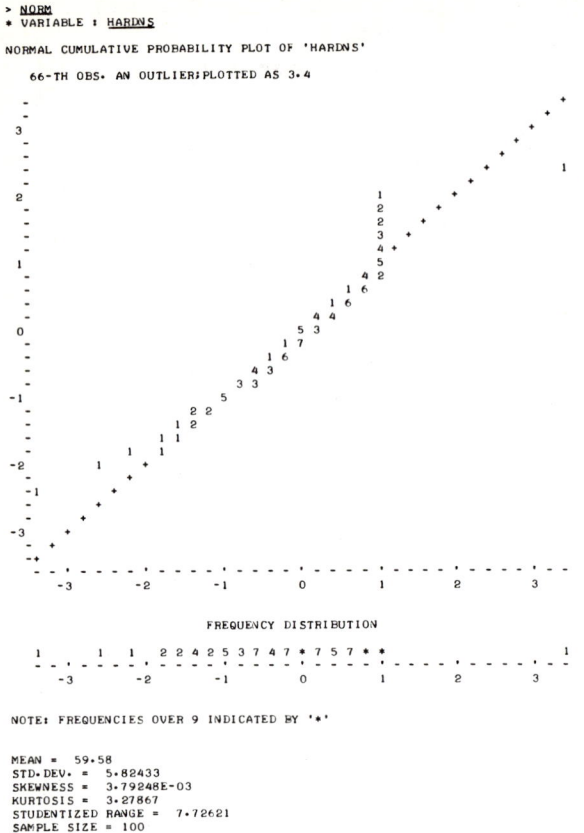

Figure 6-6. Normal Probability Plot of Rockwell Hardness Data

Suppose that the jump were smoothed out by removal of the data-fudging that we know to have occurred. What would then happen?

The sample data would still follow an elongated S-shape on the normal probability plot. Notice that the kurtosis coefficient is 3.28. Notice also that the largest reading becomes an outlier for this plot, as is shown by the warning printed at the top of the chart. From Chapter 5 we learned that this observation fell in the interval centered at 4.4 in standardized units; the normal probability plot goes only up to ± 3.5, and more extreme deviations are plotted at the edge with a printed warning at the top.

If it had been possible to print the plot without that truncation, then the picture would have been even more extreme because the extreme value would have been moved over an inch to the right?

Yes.

You said in Chapter 5 that it would be possible to delete an observation to see what the effect would be. Would you illustrate that here?

The command DELO for deleting an observation works as follows in the first level prompt:

```
> DELO
DELETE WHICH ROW (GIVE NO.) ?66
UPDATING MEAN, STD, ...
```

The normal probability plot for the remaining 99 observations is shown in Figure 6-7. I think that you will agree that the picture is substantially different, even though it is still clear that the normal model fares badly. It would be instructive for you to study the changes in detail.

THE EFFECT OF AN OUTLIER

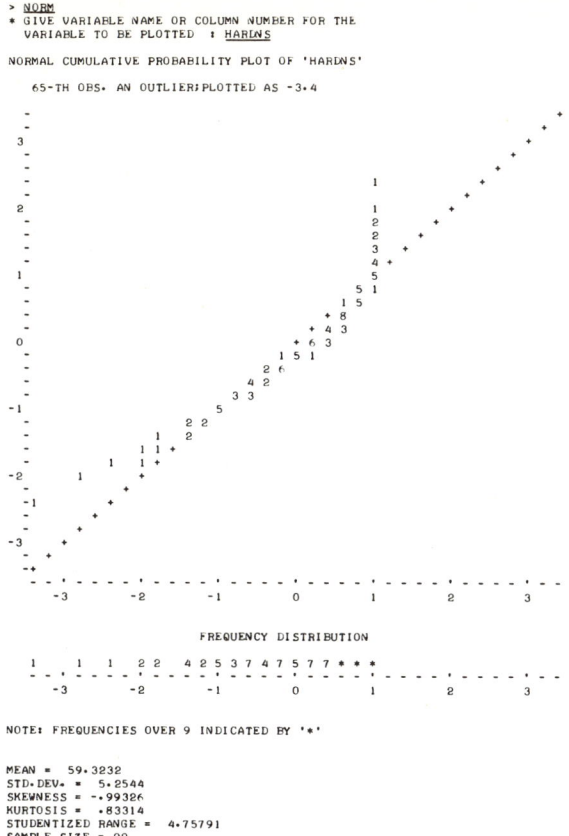

Figure 6-7. Normal Probability Plot of Hardness Data with Outlying Observation Deleted

From the way you said that, I gather that you aren't going to comment on the changes in the result of NORM. But let me ask a more general question: is it possible to get a fat-tailed or skewed appearance, or both, from basically normal data that are contaminated by occasional wild values or outliers?

It's possible. The very large value (85) in the hardness data does have a big effect, as we have just seen. It was not a measurement error. You may wish to review our discussion of this value in Chapter 5.

Note also, however, that there were some extreme values at the lower end of the distribution. At the time the study was made, there was some question as to whether these might have reflected measurement errors in the sense of readings taken from the wrong position on the coil, but this was never resolved.

Fat-tailed or skewed distributions can also arise in circumstances in which the outliers can be distinguished from the other observations in no other respect than the fact that they were outliers--no special explanations or assignable causes can be found. We shall see examples.

I get the feeling that it is vital in statistical study to worry about the observations as individuals as well as in mass.

Just as an astronomer ought to like stars and a mathematician should be fond of numbers, so a statistician ought to like his data well enough to get to know them.

A STATISTICIAN SHOULD GET TO KNOW HIS DATA

How about the normal probability plot of the stock market data?

A normal probability plot of the changes of the Dow-Jones index is shown in Figure 6-8. The normal model looks pretty good here, although close examination will pick out what appears to be almost a reverse S-shape. Corresponding to this is a substantially-negative value of the kurtosis coefficient, -1.13.

This application illustrates the need to bring in all your information, not just the sample information. Many studies have been made of stock market changes, both of individual stocks and of stock indices, and they suggest a tendency that is just the opposite of the one noted for our data. The distributions are usually more fat-tailed than the normal model would lead us to expect.

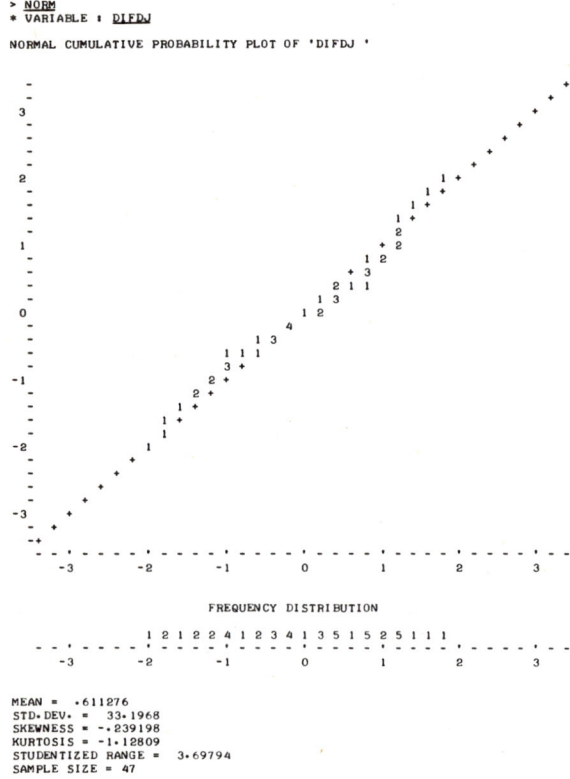

Figure 6-8. Normal Probability Plot of Dow-Jones Data

Is this because of errors in reporting the data?

Not entirely, although reporting and transcribing errors do occur. Several possible explanations for the fat-tailed tendency have been advanced and current research is attempting to clarify the question.

Let's get back to basics. Why do you attach so much importance to the normal model? After all, you said "normal" isn't necessarily "usual".

There are several reasons. Perhaps the most important for us is that the normal distribution is often a good starting point in the search for an adequate statistical model.

WHY USE THE NORMAL MODEL?

Further, when the data do not conform to the normal distribution, it is often fruitful, quite apart from the immediate task of statistical analysis, to try to figure out why they don't conform. Here are three examples:

1. A study of the ratios of actual to standard costs for a large number of items constituting a company's inventory of purchased items showed a grossly fat-tailed distribution. When extreme ratios at either end of the distribution were deleted, the remaining ratios were still grossly fat-tailed. This empirical observation led to a careful scrutiny of the accounting formulas by which standard costs were established and an overhaul of this part of the company's accounting procedures when it was found that the formulas used were seriously defective.

2. A study of the effective life of a sample of printing plates showed a distribution that was blunt-tailed, flat-topped, and, on the normal probability plot, reverse S-shaped. This led to an investigation of the causes of plate failure. It was found that there were two major modes of failure: gradual "blinding" and sudden "cracking". When evaluated separately, the normal model provided a good fit to the data for each mode of failure.

3. The persistence of findings of fat-tailed distributions of stock market price changes, and of price changes on other organized markets, has had a substantial influence on the development of the theory of finance during the last decade.

But if the data turn out to be substantially non-normal, what do you do about their statistical analysis?

One strategy that may be appropriate is to base the analysis upon an alternative distribution that may be more appropriate to the application. Several other distributions are useful. For our purposes, however, we will confine attention to the normal model, though you might think this is tying our hands unnecessarily. This isn't true, fortunately. It is often possible to transform the data to achieve a closer conformity to the normal model.

Transforming the data? That sounds bad, like cooking the data!

Don't jump to conclusions. Let's look at an example, and then you can judge for yourself. In the file $TJT66, which is printed out in Table 6-2, we have data on spoilage percentages for a particular kind of bottle cap in 56 consecutive production runs.

I suggest that you verify that the time series behavior of these data is compatible with the model of random sequence.

Now look at the "automatic" histogram, the one in which the observations are shown in standardized units at the end of the sequence plot:

TRANSFORMATION OF THE DATA

```
> FPRS
CONSECUTIVE COL. ?Y
I1,I2,J1,J2 = ?1,56,1,1
FMT = ?0,3D.2D
```

```
** 1 **   3.87       ** 29 **  4.54
** 2 **   3.13       ** 30 **  6.47
** 3 **   2.33       ** 31 **  3.91
** 4 **   2.51       ** 32 **  2.61
** 5 **   3.54       ** 33 **  3.49
** 6 **   5.17       ** 34 **  2.57
** 7 **   5.43       ** 35 **  4.43
** 8 **   3.40       ** 36 **  2.66
** 9 **   3.73       ** 37 **  3.38
** 10 **  3.45       ** 38 **  5.32
** 11 **  7.00       ** 39 **  7.29
** 12 **  3.81       ** 40 **  2.03
** 13 **  2.39       ** 41 **  4.50
** 14 **  3.49       ** 42 **  3.46
** 15 **  4.53       ** 43 **  2.55
** 16 **  5.77       ** 44 **  4.31
** 17 **  3.64       ** 45 **  3.73
** 18 **  8.71       ** 46 **  6.87
** 19 **  3.03       ** 47 **  4.28
** 20 **  3.14       ** 48 **  3.78
** 21 **  4.51       ** 49 **  4.10
** 22 **  2.50       ** 50 **  3.29
** 23 **  4.95       ** 51 **  3.18
** 24 **  3.85       ** 52 **  3.77
** 25 **  1.82       ** 53 **  2.27
** 26 **  6.04       ** 54 **  2.76
** 27 **  5.55       ** 55 **  4.06
** 28 **  2.88       ** 56 **  3.42
```

Table 6-2. Data on Bottle Cap Spoilage

```
HISTOGRAM

ABS. FREQ.
    -                     *
    -                     *
    -                     *
    -                *  * *
  5                 * *** *
    -               * **** *
    -             ** **** * *
    -             ********* *
    - *************** ****     *
     -'----'----'----'----'----'----'----'----'----'----'-
     -5   -4   -3   -2   -1   0   1   2   3   4   5

MEAN =    3.98571
STD. DEV. =   1.4281
SAMPLE SIZE =   56
```

Figure 6-9. Histogram for Bottle Cap Spoilage Data

That's a lop-sided histogram: more than half the observations are to the left of the mean, but some straggle out far to the right.

This is what I had in mind when I referred to positive skewness: the long tail pointing to the right. As you remember, this translates into a bow shape on the normal probability plot. You can see that in Figure 6-10. Both the skewness and kurtosis coefficients exceed 1.

The idea of the transformation you spoke of is to change the units in such a way that the bow will be straightened out?

Yes. And my first thought for a transformation, after I see the normal probability plot and the histogram, is to work with the logarithms of the spoilage rates rather than the percentage spoilage rates themselves.

You mean that you will take the logarithms of your data and try the plots again?

That's right.

LOGARITHMIC TRANSFORMATION

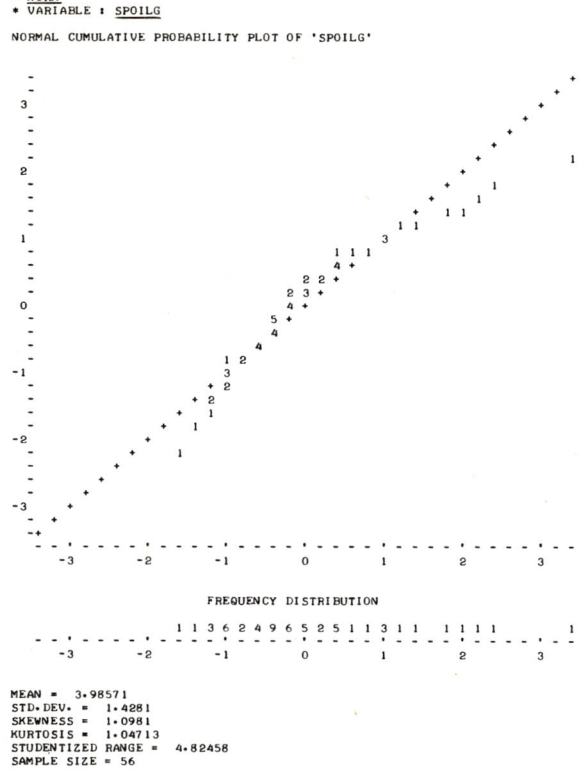

Figure 6-10. Normal Probability Plot and Frequency Distribution of Percentage Values for Bottle Cap Spoilage Data

That won't help me much, because I'm rusty on logarithms and how to look them up.

It might be useful for you to review the meaning of the logarithmic function, so that you can visualize what the transformation means in mathematical terms. However, the analysis need not await this. Just remember the command LOGE (for natural logarithms). Here's how it worked in this application:

```
> LOGE
* COL# TO PLACE VAR. : 2
* COLUMN TO BE TRANSFORMED : 1
UPDATING MEAN, STD, ..., CORR, ...

GIVE NAME OF NEW VARIABLE
VAR.  2 ?LOGSPL
```

Here I indicated that I wanted to transform to logs by typing the command LOGE. I then said that I wanted to put the transformed values into the second column of the data matrix, right next to the original values, and that I would name the transformed variable LOGSPL. The normal probability plot of LOGSPL is shown in Figure 6-11.

The normal probability plot looks good to me, and the skewness and kurtosis coefficients are close to 0. How about the histogram?

You can get a pretty good idea of what it looks like from the frequency distribution at the bottom of Figure 6-11, especially if you compare this with the corresponding distribution at the bottom of Figure 6-10. The transformation seems to have worked.

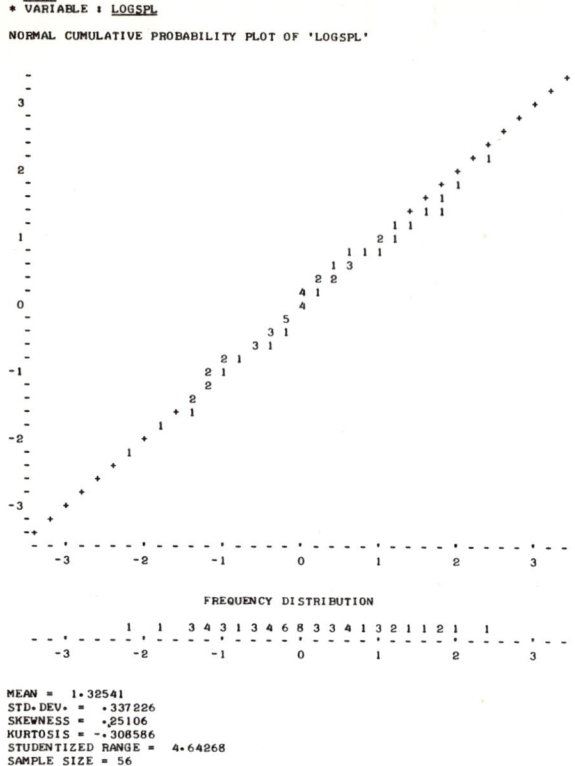

Figure 6-11. Normal Probability Plot and Frequency Distribution of Logarithmic Values for Bottle Cap Spoilage Data

But why did it work? You seemed to pull logarithms out of thin air.

First, from experience, I know that for distributions of positive data with the generally skewed-right configuration seen here, the logarithmic transformation is likely to improve considerably the normal approximation.

Second, just as we spoke earlier of conditions favorable for approximate normality of data in their original units, so here we can speak of conditions favorable for normality of the data transformed to logs. In fact, we can just translate our earlier conditions and apply them to the *logs* of the "underlying factors". If the logs of these underlying factors act independently and additively, and if no one of these factors dominates, the conditions are favorable. Another way to say this, which should be more meaningful, is to say that the factors themselves (not their logs) are *multiplicative* rather than additive in their effects. Under these conditions, then, the logarithms of the observed data themselves will tend to follow a normal distribution.

I think I understand the conclusion, but I have trouble filling in the mathematics of what you have just said.

Don't puzzle over purely mathematical points unless you are interested in them; the conclusions are what we shall need. But the key to the reasoning is that the logarithm of a product is the sum of the logarithms of the individual factors of that product.

CENTRAL LIMIT THEOREM MAY BE APPLICABLE TO LOGS OF "UNDERLYING FACTORS"

Are other transformations useful in statistical analysis?

Yes, and we shall see a few of them in due course.

Isn't it awkward to work with logarithms in a practical problem? I'd hate to tell a foreman to watch the logarithm of the spoilage percentage!

There is admittedly a problem of communication, but I don't think that it is difficult if one applies a little ingenuity and persistence:

1. There is no reason that many decisions cannot be made just as well in terms of logarithms as in terms of the original units. For example, control charts could be plotted with a vertical axis scaled logarithmically.

2. Although some subtleties may arise, it is usually possible to transform back to the original units in a satisfactory way after the statistical analysis has been made in logarithms.

CHAPTER SEVEN: INFERENCE

As I look back on what we've been doing, it seems to me that we've been spending a lot of time on the formulation of statistical models and the diagnostic-checking of the adequacy of possible models. When do we get to the statistical analysis itself?

That's a fair question, because except for one or two passing hints, I've avoided that subject in the last few chapters. Let's begin with a brief review of ideas introduced in Chapter 2:

1. In a broad sense, *inference* is the process by which we learn from experience. When we attempt to learn systematically from records or data, we speak of the process as *statistical inference*.

2. Inference is often a preliminary toward the taking of action, or, in alternative wording, the making of decisions. When decisions are made as the result of a systematic analysis of the outputs of statistical inference, we speak of the process as a *statistical decision procedure*. Medical diagnosis by computer could be an illustration.

Will we be concerned with inference or decision?

The emphasis will be on inference.

In what forms are statistical inferences expressed?

Remember first that inferences can be thought of as predictions; we are not interested in ritualistic computations but in implications of statistical analysis for what might happen in the future.

One important form of prediction is a probability assessment about some uncertain quantity. For example, we might wish to make an assessment of a probability distribution for the hardness reading of the next steel coil to be manufactured.

This is then an example of prediction expressed in terms of probabilities?

Yes. It's the same kind of prediction as a weatherman makes when he assesses the probability of rain at, say, 70%, rather than simply saying that it is going to rain.

Going back to the steel coils, didn't you say something about making a point estimate of the hardness of the next coil?

In Chapter 4, I said that the sample mean for the 100 observations would, given the assessment of a random model, be a sensible point estimate for the next reading. To get the right flavor of "point estimate", you might substitute the expression "best guess", or simply "good guess".

STATISTICAL INFERENCE AS PROBABILISTIC PREDICTION

112 □ CONVERSATIONAL STATISTICS

Thus, if 60 is the point estimate of the next hardness reading, this does not imply certainty that the next reading will actually be exactly 60.

No, it does not. A point estimate can be thought of as an incomplete probability statement. It serves to provide some idea as to the region in which our probabilities are concentrated.

But a point estimate doesn't tell whether the probability distribution is tight or spread out in the general vicinity of the point estimate.

Further information, such as an indication of dispersion, is needed for greater specificity of the inference.

I've read a little about statistics, and I've noticed a lot of emphasis on inferences about what are called parameters. How does that fit in here?

Both the terminology and the ideas underlying the terminology are important. Let's proceed by example.

Suppose that we make the assessment that the observations arise randomly in the study of the steel coils. We have a sample of 100 hardness readings for which the sample mean hardness, $\bar{x}$, has been 59.58. Now this number 59.58 is a fact about a sample of 100 particular coils already observed. Suppose, instead, that we were interested in what the process shows as mean hardness if we continued observing it indefinitely, assuming that it continued in a state of statistical control (the model of random observations).

In actual fact, the process can't go on indefinitely, nor would we even expect it to stay in the present state of statistical control for a really long time.

You're right on both counts. Still, the concept of the long run mean performance of the process can be of interest.

The mean hardness of the coils in the long run, given statistical control, is an example of a *parameter*.

If I understand the example, it would appear that a parameter cannot in fact be measured exactly. Isn't the concept a mystical one?

Philosophically it is. Yet many people find the idea intuitive, and it turns out to be a handy device for dealing with questions of statistical inference.

If you wish, you can remove the mysticism about the long run in our present example. As I suggested in Chapter 6, you can think of the long run as something that can be satisfactorily approximated by a large but finite random sample from the process. The parameter we're now discussing, the long run mean, is simply something approximated closely by the mean of the large finite sample.

You're saying, then, that long run mean hardness is a parameter, and that a parameter is different from anything we observe in samples that are obtained in practice?

PARAMETER AS A "LONG-RUN" VALUE OF A STATISTIC

PARAMETERS AND STATISTICS

INFERENCE   □   113

That's right, and the distinction between parameters and the measures actually observed for samples--which are sometimes called *statistics*--is essential to a clear understanding of the problem of inference.

To emphasize the distinction between parameters and statistics, many statisticians make a notational distinction. In our example, we shall make the following distinction:

1. $\bar{x}$ is used for the mean of a particular sample--the statistic.

2. μ (the Greek letter "mu"; pronounced "mew") is used for the long run mean--the parameter.

Are there other parameters of interest besides μ?

Yes, but we'll emphasize the parameter μ for the time being, and talk about inferences about $\tilde{\mu}$.

What's that funny little squiggle that you've drawn over the symbol μ?

That's a tilde (usually pronounced "tilda"). I sneaked that in because it serves to designate the fact that the object of inference is an *uncertain quantity*--sometimes called a *random variable*. You will see that its use is both natural and helpful.

So we'll talk about inferences about $\tilde{\mu}$?

Yes, but let's start first with inferences about a single future observation $\tilde{x}$ (note the tilde again) since it's easier to think of an inference about one future observation than about a parameter, and the same ideas are involved.

We'll use the example of the artifically generated "I.Q.'s". Suppose that I ask the computer for one more observation. How would you assess a probability distribution for this observation?

You said that one way would be to smooth this histogram of the first 100 observations, using one's best judgment.

That would be one method. But if we are satisfied that we can rely on the model of random observations from a normal distribution, we can use this assessment to aid us.

Do you mean that you would fit a normal distribution using the sample mean and standard deviation as you did in Figure 6-3 or Figure 6-4?

Not precisely that, but you're on the right track. We will base the analysis on the observed sample mean, $\bar{x} = 102.906$, and the observed sample standard deviation, $s = 15.9522$, of the sample of $n = 100$, but we will use a slightly more elaborate procedure.

Note, by the way, that the sample standard deviation, s, is another example of a statistic.

The sample values, $\bar{x} = 102.906$ and $s = 15.9522$, differ from the corresponding parameters, do they not?

APPROXIMATE INFERENCE BY SMOOTHING A HISTOGRAM

Yes, and in any application for which the normal model is an appropriate assessment, the sample statistics will differ from the corresponding parameters.

Since the I.Q. data were artifically generated, we know the parameters, since these had to be specified before the simulation could be made. In this respect, our illustration is atypical, since with real, as opposed to simulated, data the parameters are not known.

In the I.Q. simulation, you set $\mu = 100$, as I recall, and you said that the deviation between 100 and the sample mean $\bar{x} = 102.906$ was a result of sampling variation.

And the deviation between the sample standard deviation $s = 15.9522$ and the corresponding long run standard deviation 15 was also a result of sampling variation.

Do you have a special symbol for the long run standard deviation?

Yes. I'll use σ, the Greek lower case "sigma". So σ is another example of a parameter.

Check me on this: In your position as specifier of the simulation, you would assess a probability distribution for $\tilde{x}$ as normal with mean $\mu = 100$ and standard deviation $\sigma = 15$?

That's right, but in our present posture as analysts of the sample of 100, we shall pretend ignorance of how the simulation was set up. In other words, we shall pretend that we are statisticians dealing with real data.

I think I can see why fitting a normal distribution with mean $\bar{x} = 102.906$ and $s = 15.9522$ is not completely satisfactory for an inference about $\tilde{x}$. It would be correct only if $\bar{x}$ and s turned out precisely to equal μ and σ, O.K.?

O.K. And in making our inference about $\tilde{x}$, we want to make an appropriate allowance for sampling variation of the sample statistics $\bar{x}$ and s from the parameters μ and σ.

I suppose that the effect of such an allowance would be to stretch out the fitted curve of Figure 6-3.

Your intuition is sound.

If we combine our specification of the statistical model (random observations from a normal distribution) with the specification that the data constitute the bulk of our cogent evidence ("weak prior information"), there is a sharp mathematical answer to the problem of inference that we have posed.

Are you going to give the mathematics?

In line with my policy, no, since it would be a lengthy digression. When you understand the end result of the mathematical argument, you may be intrigued to see the argument itself, but let's postpone that pleasure!

ALLOWANCE FOR SAMPLING VARIATIONS OF $\tilde{\bar{X}}$ AND $\tilde{S}$

All right, then, what is the conclusion?

In our application, the probability distribution for a single future observation $\tilde{x}$ is approximated by:

1. A normal distribution,

2. centered at the sample mean $\bar{x} = 102.906$ (that is, the mean of the distribution is 102.906), and

3. dispersed according to the standard deviation

$$s\sqrt{1 + \frac{1}{n}} = 15.9522 \times \sqrt{1.01}$$

$$= 15.9522 \times 1.00499$$

$$= 16.0318$$

(that is, the standard deviation of the distribution is 16.0318).

Before I ask any other questions, please assure me that IDA will make this calculation for me!

Here it is, in first level prompts:

```
> REGR
* DEPENDENT VARIABLE (NAME OR COLUMN NUMBER) = IQ
HOW MANY INDEPENDENT VARIABLES ?0
MEAN                     = 1.02906E+02
STD. ERROR OF MEAN       = 1.59522E+00
STD. DEV.                = 1.59522E+01
STD. ERROR PRED. OBS.    = 1.60318E+01
```

For the time being, ignore "STD.ERROR OF MEAN". The other outputs are, respectively, $\bar{x}$, s, $s\sqrt{1 + \frac{1}{n}}$.

So, "STD.ERROR PRED. OBS." is $s\sqrt{1 + \frac{1}{n}}$.

Yes.

Would you explain the term "STD.ERROR"?

A standard error is really a special standard deviation, namely the standard deviation of an inferential probability distribution. Here, the probability distribution is that of $\tilde{x}$, as explained.

Then STD.DEV. is reserved for the ordinary standard deviation of the observations, $s = 15.9522$.

PROBABILITY DISTRIBUTION OF A SINGLE FUTURE OBSERVATION

In terms of Figure 6-3 or 6-4, it would be hard to distinguish this distribution from the normal distribution you drew in?

That's right. The standard deviation is only about 0.5% larger; everything else is the same.

You said that the answer is an approximation. Can you tell me what the effect of the approximation is?

The approximation is optimistic, but only slightly so for samples larger than, say, 20 or 30 observations.

1. The exact distribution is known as *Student's distribution*, or the *t-distribution*. This is a symmetrical distribution like the normal, but it is more fat-tailed.

2. The slight fat-tailed tendency by comparison with the normal distribution is scarcely noticeable except when the sample size is small.

3. The dispersion of the exact distribution as measured by the standard deviation is larger by the factor

$$\sqrt{\frac{n-1}{n-3}} = 1.01026 \ .$$

Are we going to study Student's distribution?

No, since it's really a technical refinement that we can do without.

Just remember that the normal distributions that we develop for answering problems of inference are slightly optimistic; the exact distributions are more spread out. We shall avoid treatment of very small samples for which the approximation might not be satisfactory.

I'd like to bring up a point that's been in the back of my mind for some time: Are you going to say anything more about why s, the standard deviation, seems to be so important in measuring dispersion?

It's certainly true that there's no intuitive reason why s comes in; the formula in Chapter 4 is not exactly what you'd hit on by yourself.

But if we were to show the mathematics by which results like the one above are derived (we won't actually do it!) you would see that:

1. As long as you are relying on the normal model in making inferences, the quantity

$$\Sigma (x - \bar{x})^2,$$

comes in to reflect the dispersion of the observations. The sum of the squared deviations from the mean $\bar{x}$ is involved.

NORMAL DISTRIBUTION IS USED AS AN APPROXIMATION TO STUDENT'S DISTRIBUTION

USING THE STANDARD DEVIATION TO MEASURE DISPERSION

2. In terms of formulas for dispersion in applications, such as the one we have been discussing, the most convenient form involves s, which in turn has the square root of the sum of squared deviations in the numerator.

All right, so the inference about $\tilde{x}$ is that we have approximately a normal distribution with mean 102.9 and standard deviation 16.03. How do we apply that?

Think of it as a way of packaging your information about $\tilde{x}$. Close your eyes and picture a normal distribution centered at 102.9 with a standard deviation of 16.03.

That means, for example, that you deem the chances roughly two in three that $\tilde{x}$ is within the interval 102.9 ± 16.03; that the chances are roughly 19 in 20 that $\tilde{x}$ is within the interval 102.9 ± 2 × 16.03, or 102.9 ± 32.06; that you are nearly sure that $\tilde{x}$ is within the interval 102.9 ± 3 × 16.03, or 102.9 ± 48.09.

If I wanted to make other probability calculations, could I use GAUS?

Yes. For example, suppose that you wanted to know the probability of an I.Q. between 105 and 125:

```
> GAUS
* MEAN     = ?102.906
  STD.DEV. = ?16.0318

* LOWER LIMIT L = ?105
  UPPER LIMIT U = ?125

  PR(L<X<U) =  0.36396

  MORE ?NO
```

How about the corresponding inferences about $\tilde{\mu}$?

The pattern is similar to what we have already seen. If we combine our specification of random observations from a normal distribution with the specification that the data constitute the bulk of our relevant evidence, the probability distribution for $\tilde{\mu}$ is approximated by:

1. A normal distribution,

2. centered at the sample mean $\bar{x}$ = 102.906 (its mean is 102.906), and

3. dispersed according to the standard deviation:

$$s/\sqrt{n} = 15.9522/10$$

$$= 1.595 .$$

Thus, the probability is about 2/3 that $\tilde{\mu}$ lies within the interval 102.9 ± 1.595?

That's correct.

THE MEANING OF THE INFERENCE ABOUT $\tilde{X}$

PROBABILITY DISTRIBUTION FOR THE LONG RUN MEAN

118 □ CONVERSATIONAL STATISTICS

As I look back to your illustration of the use of the command REGR, I see 1.59522 is given there, and it is called "STD.ERROR OF MEAN". So I take it that the standard deviation of the distribution of $\tilde{\mu}$ is called "standard error of the mean".

Yes, since a standard error is the standard deviation of an inferential probability distribution; in this instance, the distribution of $\tilde{\mu}$.

Is it worthwhile for me to memorize standard error formulas?

In general, no. You can get numerical values from the computer when you need them. The main thing for you to know is the interpretation of these numbers.

In a few cases, however, the formula is worth examining for lessons it can yield. The standard error of $\tilde{\mu}$ is one example:

$$s/\sqrt{n}.$$

This tells you that the dispersion of the distribution of $\tilde{\mu}$ is:

1. Directly proportional to the standard deviation, s, of the sample, and

2. indirectly proportional to the square root of the sample size n. To cut $s/\sqrt{n}$ in half, for given s, you would have to quadruple the sample size.

The standard error of the distribution for $\tilde{\mu}$ is much smaller than the standard error of the distribution for $\tilde{x}$. What is the common sense interpretation of that?

It means that you are much surer about the process mean than you are about a single observation from the process. Intuitively, the mean of a large number of observations reflects an averaging effect.

If we graphed the two distributions on a single chart, what would they look like?

Here is the graph:

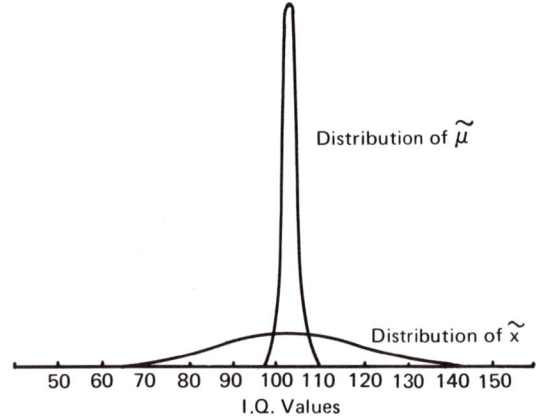

Figure 7-1. Distribution of $\tilde{\mu}$ and $\tilde{x}$ for I.Q. Data

INTERPRETATION OF S/√N

The distribution of $\tilde{\mu}$ concentrates the probability close to 102.9, the mean; the distribution of $\tilde{x}$ disperses the probability widely about that same mean.

Just a minute: I thought you said that you were drawing histograms of normal distributions. Those don't look like bells or cocked-hats.

Both curves represent normal histograms. The appearance is a consequence, to some extent, of the scales on which the curves are drawn. The standardized scales you have seen earlier are conventional.

But here we are not using standardized scales, and we must show two distributions with very different standard deviations on a single graph. If you transferred either curve to a standardized scale, it would look just like the drawings you have seen earlier.

How does all this relate to what I have heard called "confidence intervals"?

Let's take the interval 102.9 ± 1.595 for $\tilde{\mu}$ to focus the discussion.

I have been interpreting this interval directly in terms of the probability--approximately 2/3--that the long run process mean $\tilde{\mu}$ lies within it. This is the Bayesian interpretation. (See the discussion in Chapter 1 of the two statistical tribes, the sampling-theorists and the Bayesians.) We may call the interval a *Bayesian confidence interval*--or, alternatively, a *credible interval*, in order to stress that it is a direct statement of credibility or probability.

The sampling-theory tribe would call the same interval a *confidence interval*. The confidence coefficient is approximately 2/3.

What is the interpretation of "confidence coefficient"?

In repeated sampling, the method by which we arrived at the interval 102.9 ± 1.595 in the actual sample would be right about two times in three.

Then sampling-theorists and Bayesians agree in their interpretation?

No, the logic is entirely different, but it would take a long technical digression to convince you of that.

Well, I'm not going to ask for the technical digression! But while we're at terminology, would you explain the term "posterior distribution" that I've heard used?

This is the Bayesian name for the inferential distributions we have presented in this chapter.

But why "posterior"?

"Posterior" simply means "after the sample evidence has been analyzed".

CONFIDENCE INTERVALS AND CREDIBLE INTERVALS

PRIOR AND POSTERIOR DISTRIBUTIONS

Then "prior distribution" would mean "prior to the sample being analyzed"?

That's right. It would be based on all the evidence available up to, but not including, the sample being analyzed.

What prior distribution were you using in reaching the posterior distributions presented above?

Recall that I mentioned the specification that "the data constitute the bulk of our relevant evidence". This is an intuitive way of describing a prior distribution--say of $\tilde{\mu}$--that is nearly flat over wide intervals of μ. This is sometimes called a *diffuse* prior distribution. It turns out, then, that the sample data essentially determine the shape of the posterior distribution--say the distribution of $\tilde{\mu}$ shown in Figure 7-1.

Let's get back to intervals for just a moment. You said that in our example, the sampling-theory confidence interval and the Bayesian interval were the same numerically, even though there's a difference in interpretation that I didn't catch. Now, is it true that the intervals are always the same numerically?

In common applications, the intervals are exactly or approximately the same, so long as the prior distribution used in the Bayesian calculation is diffuse.

So far, we've spent most of our time talking about inferences in the form of probability distributions or probability intervals computed from probability distributions. At the start you mentioned point estimates. Can you say anything more about point estimation now?

Let me give you a rule-of-thumb. If we regard a point estimate simply as an all-purpose, condensed summary of a probability distribution, the mean of the distribution will often be satisfactory. Thus, we could use 102.9 as a point estimate of the long run process mean.

By itself, the point estimate tells nothing about the uncertainty of the estimate?

No. For example, 102.9 is a point estimate not only of the long run process mean, but also of the next individual observation. But we have seen that we are much more certain of the location of the process mean than of the next observation.

Would the standard error be a good measure of uncertainty?

In many applications, yes. In predicting the next observation, for example, we see that the standard error is 16.03, while the standard error of our inference about the process mean is only 1.595.

The all-purpose point estimate and its standard error would thus be a good way of reporting inferences in these applications. How about other applications?

INFERENCE REPORTED AS POINT ESTIMATE AND STANDARD ERROR

For a wide range of applications, as you will see, the point estimate and associated standard error do the job nicely. This is because the form of the posterior distribution is approximately normal for a wide range of applications, at least when the sample size is moderate or large and the prior distribution is diffuse.

Do you mean that in completely different applications it may be possible to interpret the point estimate and standard error in the same simple way we have learned?

It will be. For example, we shall spend a good deal of time on regression analysis, where we shall see the same pattern of inference repeated. The specific formulas used to compute the results are, of course, different, but the interpretation is the same.

Are there other forms of inferences besides the ones we have talked about?

Yes, but they will be given much less emphasis.

One important class of procedures can be called *conformity tests*. The runs procedure of Chapter 4 is an example. These are rather rough devices to aid one's judgment in selection of a statistical model.

These conformity tests are examples of a broad class of procedures called *tests of significance* or *tests of hypotheses*. Aside from conformity tests, however, I feel that tests of significance are seldom of much interest in practical problems arising in management, and I shall say almost nothing about them.

I can, however, illustrate one special test that is sometimes of interest. Consider the I.Q. data, and pretend that they are "real". Suppose you were interested in making a judgment as to whether or not the process from which the sample was drawn had a long run mean in excess of 100. The sample mean $\overline{x} = 102.9$, as you recall.

Under the circumstances, it might be of some value to assess the probability that the process mean $\widetilde{\mu}$ was, in fact, 100 or less.

That should be very easy to calculate by use of GAUS.

It is, and I have done it below:

```
> GAUS

* MEAN     = ?102.906
  STD.DEV. = ?1.59522

* LOWER LIMIT L = ?0
  UPPER LIMIT U = ?100

PR(L<X<U) =  0.03425

MORE ?NO
```

We see that we assess the probability at only 0.034 that the process mean $\widetilde{\mu}$ is less than 100.

Why did you set the lower limit L at 0?

SINGLE-TAIL PROBABILITIES

In the early part of Chapter 6, we saw that for an "upper-tail" probability, you can set U at any number greater than 4.5 standard deviations above the mean.

Similarly, for a "lower-tail" probability, as needed here, we set L at any number farther than 4.5 standard deviations to the left of the mean; 0 is easy to use, because no careful calculation is required.

How does this whole discussion tie in with random sampling from finite populations?

INFERENCES ABOUT THE MEAN OF A FINITE POPULATION

We have been talking for the last few chapters about sampling from a process that might in principle continue indefinitely under essentially the same conditions. In Chapter 2 we discussed sampling from a population for which the number of objects to be measured is finite, for example, the employees on a firm's payroll. Then we used the notation N for population size, as distinguished from n, the size of a sample.

Suppose that we are interested in an inference about the mean of a finite population of size N from which a simple random sample of size n is given. If we judge that the sample constitutes virtually all our cogent evidence, then the inference can be expressed approximately by a normal distribution for the population mean with:

1. Mean equal to the sample mean $\bar{x}$, and

2. standard deviation (standard error) equal to

$$\frac{s}{\sqrt{n}} \sqrt{\frac{N-n}{N}}$$

That's the same as what we had before except for the standard error formula.

But even that is not much different in the rather typical application in which the ratio n/N is small. Then the factor

$$\sqrt{\frac{N-n}{N}}$$

will be close to 1, and we are left with approximately $s/\sqrt{n}$. For example, if n = 100 and N = 10,000, then n/N = 0.01, and the factor displayed above is 0.995.

But insofar as there's any difference, can we say that, other things equal, the inference will be tighter for the finite population case?

Yes, since the factor

$$\sqrt{\frac{N-n}{N}}$$

is necessarily less than 1 for $n \geq 1$.

Is there any intuitive explanation for the finite population factor?

Note that

$$\frac{N - n}{N} = 1 - \frac{n}{N}.$$

There is no longer any uncertainty about the fraction n/N of the population that we have looked at, so we can subtract that out.

You mean there's no longer any uncertainty with respect to sampling error; there could still be non-sampling error, couldn't there?

Thanks for reminding me. That gives me the chance to say that none of these statistical formulas reflects non-sampling error.

I noticed that the finite population factor is 0 if n = N; what's the interpretation of that?

If n = N, the sample is a census. There is no sampling error in a census.

The inferences you have been talking about reflect the sample evidence itself rather than any more or less definite ideas you might have from other sources. Is there any way of bringing these latter in?

There are explicit Bayesian tools--non-diffuse prior distributions--for doing this, but I do not want to get into them. Just keep in mind that in a practical application you may want to make some direct judgmental allowance for non-diffuse prior information when you apply your inference; often the sample is *not* the whole story.

INTUITIVE EXPLANATION OF THE FINITE POPULATION FACTOR

124 □ CONVERSATIONAL STATISTICS

CHAPTER EIGHT: SIMPLE REGRESSION

You've emphasized the orientation of prediction, but all the examples so far have been based solely on the past performance of the variable to be predicted, as evidenced by sample data. Can't you exploit the fact that one thing may be related to another?

Often it is possible to do so. So far, we have considered examples in which there was a single variable, such as Rockwell hardness measures of steel coils or successive changes of the Dow-Jones Industrial Index. In many applications, however, you have two or more variables, and you may be able to use one or more of them to help to predict another.

I've heard the term "regression" frequently. Is that what you have in mind?

Regression is one of several statistical approaches to the problem. It is often applicable and provides a good understanding of the broader problem.

Can we work with an example?

Let's consider a study done by a company engaged in the sale of crushed blast furnace slag. Their major customer determined the iron content by a chemical test at the laboratory. The slag company developed an alternative magnetic test for iron content that could be applied quickly and cheaply before the slag was shipped out. The company wanted to determine the extent to which chemical iron content could be predicted statistically from magnetic iron content. The ability to predict would be of some interest in its own right; if this could be done, a method could be developed for controlling the chemical iron content by means of magnetic separation.

Over a period of about six months, both magnetic and chemical analyses were made for each of a sample of slag cars. It was not possible to make these paired measurements for the same material sampled from each car, but each pair refers to measurements from the same car.

The total number of pairs was 53. Entered from a file named $MCDERM, these are listed in time sequence in Table 8-1. Each pair occupies one row of this table, and each column shows one variable. Column 1 gives the chemical measurement, and column 2 gives the magnetic measurement.

A table laid out in this form--paired observations in rows, variables in columns--is an example of a data matrix; see Appendix B. It will be useful to visualize your data in this form, remembering, of course, that the numbering of the rows in the left-hand column is not strictly a part of the matrix.

PREDICTION USING REGRESSION

Table 8-1. Chemical and Magnetic Measurements of Slag for 53 Slag Cars

```
> FPRS
CONSECUTIVE COL. ?YES
I1,I2,J1,J2 = ?1,53,1,2
FMT = ?0,6D

**  1 **    24    25
**  2 **    16    22
**  3 **    24    17
**  4 **    18    21
**  5 **    18    20
**  6 **    10    13
**  7 **    14    16
**  8 **    16    14
**  9 **    18    19
** 10 **    20    10
** 11 **    21    23
** 12 **    20    20
** 13 **    21    19
** 14 **    15    15
** 15 **    16    16
** 16 **    15    16
** 17 **    17    12
** 18 **    19    15
** 19 **    16    15
** 20 **    15    15
** 21 **    15    15
** 22 **    13    17
** 23 **    24    18
** 24 **    22    16
** 25 **    21    18
** 26 **    24    22
** 27 **    15    20
** 28 **    20    21
** 29 **    20    21
** 30 **    25    21
** 31 **    27    25
** 32 **    22    22
** 33 **    20    18
** 34 **    24    21
** 35 **    24    18
** 36 **    23    20
** 37 **    29    25
** 38 **    27    20
** 39 **    23    18
** 40 **    19    19
** 41 **    25    16
** 42 **    15    16
** 43 **    16    16
** 44 **    27    26
** 45 **    27    28
** 46 **    30    28
** 47 **    29    30
** 48 **    26    32
** 49 **    25    28
** 50 **    25    36
** 51 **    32    40
** 52 **    28    33
** 53 **    25    33
```

So the column headed "24" has the chemical iron measurements, and the column headed "25" has the magnetic iron measurements.

Yes. The first observation was 24 for chemical iron and 25 for magnetic iron. The units are parts per 10000.

It looks to me that there is some tendency for the two measurements to move up and down together, but it's hard to tell much more. What's the interocular technique here?

One good way to start is to draw a scatter plot. The command is SCAT; here it is executed with first level prompts:

```
> SCAT
* GIVE VARIABLE NAME OR COLUMN NUMBER FOR THE
  VARIABLE TO BE PLOTTED ON THE VERTICAL AXIS : C IRON
* GIVE VARIABLE NAME OR COLUMN NUMBER FOR THE
  VARIABLE TO BE PLOTTED ON THE HORIZONTAL AXIS : M IRON
```

THE SCATTER PLOT

The resulting scatter plot is shown in Figure 8-1.

```
SCATTER PLOT OF STANDARDIZED VALUES OF 'C IRON' VS. 'M IRON'


  -
3 -
  -
  -
  -
  -                                                    1
2 -
  -                                 1
  -                              1     1
  -                        1       1 1     1
1 -                                           1
  -                  1        1          1       1  1
  -              1    1  2   1 1 1
  -                     1 1
  -                  1     1 1    1
0 -                          1 1 2
  -       1                 1   1
  -                  1       1
  -           1            2 1
  -
-1-                     2 2          1
  -                    3 2
  -                       1
  -                     1
-2-
  -           1
  -
  -
-3-
  -
  - - - . - - - . - - - . - - - . - - - . - - - . - - -'- - -
         -3      -2      -1       0       1       2       3


                   MEAN            STD. DEV.
VERT. VAR.        21.1321          5.04228
HORIZ. VAR.       20.7547          6.2999
SAMPLE SIZE = 53
```

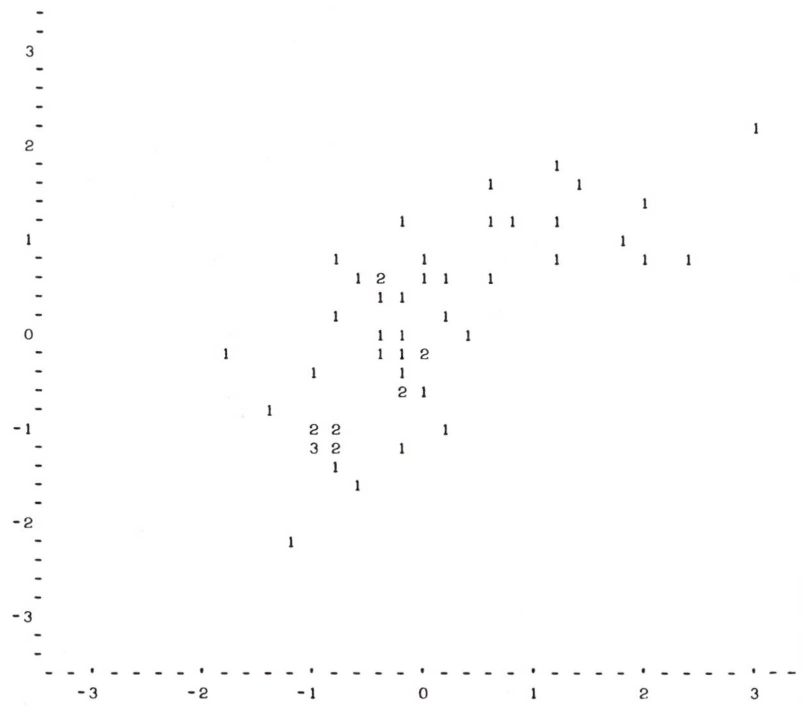

Figure 8-1. Scatter Plot of Standardized Values of 'C IRON' vs. 'M IRON'

Now tell me what SCAT has done.

Each pair of observations is represented by a point: the horizontal coordinate is M IRON expressed in standardized units, and the vertical coordinate is C IRON, also expressed in standardized units. The mean and standard deviation of each variable are printed separately beneath the plot; you can see that the means are very close, and that M IRON has a slightly greater standard deviation.

And as before, the printed digit on the graph tells how many points fell within the corresponding rectangle?

Yes. You can see that the points tend to drift upward as you move from left to right.

In other words, there is some tendency for larger values of C IRON to be associated with larger values of M IRON?

A tendency only. For any small interval of M IRON values, there is considerable dispersion of the points in the vertical direction, i.e., there is considerable dispersion of C IRON.

What next?

In this example it would make sense to see if the data conform

128 ☐ CONVERSATIONAL STATISTICS

reasonably well to what is called the simple linear regression model. This model contemplates that there is some "true" but unknown straight line, and that the observed points are distributed about it randomly and normally.

If you are up to a bit of algebraic notation, the statistical model specifies that each observed value of C IRON arose from the following scheme:

$$\beta_0 + \beta_1 x + \tilde{\epsilon} ,$$

where

1. x is any observed value of the horizontal variable; in this application, of M IRON.

2. $\beta_0 + \beta_1 x$ is the mathematical equation of a straight line.

3. β_1 is the slope of the straight line; that is, the rate of change of the line for a unit increase in the horizontal variable x.

4. β_0 is the intercept or constant term for the line; it is the height of the line at x = 0.

5. $\tilde{\epsilon}$ is a random drawing from a normal distribution with mean 0 and the same standard deviation σ_ϵ, regardless of x. $\tilde{\epsilon}$ is often called a *disturbance*.

Could you draw me a picture to represent this scheme?

We'll call the vertical variable y (C IRON in the example). Then here is the scheme for, say, the first observation (x_1, y_1):

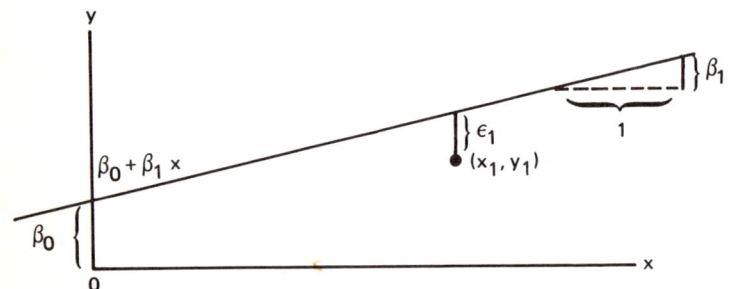

Figure 8-2. Simple Linear Regression Model

Think of the model in these steps:

1. A value, x_1, of the variable $\tilde{x}$ is observed.

2. Given $\tilde{x} = x_1$, the distribution of $\tilde{y}$ is a normal distribution with mean, $\beta_0 + \beta_1 x_1$, and standard deviation, σ_ϵ.

3. Then a value, y_1, of the variable $\tilde{y}$ is observed. See the point (x_1, y_1) on the drawing above.

THE SIMPLE LINEAR REGRESSION MODEL

GENERATION OF AN OBSERVATION IN THE REGRESSION MODEL

4. The disturbance, ϵ_1, on the first observation is $y_1 - \beta_0 - \beta_1 x_1 = \epsilon_1$. See the representation on the drawing above.

In practice, how do we find about what values are taken by β_0, β_1, ϵ so that we can draw the line?

We never know these things in practice: all we know is the observed point for each observation, such as (x_1, y_1), as shown below:

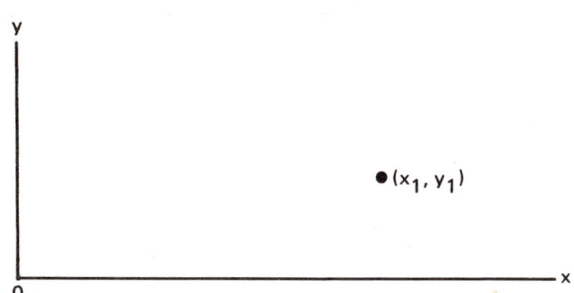

Figure 8-3. Single Known Point for Each Observation

If you never know the line, what good is it?

It is another example of a statistical model; namely, a conceptual scheme in the back of our heads that helps us to do a sensible statistical analysis of data.

In our earlier examples, we never knew the normal distribution from which we assumed the data were coming, but the model of the normal distribution guided our statistical analysis.

How does the statistical analysis proceed?

For a number of reasons, the model we have posed suggests that we fit a sample line by the method of least squares.

Why not just draw in a line by eye rather than use a mathematical method?

One practical reason is that a conscientious attempt to fit a line by eye is time-consuming by comparison to a mathematical computation done on a computer!

More basically, there are theoretical reasons--to which I have alluded above--that point to a certain mathematical procedure, the method of least squares. An exploration of these reasons, however, would take us beyond the technical level of this book. You probably won't mind avoiding this exploration!

Well, I'd like to understand anyway what "least squares" means.

To explain that, I'll introduce some more terminology and draw another picture. The line we fit by least squares will have the equation:

$$\hat{y} = b_0 + b_1 x.$$

THE MEANING OF LEAST SQUARES

130 □ CONVERSATIONAL STATISTICS

Here b_0 and b_1 are the sample estimates of β_0 and β_1; they will differ from β_0 and β_1 (miracles aside) because of random variation. Further, $\hat{y}$ is the y-value *predicted* by the sample line for any x. We shall call $\hat{y}$ a *fitted value*.

Typically, each y will differ by at least a little (with sufficiently accurate plotting) from the fitted value $\hat{y}$. The discrepancy between the actual and fitted value will be called a *residual*, denoted $\hat{\epsilon}$:

$$\hat{\epsilon} = y - \hat{y}.$$

Now for a given scatter diagram, the method of least squares produces sample estimates b_0 and b_1 such that the *sum of squared residuals*

$$\sum_{i=1}^{n} \hat{\epsilon}_i^2$$

is a minimum by comparison with any other straight line that could be fitted to the data, whether by eye or by some alternative mathematical procedure.

Show me what a residual looks like in a picture.

The picture will resemble Figure 8-2, with differences that I will explain in a moment:

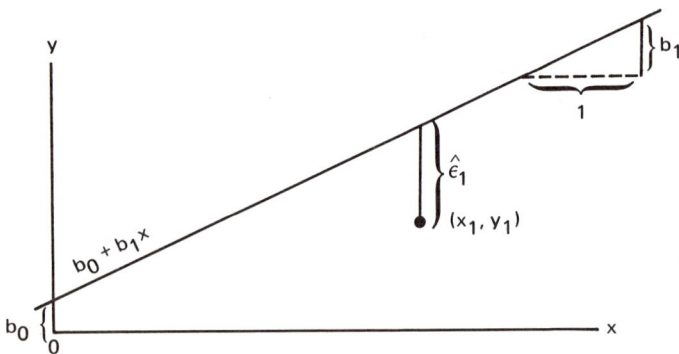

Figure 8-4. Example of a Residual

Only one point, (x_1, y_1), is shown; for comparison it is located in the same place as in Figure 8-2.

1. The point shown plus the remaining points enable the line $b_0 + b_1 x = \hat{y}$ to be fitted.

2. The line is *not* the same as the line $\beta_0 + \beta_1 x$ of Figure 8-2; the latter is conceptual and never observed in practice, while the former is actually computed from the data of a particular sample.

3. The residual, $\hat{\epsilon}_1$, is the *vertical* deviation of the point from the line. In this example, the residual is negative since it's below the line.

COMPUTING b_0, b_1, AND THE $\hat{\epsilon}$'s; THESE ARE POINT ESTIMATES OF $\tilde{\beta}_0$, $\tilde{\beta}_1$, AND THE $\tilde{\epsilon}$'s

The sample line $b_0 + b_1 x$ is different from the line specified by the model and shown in Figure 8-2.

Yes. The sample line is steeper ($b_1 > \beta_1$), its intercept is smaller ($b_0 < \beta_0$), and the residual, $\hat{\epsilon}_1$, is smaller than the disturbance, ϵ_1 (in absolute value, $|\hat{\epsilon}_1| > |\epsilon_1|$).

But we actually can observe only b_0, b_1, and $\hat{\epsilon}_1$, not β_0, β_1, and ϵ_1?

That is right, at least in practical applications. For teaching purposes, of course, we can construct simulations for which β_0 and β_1 are known, and ϵ is observed. (Just as we used subcommand NORM of RAND to generate the artificial I.Q. data in Chapter 5, so we can use the computer to create a simulation of the regression model. This will be illustrated at the end of this chapter.)

All right, how do you actually compute b_0 and b_1?

You use the command REGR followed by the command COEF. Here's the printout:

```
> REGR
* DEPENDENT VARIABLE (NAME OR COLUMN NUMBER) = C IRON
HOW MANY INDEPENDENT VARIABLES ?1
INDEP. VAR.  1 = ?M IRON
ANALYZING RESIDUALS ...

WARNING: RESIDUALS MAY BE EXCESSIVELY CORRELATED .
    OBSERVED NUMBER OF RUNS= 18
    EXPECTED NUMBER OF RUNS= 27.4151
    STANDARD DEVIATION OF RUNS= 3.59311
        (OBS.-EXP.)/(STD.DEV.)=-2.62032
```

As a response to the command REGR, IDA automatically makes certain diagnostic checks and, if necessary, issues certain warnings about the appropriateness of the model you are using. All these messages will be explained in due course.

```
> COEF

VARIABLE  B(STD.V)    B          STD.ERROR(B)   T

M IRON    0.7330    5.8664E-01   7.6242E-02     7.695
CONSTANT            8.9565E+00   1.6523E+00     5.420
```

What are you going to do about the warning given in IDA?

For the moment, nothing, since I am now using the data only as an illustration. Look next at the output given by COEF: the column headed B gives b_0 and b_1. In this printout, the line starting with CONSTANT gives $b_0 = 8.9565$; the line starting with M IRON gives $b_1 = 0.58664$.

Hence

$$\hat{y} = b_0 + b_1 x$$
$$= 8.9565 + 0.58664 x .$$

USE OF *REGR* COMMAND

"Dependent variable" is y; the variable plotted on the vertical axis of a scatter diagram; the variable to be predicted.

"Independent variable" is x; later we'll see examples with more than one independent variable.

THE REGRESSION LINE SHOWN ON THE SCATTER PLOT

How would the line appear on the actual scatter diagram of the original observations?

The computer plot is in standardized units. Figure 8-5 shows the original data in a redrawing of a plot made by the author of the original study.

I have modified his diagram in only one essential respect; I have shown that the least squares regression line must pass through the point $(\bar{x}, \bar{y})$, that is, the mean of each of the variables. That's a handy fact to remember.

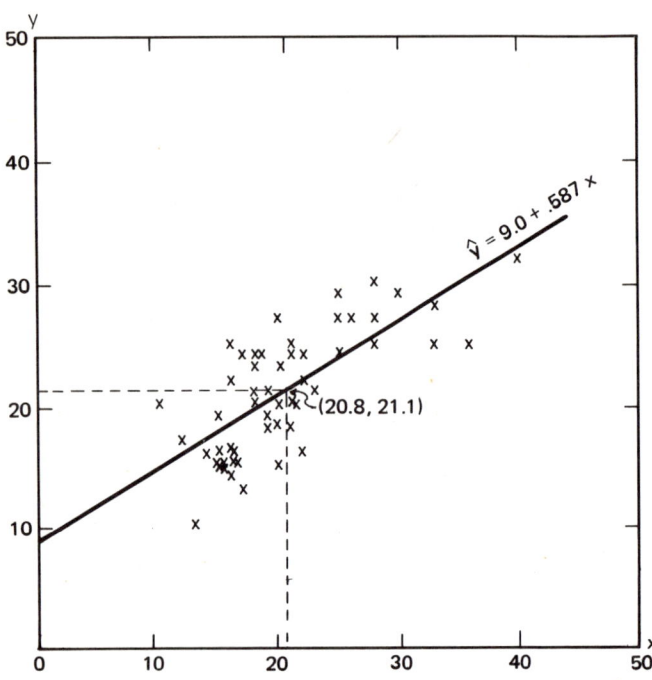

Figure 8-5. Original Data Plotted by Hand

The intercept is 8.9565. That would seem to say that we would predict chemical iron as 8.9565 if magnetic iron is 0. That doesn't make sense.

Arithmetically, the prediction is just what you say, and no, it doesn't make sense. If you reflect carefully on *why* it doesn't make sense, you will be well rewarded by a lesson about statistical models that applies in *any* application. In entertaining the simple linear regression model, at most we were saying that we thought this model might be consistent with data not too far from the mean of the set of data to be analyzed. We made no assertion that it would hold for any values of x and y. Later, we shall see how to give the data careful scrutiny to see whether they conform to the model we had in mind.

Now that you mention it, it looks to me like the dots have a tendency to curl around the line.

We'll talk about that in due course.

Now let's examine the plot of the regression line on our computer plot, shown in Figure 8-6.

1. Just as the plot above has the line passing through ($\bar{x}, \bar{y}$), so the computer plot of standardized variables will have the line passing through (0, 0).

2. In standardized units, the slope of the line is given by the number called B(STD. V.) in the printout on an earlier page: 0.7330. So the slope of the line in the computer plot of Figure 8-6 is 0.7330. Note that the CONSTANT or intercept in standardized units is zero.

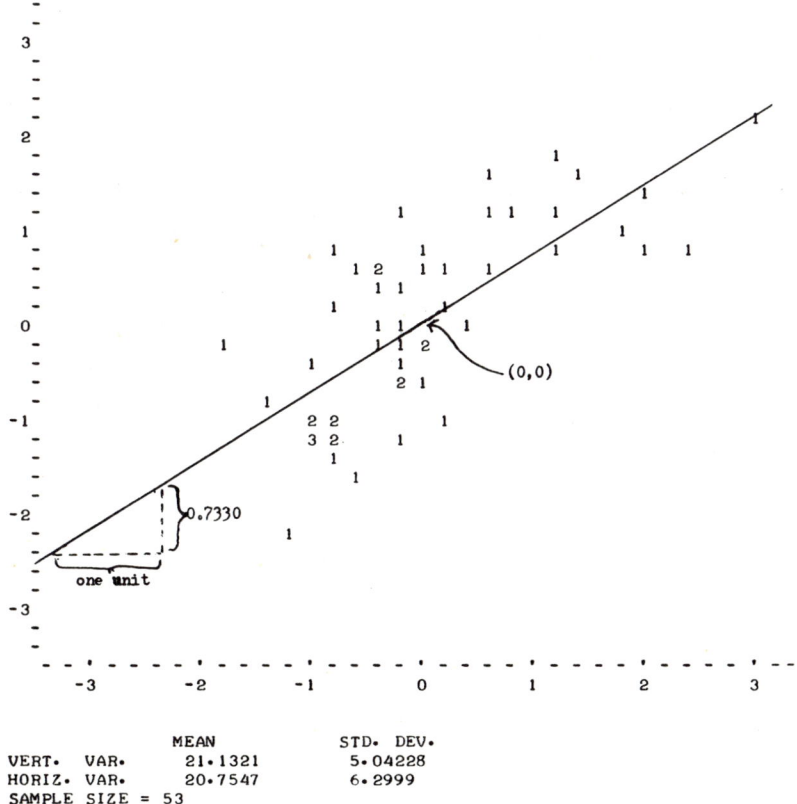

```
> SCAT
* VERT. VAR. : C IRON
* HORIZ. VAR. : M IRON
SCATTER PLOT OF STANDARDIZED VALUES OF 'C IRON' VS. 'M IRON'
```

	MEAN	STD. DEV.
VERT. VAR.	21.1321	5.04228
HORIZ. VAR.	20.7547	6.2999

SAMPLE SIZE = 53

Figure 8-6. Computer Plot of Data with Regression Line Superimposed

What's the purpose of giving B(STD.V.) in COEF?

For our usual purposes, we shall not be interested in B(STD.V.) as such. It is printed out only because it is of interest in certain specialized applications. *I shall ask you simply to ignore it when you look at the results of COEF.*

In the special case of simple regression, the model of this chapter, B(STD.V.) is of interest because it can be interpreted as *the coefficient of correlation* between y and x.

Do you mean that the coefficient of correlation is the slope of a special least squares line?

That's right. If x and y are expressed in standardized units, then

THE COEFFICIENT OF CORRELATION

the slope of the least squares regression line of y on x is numerically equal to the coefficient of correlation.

Can I get the coefficient of correlation directly from IDA?

It is one of the values given by SUMM, as shown below. In particular, you see that "UNADJUSTED MULTIPLE R" is given as 0.7330.

```
> SUMM
                  MULTIPLE R    R-SQUARE
    UNADJUSTED       0.7330       0.5372
    ADJUSTED         0.7267       0.5282
    STD. DEV. OF RESIDUALS   =      3.46359
```

Why "MULTIPLE"?

The same command SUMM applies also when there is more than one independent variable ("multiple regression"), where the important correlation concept is that of *multiple* correlation. When there is just one independent variable, as here, the multiple correlation coefficient is also the simple correlation coefficient, usually denoted r.

Will we discuss the multiple correlation coefficient when we get to multiple regression?

Yes.

Can I think of the sample correlation coefficient 0.7330 as an estimate of some "true" correlation coefficient?

Just as $b_1 = 0.58664$ is a point estimate of the parameter β_1 in the statistical model, so 0.7330 is an estimate of the "true" correlation coefficient ρ, a parameter implied by the model. For the most part, however, we shall be more interested in regression coefficients than in correlation coefficients.

But I can think of the simple correlation coefficient of a sample as the slope of the least squares line fitted to the standardized values of y and x?

Yes.

What is the meaning of "adjusted" R in the output of SUMM?

For technical reasons, the "adjusted" R is usually a preferred point estimate of the "true" correlation coefficient. In absolute value, it is somewhat smaller than the unadjusted R.

And what is R-square?

The square of R, that is, $R \times R = R^2$. R^2 is sometimes called the "proportion of variance explained by regression", but this interpretation lends itself so easily to misunderstanding that I shall not use it.

And what is STD. DEV. of RESIDUALS?

Denoted s_ϵ, it is a sample point estimate of σ_ϵ, the standard

STANDARD DEVIATION OF RESIDUALS

deviation of $\tilde{\epsilon}$. The formula is almost identical to the standard deviation formula for s in Chapter 4, applied now to the residuals $\hat{\epsilon} = y - \hat{y}$ from the least square sample line $\hat{y} = b_0 + b_1 x$:

$$s_{\hat{\epsilon}} = \sqrt{\frac{\sum \hat{\epsilon}^2}{n - 2}}$$

Why did you divide the sum of squared residuals by (n - 2) instead of n or (n - 1)?

It seems intuitively natural to divide by n. Even for the simple standard deviation of a sample of observations on a single variable, it is intuitive to divide the sum of squared deviations about the mean by n rather than by (n - 1).

The best answer that I can give may not satisfy you because it is not intuitive; the various standard error formulas will be simpler in appearance if we define the standard deviations in the way we have. It just comes out of the mathematics in that way. It might also help to remark that the "-1" in the simple standard deviation reflects the fact that the deviations are about the single fitted value $\bar{x}$; that the "-2" above reflects the fact that the deviations $\hat{\epsilon}$ are taken about the fitted line $\hat{y}$, which is a function of the two fitted values b_0 and b_1 (recall that $y = b_0 + b_1 x$), and that this same counting rule works for more complicated models to be taken up later.

Sorry I brought it up! Let's get back on the main track. Where do we go from here?

The next question is whether it was sensible to apply our regression model to this set of data.

The initial line of attack will be similar to the one we are already familiar with; we'll begin with PLTS, but we'll apply it to *residuals* from the regression just computed. The result is Figure 8-7.

How did you ask for Figure 8-7?

Residuals from the command REGR are stored by IDA under the name RESIDU, so I applied PLTS as follows.

```
> PLTS
* VARIABLE : RESIDU
ALL ROWS ?Y
```

My interocular procedure suggests that there is some tendency for residuals close together in the time sequence to resemble each other more closely than residuals that are farther apart. Can I check this impression by a runs count?

Since the runs count is in fact suspicious, IDA warned you of this at the time REGR was executed. But you can get the RUNS analysis of residuals in the usual way:

DIAGNOSTIC CHECKS OF RESIDUALS

136 □ CONVERSATIONAL STATISTICS

```
>  RUNS
*  VARIABLE : RESIDU

    OBSERVED NUMBER OF RUNS= 18
    EXPECTED NUMBER OF RUNS= 27.4151
  STANDARD DEVIATION OF RUNS= 3.59311
       (OBS.-EXP.)/(STD.DEV.)=-2.62032
```

Note that you can always call on residuals from the current regression by the name RESIDU. If you do another regression, then new residuals are computed by IDA, and these displace the old ones. But you can always put the current residuals directly into your data matrix by the command SAVR.

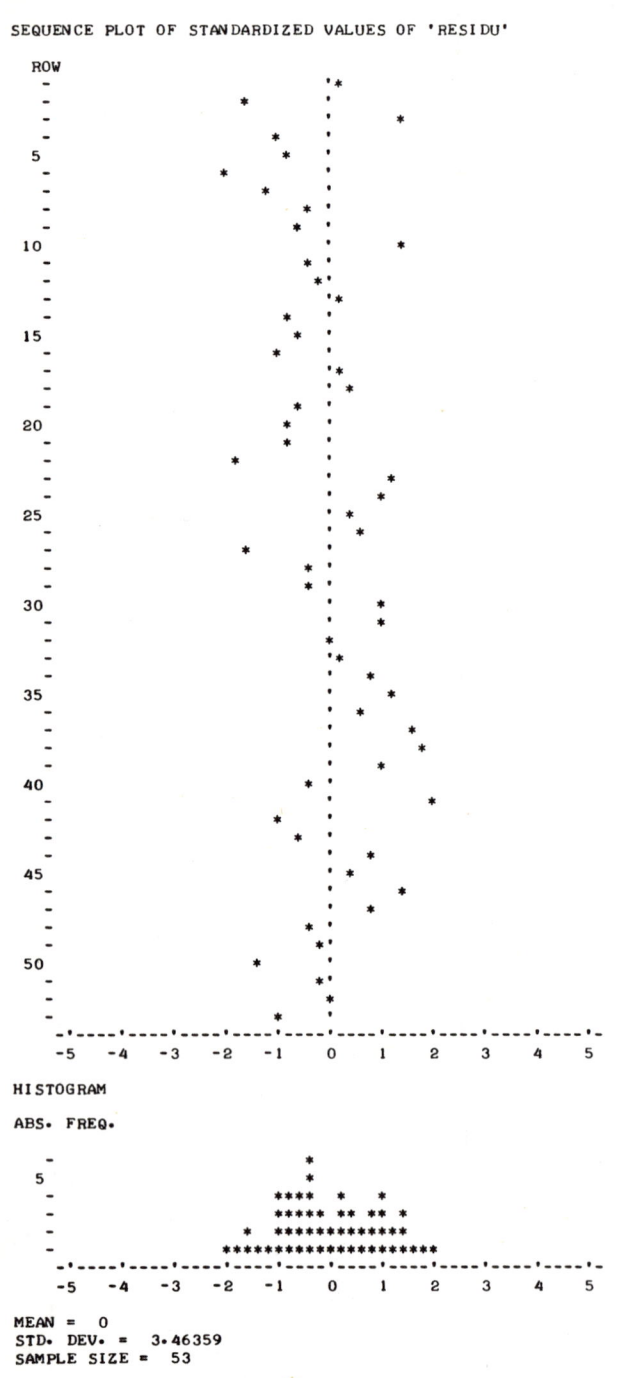

Figure 8-7. Sequence Plot and Histogram of Residuals of 'C IRON' on 'M IRON'

Do we stop at this point and see if we can find a better model?

That would be the natural thing to do. But let's defer the search for a better model until later, so that we can explore the other aspects of the problem. Our present purpose is pedagogical, and the simple linear regression model is only mildly, rather than severely, inappropriate to these data.

First, let's look at the normal probability plot of residuals in Figure 8-8.

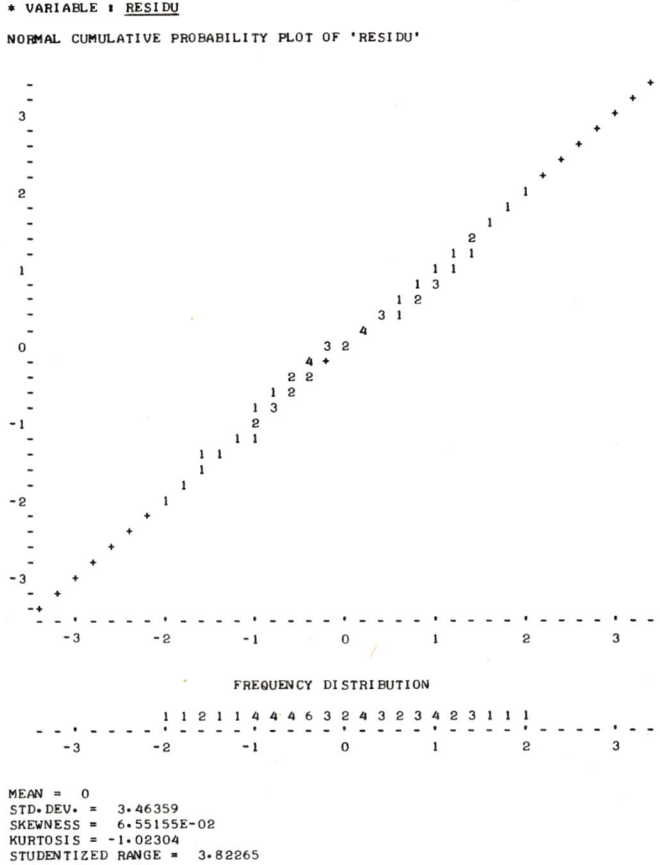

Figure 8-8. Normal Probability Plot of Residuals of 'C IRON' on 'M IRON'

Although the normal probability plot is not egregiously bad, it does hint at a reverse S-shape, and this is reflected in the kurtosis coefficient of -1.02.

How about the hint of curvilinearity we noted in the original scatter diagram?

Now that the regression has been carried out, we can get a slightly more informative plot bearing on this, namely a scatter diagram of *residuals* versus *fitted*; see Figure 8-9. *Fitted* means values of the sample line $\hat{y} = b_0 + b_1 x$ for the x's of the sample.

Note that I have drawn a horizontal line at height 0. If the linearity aspect of the regression model is appropriate, then the horizontal line should appear to the eye to be a reasonable fit to the points.

FURTHER DIAGNOSTIC CHECKS

138 □ CONVERSATIONAL STATISTICS

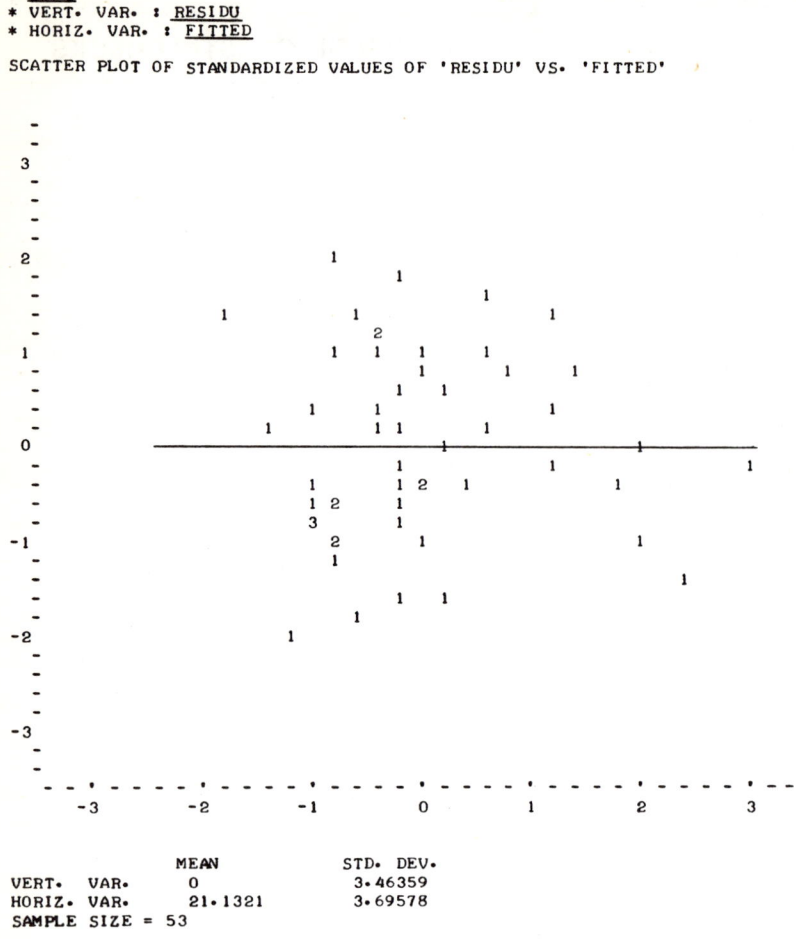

Figure 8-9. Scatter Plot of Residuals Versus Fitted Values of 'C IRON' on 'M IRON'

What is the advantage of this representation over the scatter diagram of y versus x, as shown in Figure 8-1 or Figure 8-6?

Essentially the same information about adequacy of the model is given in both representations, but it is easier to compare the points to a horizontal line than to a line sloping either positively or negatively.

Just a minute: do you mean to say it's possible to have a negative slope in a regression?

Yes, if the sign of b_1 is negative, the least squares line will slope downward. Note that a negative slope can be useful in prediction.

You can also see the deviations of the residuals somewhat more clearly in Figure 8-9 because of the way in which the scaling works. This is especially valuable when the correlation coefficient is close to +1 or -1.

I notice that the last few observations at the right of Figure 8-9 are at or below the horizontal line. The next few to the left are mostly above the line. That suggests the possibility of curvilinearity, doesn't it?

It does, but unfortunately the suggestion is weak so that the traumatic test is not traumatic in this example. We shall pursue the question in Chapter 10 by another route.

REGRESSION LINE MAY HAVE NEGATIVE SLOPE

Let me mention another command, RVSF (residuals versus fitted), that gives a quick mini-plot of Figure 8-9.

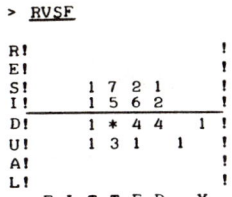

I have drawn in the appropriate horizontal line for you. With such broad intervals, most of the hint is lost, but we see that the two right-hand points show up beneath the line. If there were a strong hint of curvilinearity, the RVSF plot would have a good chance of picking it up.

Besides curvilinearity, are there any other kinds of model inadequacy that can be exposed by the plot of residuals versus fitted?

Yes. The regression model says that the disturbances come from a distribution with the same standard deviation regardless of x. This is the *constant scatter* assumption, for it implies that the dispersion of the residuals around the line should be about the same everywhere, after allowance for the fluctuations of sampling. One common violation of this assumption is for the disturbances to be more dispersed for large values of x than for small ones. This shows up in a tendency for the columns of residuals to show greater dispersion as you move to the right-hand side of the diagram.

Caution: Do not judge dispersion by the *height* of the columns. To do so is essentially to use the *range* (absolute difference between the largest and smallest observation) as a measure of dispersion. Unfortunately, the range tends to increase with sample size, so that a column with more residuals will tend to be more dispersed as measured by the range because of this deficiency of the range as a measure of dispersion. One way to get around this is to count the fraction of observations within each column that lie, say, within ± one standard deviation of the vertical variable from the mean of that variable. (RVSF facilitates this count, since it is scaled in units of one standard deviation.) After rough allowance for sampling variation, you should decide whether or not the fraction is about the same in all columns.

Figure 8-9 shows no substantial evidence against the constant scatter assumption.

It seems to me that I've heard a fancy Greek expression for what you have called constant scatter.

The term is *homoskedasticity*. Nonconstant scatter is *heteroskedasticity*. Try those on your friends.

As I have promised, we shall turn in Chapter 10 to modification of the model in order to remove some of the blemishes that we have observed: the hint of non-normality, the persistence of residuals by comparison to what would be expected under randomness, and the slight hint of curvilinearity.

THE CONSTANT SCATTER SPECIFICATION

140 ☐ CONVERSATIONAL STATISTICS

Now, purely for pedagogical purposes, we shall illustrate the inferences that would be drawn if we were content with the model. This will be an extension of the approach to inference given in Chapter 7.

Why don't you start with the prediction of a new observation?

All right. In predicting the next y--chemical iron--we will have an x--magnetic iron--to help us. For illustration, suppose that the magnetic iron reading is 23.

Before you consider that, how would you make the prediction if you didn't know the magnetic iron reading?

That question involves subtleties that you may not have thought of. We didn't study the chemical iron readings separately, in the manner of earlier chapters. Let me leave that to you as an exercise, but remember that statistics involves common sense as well as calculation! We shall see more about this question in Chapter 9.

To get a point prediction of chemical iron given a magnetic iron reading of 23, we will simply substitute 23 for x in the regression equation, as follows:

$$\begin{aligned}\hat{y} &= b_0 + b_1 x \\ &= 8.9565 + 0.58664 \times 23 \\ &= 22.449 \ .\end{aligned}$$

In terms of a scatter plot with the sample line plotted, this amounts to reading off the height ($\hat{y}$) of the line at the x in question.

This number can also be interpreted as the point estimate of the mean of a very large number of y's at x = 23--that is, the height of the "true" line (see Figure 8-2) at x = 23.

PREDICTION OF A NEW OBSERVATION

Can you use a normal approximation as before in assessing the distribution of a future $\tilde{y}$ at x = 23?

Yes, once you know the point estimates and the standard error. And, as before, the normal distribution that you use as an approximation will be slightly too optimistic. The exact distribution is somewhat more dispersed.

You can get both point estimate and standard error by the command SEPR, as illustrated on the first line of the following printout. The printout also gives a standard error of FITTED that will be explained below.

```
> SEPR
WANT TO USE THE VALUES OF THE INDEPENDENT
VARIABLES GIVEN IN THE DATA MATRIX ?NO
NUMBER OF Y'S TO COMPUTE S.E. = ?3

                  Y         S.E.(FITTED) S.E.PRED.(Y)

GIVE VALUES OF INDEP. VARS., SEPARATED BY COMMAS :
?23
           2.2449E+01    5.0562E-01    3.5003E+00

GIVE VALUES OF INDEP. VARS., SEPARATED BY COMMAS :
?20.7547
           2.1132E+01    4.7576E-01    3.4961E+00

GIVE VALUES OF INDEP. VARS., SEPARATED BY COMMAS :
?0
           8.9565E+00    1.6523E+00    3.8375E+00
```

I entered x = 23, as you see on the first line, and I got three numbers:

1. The value for Y is 2.2449E + 01; this is the point estimate just discussed.

2. The value for S.E.(FITTED) is the standard deviation of the normal distribution that expresses our approximate inference about the mean of a very large number of observations of $\tilde{y}$ at x = 23. Or, if you prefer, it is our approximate inference for the height of the "true" regression line at x = 23. This approximating distribution is thus normal with mean 22.449 and standard deviation 0.50562.

3. The value for S.E. PRED.(Y) is the standard deviation of the normal distribution that expresses our approximate inference about a single future $\tilde{y}$ given x = 23. This approximating distribution is normal with mean 22.449 and standard deviation 3.5003.

You are thus much more certain about the location of the "true" line than of a single future observation?

Yes, just as in Chapter 7 we were more certain about $\tilde{\mu}$ than about a single future $\tilde{x}$.

ALLOWANCE FOR UNCERTAINTY IN REGRESSION PREDICTIONS FOLLOWS THE OUTLINE DEVELOPED IN CHAPTER 7.

How about lines 2 and 3 of the printout?

In line 2, I got the answers to the same questions for an x = 20.7547, which happens to be the mean of the x's observed in the sample. Note that both standard deviations are smaller than the corresponding standard deviations for the inferences given x = 23.

In line 3, I got the answers for x = 0. Note that the standard deviations are substantially larger than the ones on either of the first two lines.

Judging from these examples, I would guess that the uncertainty of our inferences increases the farther x is from the mean of the sample just analyzed.

That's a good guess. The farther away x is from $\bar{x}$, the greater the effect of the uncertainty about the "true" slope $\tilde{\beta}_1$ on our inference about the height of the line or the value of an individual $\tilde{y}$.

But still the inference for x = 0 is unrealistic?

Yes. Just as we said before, x = 0 is far outside the range of the x's observed in the sample; we don't put trust in the model that far, even if we are satisfied with it for x's near the range of observed x's.

Can you give some indication of how these standard errors are computed?

In Chapter 7, we saw that the standard error of the distribution for a single future observation was

$$s\sqrt{1 + \frac{1}{n}}$$

while the standard error of the distribution of $\tilde{\mu}$ (the mean of a very large future sample) was

$$s/\sqrt{n} = s\sqrt{\frac{1}{n}}$$

where s was the standard deviation of the observations in the sample.

The formulas used in SEPR are somewhat more complex, but we can understand them in relation to the ones above if we examine the case of an inference conditional on an x equal to the mean $\bar{x}$ of the sample just analyzed. Above, s was the standard deviation of the sample observations. In the new formulas we use the standard deviation *of the residuals* $\hat{\epsilon} = y - \hat{y}$ from the least squares regression line, which we call $s_{\hat{\epsilon}}$, defined earlier in this chapter.

In this example, $s_{\hat{\epsilon}}$ = 3.46359, as shown by SUMM?

Yes, and also under the various plots of residuals.

INTERPRETING STANDARD ERRORS IN REGRESSION

Please continue your explanation of SEPR.

Let's look at the standard error of the inference about $\tilde{y}$ given that $x = \bar{x}$. The formula is this:

$$s_{\hat{\epsilon}}\sqrt{1 + \frac{1}{n}}$$

which parallels the first formula on the previous page.

What if x is not equal to $\bar{x}$?

Then there is a positive term under the square root sign, proportional to the squared distance from x to $\bar{x}$, which reflects our uncertainty about the slope of the "true" line. The standard error of $\tilde{y}$ is larger for x's more distant from $\bar{x}$, as we noted in the numerical illustrations, since the effect of uncertainty about the slope becomes more pronounced the farther x is from $\bar{x}$.

Similarly, the formula for the inference about the "true" line given $x = \bar{x}$ shows that the standard error is

$$s_{\hat{\epsilon}}\sqrt{\frac{1}{n}}$$

For x's not equal to $\bar{x}$, the additional positive term noted for the standard error of $\tilde{y}$ appears under the radical sign; the standard error is again larger for x's more distant from $\bar{x}$, again reflecting uncertainty about the slope of the "true" line.

So the standard errors of a single predicted $\tilde{y}$ are slightly larger than the standard deviation of residuals, 3.464?

Yes, even for $x = \bar{x}$, we saw that S.E.PRED.(Y) was 3.496.

Is there ever any interest in inferences about the "true" slope $\tilde{\beta}_1$?

Often this is the major purpose of a regression investigation. You want to estimate the average change in $\tilde{y}$ for a unit increase in x.

The inference follows the pattern that should be familiar by now. The approximating distribution for $\tilde{\beta}_1$ is normal with mean equal to b_1 and a standard deviation given in the printout following COEF. In the earlier explanation of COEF, you see that this distribution has a mean of 0.58664 and a standard deviation of 0.076242, or roughly 0.589 and 0.076.

You can visualize the inference in your mind's eye by picturing a normal distribution centered at the observed slope of 0.589 with a standard deviation of 0.076.

What is the meaning of "T" in that printout?

That is simply the ratio of the estimated slope to its standard error; thus $0.58664/0.076242 = 7.69$.

FURTHER DISCUSSION OF STANDARD ERRORS

INFERENCES ABOUT $\tilde{\beta}_1$ AND $\tilde{\beta}_0$

144 □ CONVERSATIONAL STATISTICS

What does that tell me?

Our interpretation is this: suppose that you were interested in whether the "true" slope $\tilde{\beta}_1$ were positive or negative. You see that your distribution of $\tilde{\beta}_1$ is centered at 0.58664, and that this center point is nearly eight standard deviations from zero. Therefore practically all your probability is on positive values of $\tilde{\beta}_1$, thus justifying the inference that $\tilde{\beta}_1$ is almost surely positive.

And is the value of T for CONSTANT to be interpreted in the same way?

Yes, from the T of 5.42, you see that $\tilde{\beta}_0$ is also almost surely positive.

This must be taken with a grain of salt, however, for reasons already mentioned; we don't have much confidence in our model itself for values of x this far removed from $\bar{x}$.

If it had been necessary to make explicit calculations for the probability $\hat{\beta}_1 > 0$ or the probability $\tilde{\beta}_0 > 0$, could we have used GAUS?

Yes, but remember that the normal distribution gives an approximation rather than an exact result.

We've spent a good deal of time in this chapter talking about one aspect or another of scatter diagrams, but we've studied only one example. It would help me to see some other examples for comparison, especially so that I would know better what to look for.

You may find it helpful to examine a series of scatter diagrams from simulated samples. I have prepared a series of simulations of samples of size n = 100 according to the following model:

1. $\tilde{x}$ is drawn from a normal distribution with mean 0 and standard deviation 1.

2. Given $\tilde{x}$ = x, a $\tilde{y}$ is generated according to the linear regression model of this chapter with varying regression coefficients. If you look at the distribution of $\tilde{y}$ separately, it also follows a normal distribution with mean 0 and standard deviation 1. Therefore, the regression coefficient specified by the model is equal to the correlation coefficient, and we shall discuss the plots in terms of the correlation coefficient.

In the first simulation, the "true" correlation ρ was 0; in the sample of 100 observations, the coefficient of correlation is r = 0.055. The difference between 0 and 0.055 is, of course, attributable to sampling variation. Similarly, the sample values of means and standard deviations differ from the "true" values of 0 and 1.

A REGRESSION SIMULATION

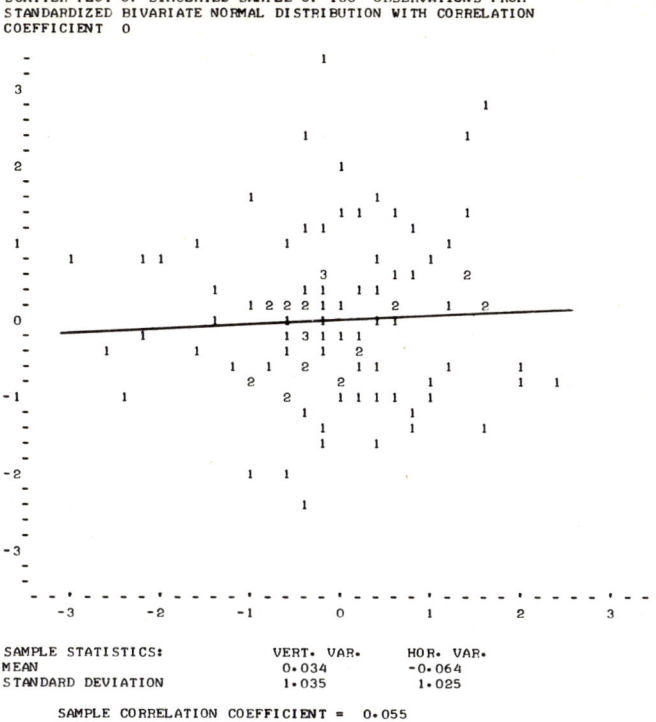

Figure 8-10. Scatter Plot of Simulated Sample of 100 Observations from Standardized Bivariate Normal Distribution with Correlation Coefficient 0

If I had just looked at the scatter plot without the line, I wouldn't have been able to tell whether the line should slope up or down.

In the next example, you can see the direction of the slope, but the relationship is certainly not striking.

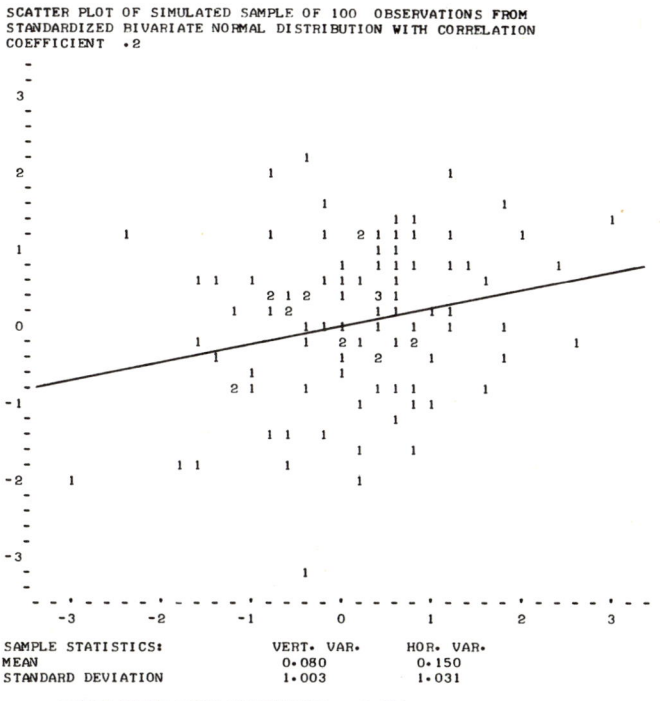

Figure 8-11. Scatter Plot of Simulated Sample of 100 Observations from Standardized Bivariate Normal Distribution with Correlation Coefficient .2

146 □ CONVERSATIONAL STATISTICS

Well, with r = 0.246, I can see from the plot that the line should slope upwards, but I'm still not much impressed by the association.

Even so, the sample correlation is somewhat larger than the "true" correlation of 0.20; sampling variation again. Now look at a sample r of 0.498 in a model in which the "true" correlation was 0.50:

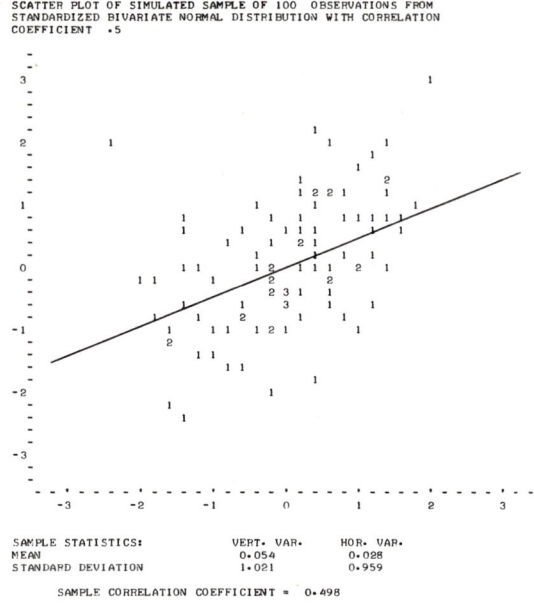

Figure 8-12. Scatter Plot of Simulated Sample of 100 Observations from Standardized Bivariate Normal Distribution with Correlation Coefficient .5

I'm surprised to see so much scatter about the least squares line; I would have expected a much greater reduction, especially since 0.5 is halfway from 0 to 1.

It can be shown that $s_{\hat{e}}$ is only about 13% less for a correlation coefficient of 0.5 than for one of 0, other things being equal. In Figure 8-13, r = 0.852 (for a "true" correlation coefficient of 0.80); $s_{\hat{e}}$ is more than 50% less than for r = 0.

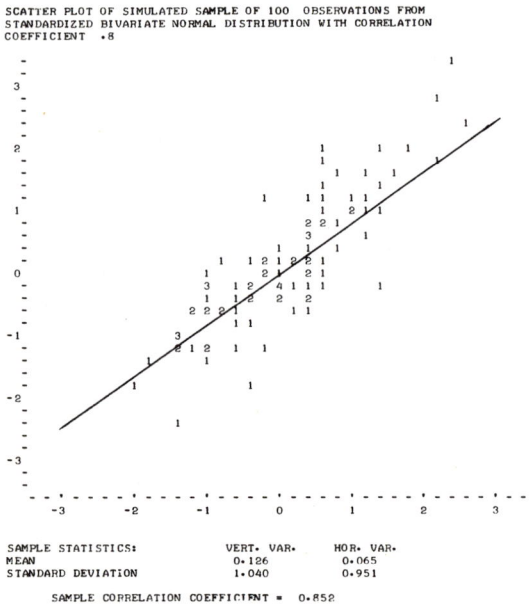

Figure 8-13. Scatter Plot of Simulated Sample of 100 Observations from Standardized Bivariate Normal Distribution with Correlation Coefficient .8

MORE SIMULATIONS

In Figure 8-13, I notice a tendency for the points to curl upward in the upper right-hand corner. A hint of curvilinearity?

Yes, there is a hint. But here, as compared with the slag measurements, we know that the linearity of the model is correct--unless something went wrong with the computer simulation. Interocular examination of data must be tempered by a healthy regard for the effects of sampling variation.

In the next example, we illustrate a negative correlation coefficient:

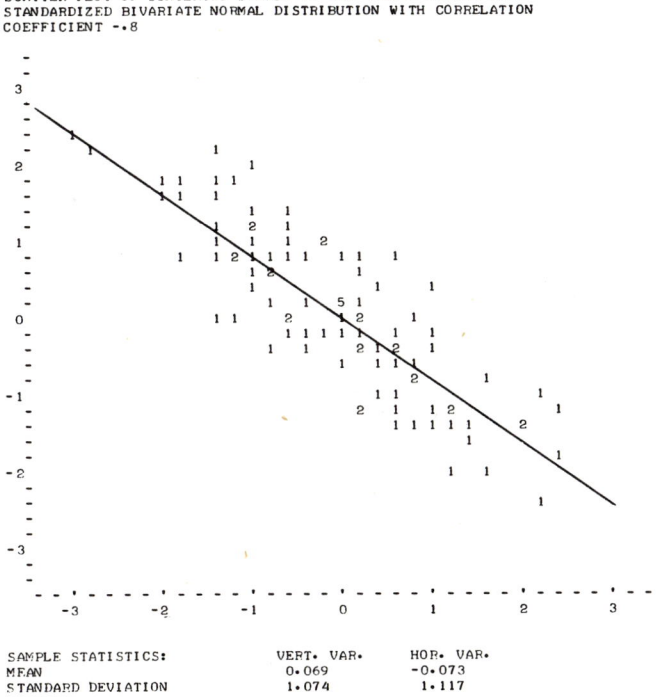

Figure 8-14. Scatter Plot of Simulated Sample of 100 Observations from Standardized Bivariate Normal Distribution with Correlation Coefficient -.8

My general impression is that Figure 8-14 looks like the mirror image of Figure 8-13.

That's a good way to describe the relationship between positive and negative correlation.

Our final simulation is that of an r of 0.955, given a "true" correlation of 0.95.

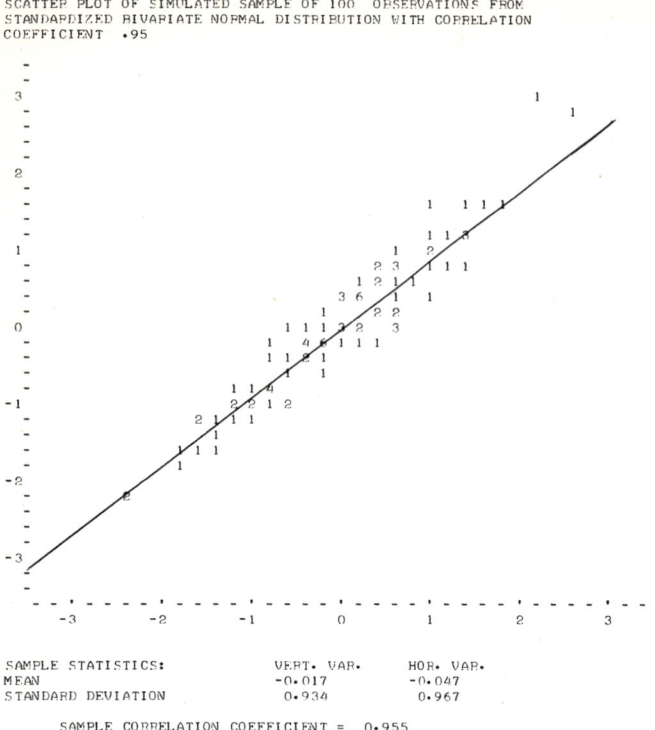

Figure 8-15. Scatter Plot of Simulated Sample of 100 Observations from Standardized Bivariate Normal Distribution with Correlation Coefficient .95

How could IDA be used to do simulations like this?

The key command is RAND, which we met back in Chapter 4. This command permits you to do simulations for practice in looking at random data, and also to review the key features of the regression model.

To begin with, I'll use RAND, and subcommand NORM, to produce a sample of n = 100 observations from a normal distribution with mean 0 and standard deviation 1. Since I started with an empty data matrix, I can specify 100 observations and RAND automatically stores these 100 observations in column 1; I named this variable X.

```
> RAND
*DIST = NORM
* N = 100
M,S = ?0,1
WANT TO SUPPLY NAMES TO SAMPLE?YES
GIVE NAME OF NEW VARIABLE
VAR. 1 ?X
COMPUTING...
```

Now at this stage I could, if I wished, do sequence plots, normal probability plots, or whatever, on X?

Yes, you could study X in any way you wished, just as if it represented actual data. And it's good practice to do so!

Next, I use RAND-NORM a second time to produce a second variable Y, also drawn from a normal distribution with mean 0 and standard deviation 1, and store Y in column 2. (N = 100 since 100 observations have already been entered in column 1 of the data matrix.)

USE OF *IDA* FOR REGRESSION SIMULATIONS

```
> RAND
*DIST = NORM
* COL. # TO PLACE SAMPLE : 2
M,S = ?0,1
GIVE NAME OF NEW VARIABLE
VAR.  2 ?Y
COMPUTING...
```

Then the "true" correlation between X and Y should be 0?

That's right, since the two samples were independently generated. Here's what the scatter diagram looked like:

```
> SCAT
* VERT. VAR. : Y
* HORIZ. VAR. : X
SCATTER PLOT OF STANDARDIZED VALUES OF 'Y   ' VS. 'X   '
```

[scatter plot]

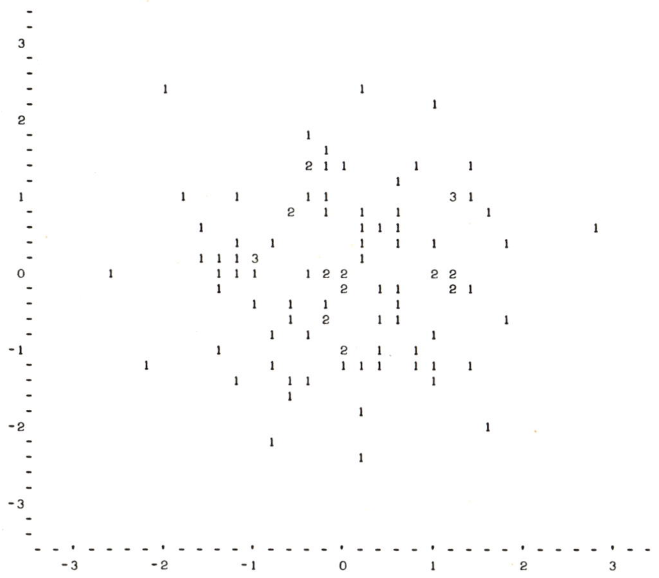

```
                MEAN           STD. DEV.
VERT. VAR.   1.09279E-02       1.03249
HORIZ. VAR.  3.46577E-02       .85456
SAMPLE SIZE = 100
```

Figure 8-16. Scatter Plot of Standardized Values of 'Y' vs 'X'

SCAT doesn't give the sample correlation coefficient. How can I get that?

Use the command CORR, shown below in third level prompts:

```
> CORR
HOW MANY VARIABLES ?2
COL. #'S: ?1,2
*  # DECIMALS = 3

            Y         X
    Y     1.000
    X    -0.002    1.000
```

Thus the sample correlation of X and Y was -0.002, very close to the "true" correlation of 0?

Right.

Suppose that I wanted to simulate a regression model, with any β_0 and β_1. How would that work?

First, you have to have the X variable. You could get this in any way you wished, including simulation, actual data, or even numbers you enter yourself.

Do you mean that the independent variable doesn't have to be normally distributed according to the regression model?

No, it does not have to be. Review carefully the exposition of the regression model at the beginning of this chapter.

Since, however, I am free to use the normal distribution if I wish, that is how I started out the simulation: for n = 100, I specified that $\tilde{x}$ be normal, mean 0, standard deviation 1.

```
> RAND
*DIST = NORM
* N = 100
M,S = ?0,1
WANT TO SUPPLY NAMES TO SAMPLE?Y
GIVE NAME OF NEW VARIABLE
VAR.  1 ?X
COMPUTING...
```

REQUIREMENTS FOR THE INDEPENDENT VARIABLE

Now you have your x's in column 1. What next?

Next, I generate the disturbances $\tilde{\epsilon}$. The regression model requires that these be normal with mean 0, and a standard deviation σ_ϵ to be supplied by me. I generated these disturbances, specifying σ_ϵ = 0.866025 and placing the 100 numbers in column 2 under the temporary name Y.

```
> RAND
*DIST = NORM
* COL. # TO PLACE SAMPLE : 2
M,S = ?0,.866025
GIVE NAME OF NEW VARIABLE
VAR.  2 ?Y
COMPUTING...
```

When does β_1 get into the act?

I next specified β_1 = 0.5, and computed $\beta_1 x_i$ for i = 1, 2, ..., 100 by multiplying all the values X in column 1 by 0.5, and placing the result in column 3 under the name Z.   The IDA command that does this is MULC--

Which means "multiply by a constant", I'll bet!

You're right.

```
> MULC
* COL# TO PLACE VAR. : 3
* COLUMN TO BE TRANSFORMED :1
* C = .5
UPDATING MEAN, STD, ..., CORR, ...
GIVE NAME OF NEW VARIABLE
VAR.  3 ?Z
```

BUILDING THE REGRESSION SIMULATION WITH *IDA* COMMANDS

Then you could get β_0 in the picture by adding a designated value for β_0 to each entry in column 3?

The IDA command that does that is ADDC ("add constant"), but I'll be lazy and set $\beta_0 = 0$ and skip this.

I'm getting just a little mixed up in following this. How are we going to put things together?

$$\tilde{y} = \beta_0 + \beta_1 x + \tilde{\epsilon}.$$

Now we have the ϵ's in column 2, the $\beta_0 + \beta_1 x$'s in column 3, so we can get the y's by adding columns 2 and 3. That is done below with ADDV ("add variable" or "add vector"), and the final result is stored in column 2 with the name Y:

```
> ADDV
* COL# TO PLACE VAR. : 2
I,J = ?2,3
UPDATING MEAN, STD, ..., CORR, ...
GIVE NAME OF NEW VARIABLE
VAR. 2 ?Y
```

So now you have the X's in column 1 and the Y's in column 2?

Yes, and the simulation is accomplished. Look at the scatter plot in Figure 8-17.

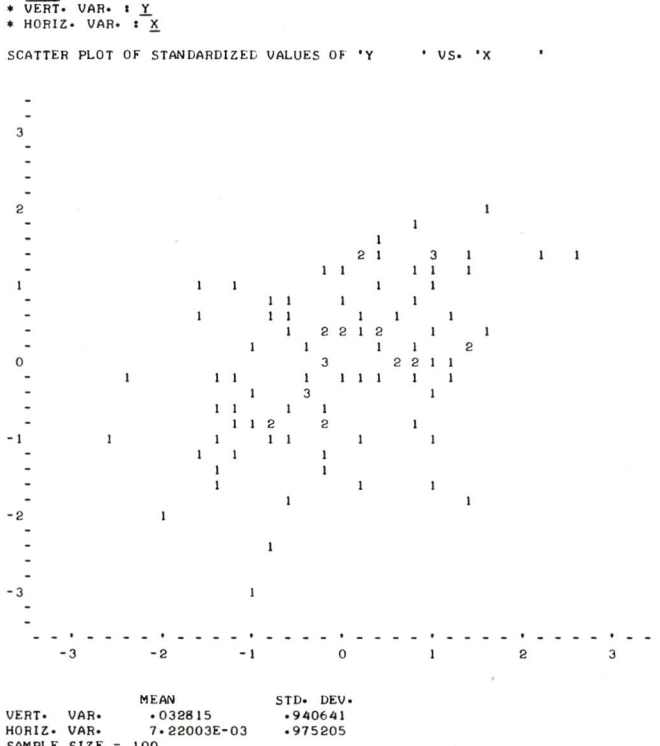

Figure 8-17. Scatter Plot of Standardized Values of 'Y' vs. 'X'

CONVERSATIONAL STATISTICS

What is the sample correlation coefficient?

```
> CORR
HOW MANY VARIABLES ?2
COL. #'S: ?1,2
*  # DECIMALS = 3

         X       Y
  X     1.000
  Y     0.453   1.000
```

Can you tell me what "true" correlation coefficient was implied by your specification?

The "true" correlation was 0.5.

Please give me the procedure by which you computed this.

Notation: (1) σ_x = standard deviation of $\tilde{x}$; (2) σ_y = standard deviation of $\tilde{y}$; (3) ρ = "true" correlation coefficient.

1. $\sigma_y = \sqrt{\beta_1^2 \sigma_x^2 + \sigma_\epsilon^2}$

    $= \sqrt{(0.5)^2 1 + (0.866025)^2}$

    $= \sqrt{0.25 + 0.75}$

    $= 1.$

2. Then the "true" correlation coefficient is

    $\rho = \beta_1 \dfrac{\sigma_x}{\sigma_y} = 0.5\left(\dfrac{1}{1}\right) = 0.5$.

Could I apply any of the other IDA commands of this chapter to the simulation?

Certainly. But if you want to spend more time on the data later, be sure to save your simulation in a file, using SAVF (see Computer Notes, Appendix B).

CORRELATION COEFFICIENTS IN THE SIMULATION

CHAPTER NINE: AUTOREGRESSION

I think that I've been following the development step by step, but I feel a little uneasy about my grasp of the overall strategy. Could we review a little before moving ahead?

That sounds like a good idea. Where do we start?

Well, I've just finished Assignment 3 of the first application in Appendix A. Even though I think I've done the right computations, I'm not sure about the meaning of my results.

That's the "market model" problem, in which you're asked to do a simple regression of monthly returns on AT&T common stock (dependent variable) on returns of a market index called NYSE. The time period is the 84 months from January 1961 through December 1967. (Detailed description in Application 1 of Appendix A.)

I got a correlation coefficient of 0.50, and I found myself wondering whether that was good or bad.

It's not very helpful to think of correlation coefficients as "good" or "bad". It's possible for a very small correlation coefficient to accompany a model that has very useful applications, and for a large coefficient of correlation to accompany an inappropriate, useless model.

How, then, do we judge the success of a statistical analysis?

There are two components:

1. Have we found an adequate statistical model—one that does justice to what seems to be going on in the data?

2. Given an adequate model, does it have useful applications?

Would you review the considerations of model adequacy?

When we contemplate a possible statistical model, there are two broad kinds of checks on its adequacy. The first is how well it fits in with our *a priori* knowledge, that is, with all information we have other than that of the sample being analyzed. The second comprises the various diagnostic checks that we have been studying.

Would you go over the diagnostic checks?

We started with applications in which there was just a single variable, such as Rockwell hardness measures or bottle cap spoilage. We considered a simple model that specified randomness and normality. Randomness meant essentially that there is no way to exploit the sequence of past observations in order to make predictions or inferences about the future.

And here we can use the sequence plot and a measure of conformity based on runs?

INTERPRETING CORRELATION COEFFICIENTS AND JUDGING THE SUCCESS OF A STATISTICAL ANALYSIS

154 □ CONVERSATIONAL STATISTICS

Yes, and for normality, our principal reliance is on the normal probability plot supplemented by conformity measures of skewness and kurtosis.

And how do these diagnostic checks extend to regression?

The regression specification includes the specification that the disturbances from the regression model are random and normal. Hence we can use the same checks as before, but we apply them to the residuals computed from the regression analysis. These residuals can be thought of as point estimates of the disturbances.

In addition, however, the regression model introduces the specification that the expected value of the dependent variable is a linear function of the independent variable. The scatter plot is an excellent graphical check on this specification. You can either plot the dependent variable against the independent variable or, for slightly better visibility, the residuals from regression versus the fitted values.

The scatter plot of residuals versus fitted values serves to check the linearity?

Yes. It is also useful to check the constant-scatter assumption. The mini-plot RVSF is especially useful for this latter check, because you can see easily whether the proportion of residuals within, say, one standard deviation of zero is about the same for low, medium, and high levels of the fitted values.

Are there other important diagnostic checks?

Certainly, and we shall see some of them. In this chapter, for example, we shall learn about *autocorrelations*.

Do users of statistics usually do the kinds of diagnostic checks that we have learned?

Often, unfortunately, they do not.

What happens to them then? Are they struck by lightning?

If they try to use the results of the statistical analysis for serious purposes, they do risk some such dire consequence. Often, however, the statistical analysis is just window-dressing that is not taken seriously, so grossly inadequate models are not discovered.

That leads me to a different line of questioning. You have said that finding a reasonably adequate model is one criterion of success. The other, you said, was whether or not the model finds useful applications. Could you discuss the kinds of applications served by regression models?

The most obvious and vivid is the prediction, in probabilistic form, of future values of $\tilde{y}$ given some specific value x of $\tilde{x}$. In the example of Chapter 8, for example, you might well be interested in predicting a value of C IRON given a value of M IRON. This prediction, in turn, may serve various decision objectives, such as quality control and process adjustment, either at the supplier's plant or in applications by the customer.

EXTENDING DIAGNOSTIC CHECKS TO RESIDUALS AND FITTED VALUES

That application assumes that you know M IRON before C IRON, does it not?

In the form we have discussed the problem, yes. We have discussed prediction of $\tilde{y}$ given $\tilde{x} = x$.

In the example of AT&T versus the NYSE index, we would actually learn the values of both variables at about the same time. Is there any interest in prediction of AT&T given NYSE in that case?

Potentially, there are two possible applications.

First, we might like to judge whether a future return of AT&T was better or worse than what would be expected given the return observed for the market in that same future period. We could answer this purpose by comparing the actual return with that predicted by the regression point estimate.

If I may interrupt you at that point, could you use the residuals from the regression analysis of the past data in the same way?

Yes, we could. We might further try to find other variables that might be associated with better-than-market performance, as evidenced by positive residuals, and then build these variables into a more complex model.

The second possible application concerns the fact that we may be able to predict the independent variable, such as NYSE, on the basis of some other analysis, and then combine the two predictions in order to forecast future values of AT&T. Unfortunately, in the particular example, the returns on NYSE are themselves as difficult to forecast as would be AT&T, so we don't gain much here. In other applications, we might gain something.

I've heard that some companies forecast sales in a two-step procedure suggested by what you just said. They use statistical regression to predict company sales given Gross National Product (GNP); then they get a GNP forecast from the company economist. Is that the sort of thing you had in mind?

It is.

What are some of the other applications of a regression analysis?

A second type of application is a prediction of a long run average of $\tilde{y}$ at some specified value $\tilde{x} = x$. For example, $\tilde{y}$ might be a process yield and $\tilde{x}$ might be a process input under the control of management. It might be natural, then, to want to assess the process capability if the input is set at some proposed level x.

The point estimate for that is the height of the sample line at x?

And the process capability, or long run mean, is the height of the "true" regression at x.

What about the slope coefficient β_1 and its sample point estimate b_1? The stock market regression of AT&T on NYSE, for example, gave me $b_1 = 0.46$.

USING REGRESSION TO EVALUATE PERFORMANCE

THE "BETA" OF PORTFOLIO ANALYSIS IS AN ESTIMATED REGRESSION SLOPE COEFFICIENT

In some applications there may be considerable interest in b_1 for its own sake. This is probably not true in the application of Chapter 8, but it is true in your example. The $b_1 = 0.46$ that you cite, for example, means that AT&T tends to fluctuate less than does the market: the predicted AT&T return is only 46% of the NYSE return, aside from the contribution of b_0.

That means that when the market goes up, AT&T tends to go up, but not so fast, and when the market goes down, AT&T tends to go down, but again, not so fast? (I'm abstracting from b_0, of course.)

Right. For this reason, the b_1 for a stock (often called its "beta") is an estimate of riskiness. A stock with a $b_1 > 1$ tends to swing more widely than the market, and is thus more risky than a stock with $b_1 < 1$, like AT&T.

If you were sure that the market was going up, then you'd want to buy a portfolio of stocks with high b_1?

That's the idea. Of course, it's also nice to have high b_0!

Let me now ask about the correlation coefficient of 0.50 that I got in the market problem You've said that I shouldn't think of it as good or bad, so I'll just accept it for what it is. But what does it tell me?

In that application, the scatter diagram looks pretty much like that of the simulation of Figure 8-12, except for the presence of a couple of outlying observations. Thus, in standardized units, the slope of the least-squares line is about 0.50; that leaves plenty of scatter about the line.

Could I say, then, that AT&T shows a moderate systematic relationship to the NYSE, with a lot of variability that is not associated with the NYSE?

That's the right idea. The systematic relationship captured by the sample correlation coefficient $r = 0.50$ is sometimes called the *systematic risk* in finance. To put the systematic risk in the right numerical perspective, however, it would be well to compare the standard deviation of residuals about the regression line with the standard deviation of the original values of **AT&T**. In that application, this gives $0.0342127/0.0392078 = 0.873$. As measured by the standard deviation, the dispersion of the residuals is fully 87.3% of that of the dependent variable. Thus, after allowance for marketwide forces that impact on NYSE, the standard deviation of residuals is only about 12.7% less than the standard deviation of the individual **AT&T** returns.

Hence the systematic risk is relatively small?

Yes.

Let me turn to another kind of question. As we try to find a regression model, what is the strategy we are following?

So far, we've seen the strategy only incompletely. The broad idea, however, is to introduce independent variables, transformations, and other building blocks of regression models to achieve two objectives:

THE STRATEGY OF SEARCH FOR A REGRESSION MODEL

1. To capture systematic influences on the dependent variable in the deterministic part of the regression model — that is, the part of the model other than the disturbance $\tilde{\epsilon}$.

2. To formulate the model in a way that the disturbances will conform to the random, normal specification.

And you judge your success from the behavior of the residuals $\hat{\epsilon}$?

Yes, since these are the sample point estimates of the disturbances. To the extent that we reduce the dispersion of residuals by modifying the specification of the systematic part of the model, we feel that we have made some progress toward the first objective. To the extent that the diagnostic checks on residuals are improved, we feel that we have made some progress toward the second objective. This is a bit of an oversimplification, but it will give you some useful perspective for the time being.

What do we do when we fail to find a good model?

Well, first we keep working until we exhaust our statistical "tricks" for improving model-specification. I suppose, however, that you want to know what we do then.

As an example, take the stock-market regression problem that you were working on. The regression model comes through pretty well there except for a hint of fat-tailedness in the normal probability plot of the residuals. Now that kind of model-blemish is not easy to deal with, at least with relatively elementary methods. My feeling is this: when you reach the limits of what you can do technically with formal statistical analysis, use your common sense and intuition to extend the analysis informally. It is plausible, for example, that the uncertainty of your inferences will be somewhat enhanced by the fat-tailed tendency. But your formal analysis gives you at least a lower bound on the uncertainty.

WHEN AN ADEQUATE REGRESSION MODEL CAN'T BE FOUND

Failure to find a good statistical model, or to improve a model previously obtained, may have good by-products. We learn what we *don't* know about the problem at hand. This can help us to avoid fruitless kinds of analysis, and can guide our search for further understanding. For example, the failure to find statistical models to predict the stock market beyond the very simple approach of Chapter 4 suggests two things:

1. It may not be wise to spend money or time on services that claim to predict the stock market by examination of past sequences of prices — the "chartists" and "technical analysts".

2. One may be rewarded by a search for theories consistent with the seemingly meagre statistical findings. Various versions of the "Efficient Markets Hypothesis", for example, have been stimulated in part by the "negative" statistical findings.

To paraphrase a wise economist, Frank H. Knight, it isn't so much what we don't know that hurts us, but the fact that we know so darn much that isn't so.

Well, that gives me a little perspective that I can mull over. Can we move ahead now?

I'm ready if you are.

Suppose that a sequence plot suggests nonrandom behavior; in particular, each observation tends to be closer to the preceding observation than to the mean of all preceding observations. Can we use regression methods to exploit this tendency?

One possibility is to treat the immediately preceding observation as an independent variable in a regression. Since this preceding observation is simply an earlier realization of the variable of interest, this approach can be called *auto*regression.

Suppose that, in the example of the last chapter, we had measurements only on C IRON to work with: no M IRON readings available to aid in prediction.

Then you'd start out with a sequence plot?

Yes, as shown in Figure 9-1. The RUNS analysis is shown below: too few runs are present by comparison with the random expectation.

```
>  RUNS

*  VARIABLE :  C IRON


      OBSERVED NUMBER OF RUNS= 15
      EXPECTED NUMBER OF RUNS= 27.4151
   STANDARD DEVIATION OF RUNS= 3.59311
        (OBS.-EXP.)/(STD.DEV.)=-3.45525
```

The time sequence effects seem strong. I notice a persistence from observation to observation and an apparent upward drift of the observations.

Good. Now perhaps you can suggest a way to exploit the apparently nonrandom behavior to our advantage in prediction.

I don't understand what you mean.

Well, you said, for example, that you noticed a persistence from observation to observation. Is that of any value?

You mean that we would pay special attention to the most recent observations?

You're on the right track. But *how* do we do this?

Oh, I see: you might, for example, treat the immediately preceding variable as an independent variable in regression! That's what you were saying before.

That's the idea.

LAGGING TO SET THE STAGE FOR AUTOREGRESSION

```
> PLTS
* VARIABLE : C IRON
ALL ROWS ?YES

SEQUENCE PLOT OF STANDARDIZED VALUES OF 'C IRON'
```

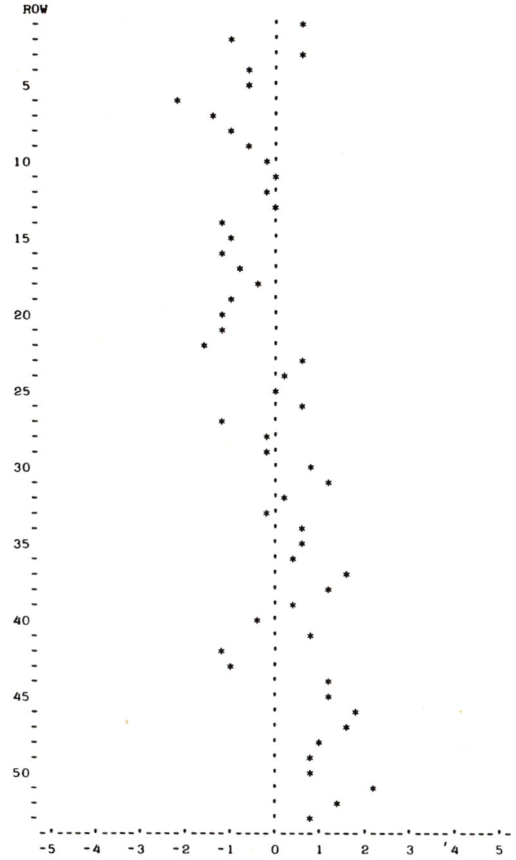

(sequence plot of standardized values of 'C IRON', rows 1–53, plotted on axis from -5 to 5)

```
HISTOGRAM
ABS. FREQ.
```

(histogram of standardized values on axis from -5 to 5)

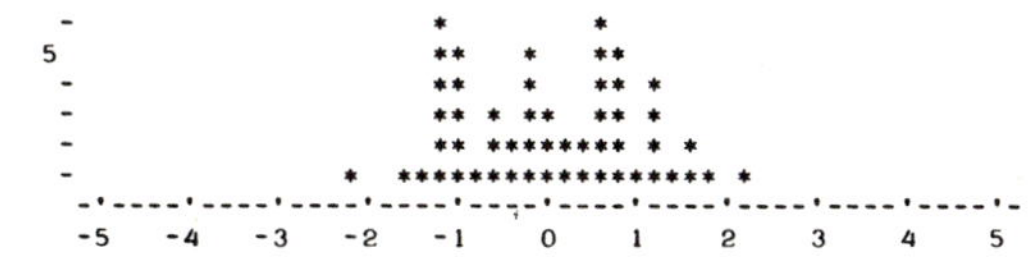

```
MEAN       = 21.1321
STD. DEV.  =  5.04228
SAMPLE SIZE = 53
```

Figure 9-1. Sequence Plot and Histogram of Standardized Value of 'C IRON'

Is it statistically correct to do this?

In many cases, yes. One dangerous situation arises when adjacent observations resemble each other *very* closely. But this set of data is safe.

I suppose you'd start with a scatter diagram. But how do you set up the data to do this?

The command is LAGG, which creates a new variable that is always a certain earlier value of the variable we start with.

If we want the value exactly *one* period back in time, we would answer 1 to the prompt "GAP = ?", as shown below.

```
> LAGG

* COL# TO PLACE VAR. : 2
* COLUMN TO BE TRANSFORMED : 1
GAP = ?1
UPDATING MEAN, STD, ...
GIVE NAME OF NEW VARIABLE
VAR. 2 ?CIRO-1
```

160 □ CONVERSATIONAL STATISTICS

What does the data matrix look like now?

The FPRS command gives this in Table 9-1. The first variable is C IRON, the second is CIRO-1.

Table 9-1. Data Matrix for Revised C IRON on M IRON Data

```
> FPRS
CONSECUTIVE COL. ?YES
I1,I2,J1,J2 = ?1,53,1,2
FMT = ?#,6D

**   2 **     16     24
**   3 **     24     16
**   4 **     18     24
**   5 **     18     18
**   6 **     10     18
**   7 **     14     10
**   8 **     16     14
**   9 **     18     16
**  10 **     20     18
**  11 **     21     20
**  12 **     20     21
**  13 **     21     20
**  14 **     15     21
**  15 **     16     15
**  16 **     15     16
**  17 **     17     15
**  18 **     19     17
**  19 **     16     19
**  20 **     15     16
**  21 **     15     15
**  22 **     13     15
**  23 **     24     13
**  24 **     22     24
**  25 **     21     22
**  26 **     24     21
**  27 **     15     24
**  28 **     20     15
**  29 **     20     20
**  30 **     25     20
**  31 **     27     25
**  32 **     22     27
**  33 **     20     22
**  34 **     24     20
**  35 **     24     24
**  36 **     23     24
**  37 **     29     23
**  38 **     27     29
**  39 **     23     27
**  40 **     19     23
**  41 **     25     19
**  42 **     15     25
**  43 **     16     15
**  44 **     27     16
**  45 **     27     27
**  46 **     30     27
**  47 **     29     30
**  48 **     26     29
**  49 **     25     26
**  50 **     25     25
**  51 **     32     25
**  52 **     28     32
**  53 **     25     28
```

What happened to the first row of the data matrix? I see from looking back at Chapter 8 that the first value of C IRON was 24.

The 24 is moved over and down to the second column of row 2. But what would we put in the second column of row *1?*

We don't know what that lagged observation was, because it preceded the data we're working with.

And so IDA simply masks row 1, not only for CIRO-1 but also for C IRON. We have to work with paired observations.

And if there had originally been another variable in the data matrix, its first value would have been chopped off also when LAGG with GAP = 1 was executed?

TREATING THE LAGGED VARIABLE AS AN INDEPENDENT VARIABLE IN AN ORDINARY REGRESSION ANALYSIS

That's right. So now we're ready for SCAT, used with respect to C IRON and CIRO-1 (see Figure 9-2)

```
> SCAT
* VERT. VAR. : C IRON
* HORIZ. VAR. : CIRO-1

SCATTER PLOT OF STANDARDIZED VALUES OF 'C IRON' VS. 'CIRO-1'
```

[scatter plot]

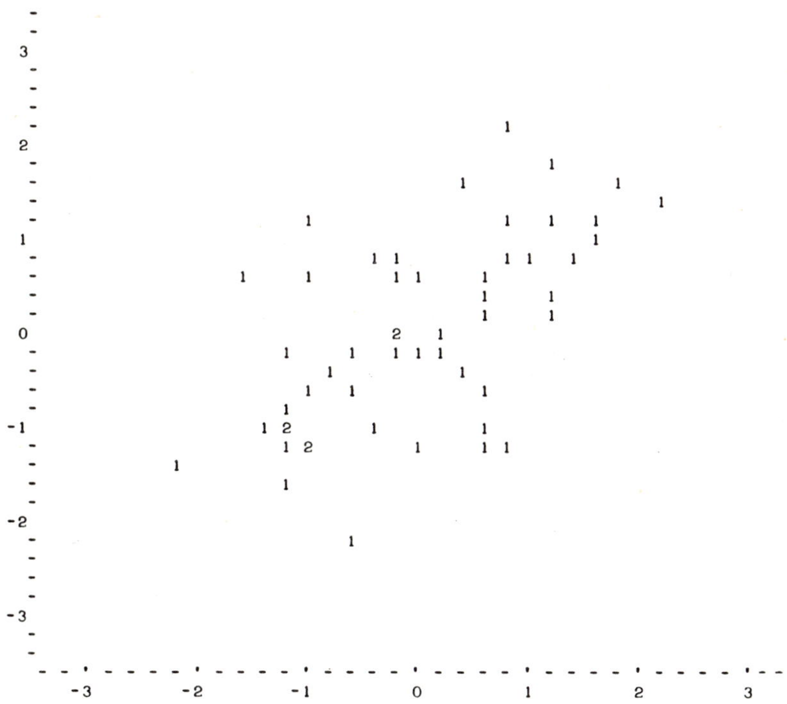

```
              MEAN        STD. DEV.
VERT. VAR.    21.0769     5.0753
HORIZ. VAR.   21.0577     5.06202
SAMPLE SIZE = 52
```

Figure 9-2. Scatter Plot of Standardized Values of 'C IRON' vs. 'CIRO-1'

So far as I can tell, there's a moderate, positive association between C IRON and CIRO-1. And the regression model looks good.

I must warn you that sampling-theory statisticians have reservations about using ordinary linear regression without modification in such applications. Bayesian statisticians feel that autoregression is essentially like ordinary regression, with only this qualification: we must beware of cases in which the scatter is very close, say a sample correlation coefficient in excess of 0.7, and also where the sample size is only moderate — say 50 or 100.

These cases require special treatment?

Yes. Often, differencing, as in the Dow-Jones data at the end of Chapter 4, is all we need. Sometimes other methods are required.

Here, however, we're pretty safe in using REGR, so I'll go ahead.

```
> REGR
UPDATING CORR. MATRIX...
* DEP. VAR. = C IRON
HOW MANY INDEP. VAR. ?1
INDEP. VAR.   1 = ?CIRO-1
ANALYZING RESIDUALS...

> COEF

VARIABLE    B(STD.V)      B           STD.ERROR(B)    T

CIRO-1      0.5959     5.9745E-01     1.1387E-01      5.247
CONSTANT               8.4959E+00     2.4648E+00      3.447

> SUMM

              MULTIPLE R    R-SQUARE
UNADJUSTED      0.5959       0.3551
 ADJUSTED       0.5850       0.3422
  STD.DEV. OF RESIDUALS  =   4.11636

> MEAN

VARIABLE        MEAN          STD. DEV.

C IRON       2.10769E+01    5.07530E+00
CIRO-1       2.10577E+01    5.06202E+00
```

Why did you do COEF, SUMM, and MEAN before looking at model adequacy commands, such as PLTS for residuals?

There's no fixed sequence, so long as we look at model adequacy before we're done.

So a point prediction of a C IRON reading would be made by adding 8.496 to the product of the previous C IRON (CIRO-1) and 0.597?

Yes, but remember that this is the point estimate based only on the immediately preceding value of the dependent variable. We're not using M IRON at all in this chapter.

I notice that the standard deviation of residuals is 4.116, which is larger than the 3.464 that we obtained using M IRON as the independent variable in Chapter 8.

You're perfectly right, but we'll see in Chapter 10 that we're not faced with an either/or choice; we'll see how to use both variables.

But for the present, we'll concentrate on a discussion of the autoregression just run. And note that the standard deviation of residuals, 4.116, is substantially less than the standard deviation of C IRON of 5.075 given by MEAN.

Next, note also that the correlation coefficient of this regression is 0.596. That means that when C IRON and CIRO-1 are plotted as in the scatter diagram of Figure 9-2, 0.596 is the slope of the least squares line.

Why not use the autocorrelation coefficient as another measure of randomness?

A zero "true" (long run) autocorrelation coefficient would be consistent with the notion of a random process producing the data. However, there's a little more to the question than the autocorrelation coefficient between the current and previous observation.

We can also define an autocorrelation coefficient between, say, the current observation and the observation *two* periods back. Just as

AUTOREGRESSION MEASURES

the autocorrelation already studied is often called the *first* autocorrelation coefficient, so this one is called the *second* autocorrelation coefficient.

The scatter diagram corresponding with the second autocorrelation coefficient would be made by first creating a variable lagged two periods?

Yes. We would simply modify the procedure under LAGG by calling for a gap of 2 rather than 1. You might try that yourself.

There would then be (n - 2) points in the scatter diagram?

Just as there are (n - 1) in Figure 9-2.

And you could go on to define a third, fourth, etc. autocorrelation coefficient?

Yes. The "true" or long run values for a process that could in principle go on forever can be defined as far back as you wish.

And all should be zero if the process is random?

Precisely. Of course, the sample coefficients we compute will show random or sampling variation even if the process is random.

How can we evaluate the role of sampling variability?

If all the long run autocorrelations are zero, then the standard error of any sample autocorrelation is approximately $1/\sqrt{n}$, where n is the sample size.

In our example, this means a standard error of $1/\sqrt{53} = 0.137$.

What is the IDA command for sample autocorrelation coefficients at different lags?

AUTO. Here's how it works for our example for the entire 53 observations of C IRON.

```
> AUTO
* VARIABLE : C IRON

* MAX ORDER : 13

  ORDER   AUTOCORR.
    1      +0.591
    2      +0.452
    3      +0.302
    4      +0.213
    5      +0.250
    6      +0.306
    7      +0.332
    8      +0.229
    9      +0.099
   10      +0.157
   11      +0.133
   12      +0.167
   13      +0.252

  S.E. OF EACH COEF. GIVEN RANDOM MODEL=0.137

  BOX-PIERCE STATISTIC= 61.0683
  EXPECTATION GIVEN RANDOM MODEL= 13
      (OBS.-EXP.)/(STD.DEV.) = 5.29181
```

Why did you compute only the first 13 autocorrelation coefficients?

AUTOCORRELATION COEFFICIENTS AT DIFFERENT LAGS

164 CONVERSATIONAL STATISTICS

Because of sampling variability, it is usually not a good idea to go back to lags more distant than about one-fifth or one-fourth the sample size. IDA will not permit you to go back further than lag 20, which is one-fifth the standard maximum sample size of 100.

I notice that the first autocorrelation is 0.591 rather than 0. 596. Why is that?

For technical reasons, a slightly different formula (the one recommended by Box and Jenkins, whom we shall encounter later) is used in AUTO.

All the correlations are positive; many of them are substantially larger than the "standard error given random model".

Clearly, the autocorrelation coefficients confirm our earlier visual analysis of the sequence plot of C IRON, and also the runs count at the beginning of the chapter.

Is this pattern the one to expect if the process is nonrandom?

Not necessarily. There are many nonrandom patterns. This particular one, however, suggests that the simple autoregression model is an intelligent model to try. We will see in a moment how well the model has succeeded in representing what seems to be going on.

Would you first explain the Box-Pierce statistic?

It is a rough overall measure for the set of autocorrelations above. Actually it is: (the sum of the squares of the autocorrelation coefficients) times (sample size). The larger the Box-Pierce statistic, the poorer the correspondence to what is expected in a random model, namely a number equal to the number of autocorrelations asked for (13, in the example). The deviation of the actual from expected in standard deviation units is printed out on the third line; here the observed is more than 5 standard deviations from expected. That represents a gross discrepancy.

Can you apply AUTO to residuals from regression or autoregression?

As a rough guide (shown below). The RESIDU there comes from the regression (actually, autoregression) executed back several pages.

```
> AUTO
* VARIABLE : RESIDU

* MAX ORDER : 13

ORDER   AUTOCORR.
  1      -0.081
  2      +0.107
  3      +0.059
  4      -0.056
  5      +0.080
  6      +0.108
  7      +0.213
  8      +0.098
  9      -0.146
 10      +0.139
 11      -0.022
 12      -0.030
 13      +0.187

S.E. OF EACH COEF. GIVEN RANDOM MODEL=0.139

BOX-PIERCE STATISTIC= 9.0946
EXPECTATION GIVEN RANDOM MODEL= 13
     (OBS.-EXP.)/(STD.DEV.)=-.727967
```

THE BOX-PIERCE STATISTIC AS AN OVERALL CONFORMITY MEASURE FOR RANDOMNESS

The autocorrelation coefficients are generally small and have well-mixed signs. Only two exceed their standard errors in absolute value.

Also, the Box-Pierce statistic is small.

Can you interpret the Box-Pierce statistic for residuals from an autoregression in the same way as for original data, such as C IRON?

Generally the same, but we should deduct one from the random expectation of 13, so we have 13 - 1 = 12. But the observed value of 9.095 is still less than this.

Can you use autocorrelations for residuals from other regression models in order to check model adequacy?

Yes, but generally only as a very rough guide. Other things equal, we should expect the autocorrelations and expectation of the Box-Pierce statistic to be smaller as additional independent variables are included in the model; see my answer to the previous question. Here, of course, I'm anticipating the next chapter on multiple regression. If there is more than one lagged variable, we shall subtract one for each such variable.

Do you have rules-of-thumb in evaluating the Box-Pierce statistic?

My suspicions start arising when the standardized value exceeds one, and increase steadily for larger positive deviations from the random expectation.

How about cases in which the Box-Pierce statistic is much smaller than the random expectation?

Ordinarily, this would not be important. It suggests that sampling variation happened to be less than expected under a random model. The only catch is that the data may have been fudged or doctored.

Can you give me a formula for the standard deviation associated with the Box-Pierce statistic?

I'll have to confess a little technical problem here. The approximate distribution, given randomness, of the Box-Pierce statistic is non-normal. So that you would have a result interpretable in the usual way, I had to transform this statistic to get the "(OBS.-EXP.)/(STD.DEV)" shown in the IDA printout after AUTO. So I can't answer your question.

How about the other model adequacy checks for our autoregression?

Pretty good, as you can see:

```
> RUNS
* VARIABLE : RESIDU

    OBSERVED NUMBER OF RUNS= 23
    EXPECTED NUMBER OF RUNS= 26.9615
   STANDARD DEVIATION OF RUNS= 3.56464
       (OBS.-EXP.)/(STD.DEV.)=-1.11134
```

166 □ CONVERSATIONAL STATISTICS

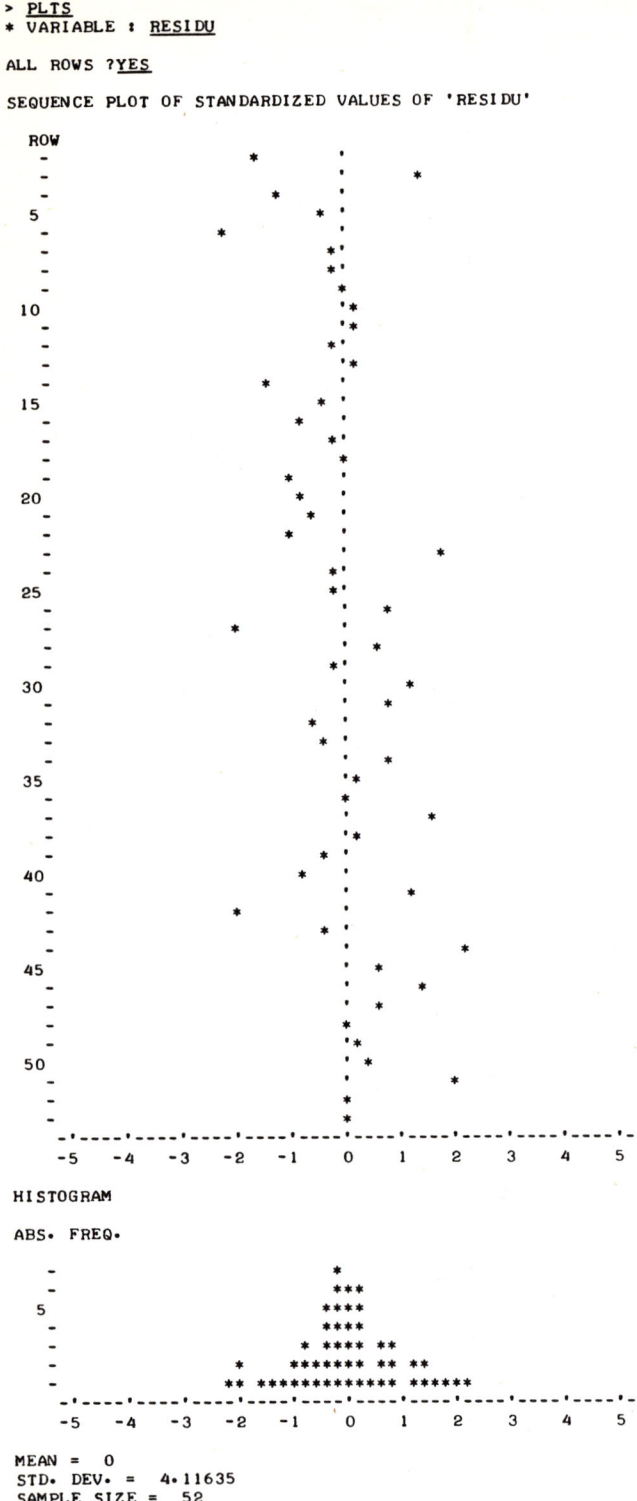

Figure 9-3. Sequence Plot and Histogram of Standardized Values of 'RESIDU'

```
> NORM
* VARIABLE : RESIDU
NORMAL CUMULATIVE PROBABILITY PLOT OF 'RESIDU'
```

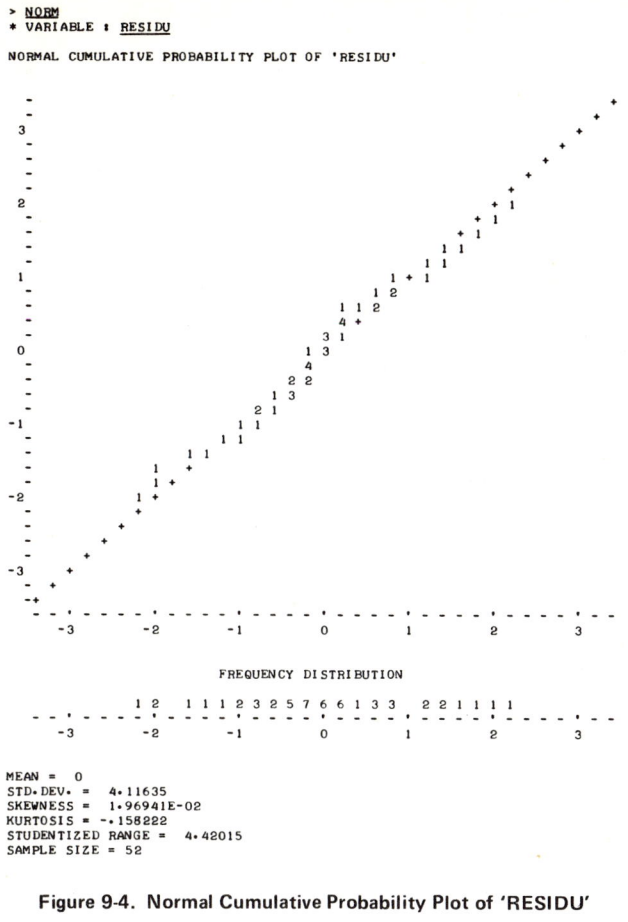

```
MEAN = 0
STD. DEV. =    4.11635
SKEWNESS =    1.96941E-02
KURTOSIS = -.158222
STUDENTIZED RANGE =    4.42015
SAMPLE SIZE = 52
```

Figure 9-4. Normal Cumulative Probability Plot of 'RESIDU'

```
> SCAT
* VERT. VAR. : RESIDU
* HORIZ. VAR. : FITTED
SCATTER PLOT OF STANDARDIZED VALUES OF 'RESIDU' VS. 'FITTED'
```

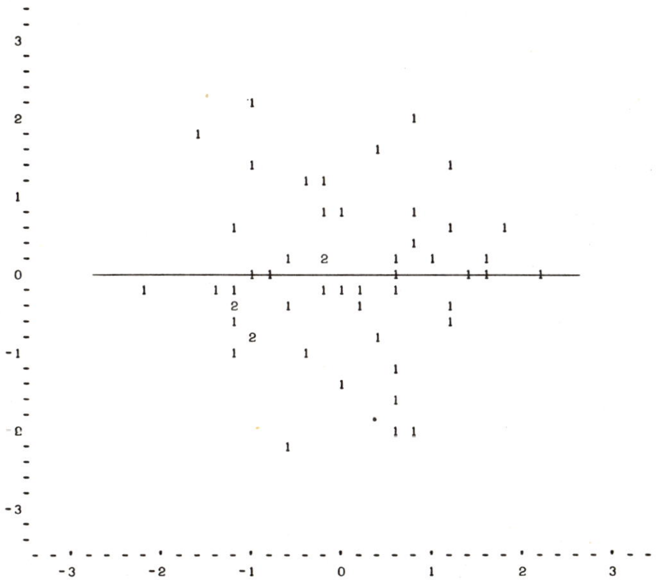

```
                MEAN            STD. DEV.
VERT.  VAR.      0                4.11635
HORIZ. VAR.    21.0769            3.02432
SAMPLE SIZE = 52
```

Figure 9-5. Scatter Plot of Standardized Values of 'RESIDU' vs. 'FITTED'

168 □ CONVERSATIONAL STATISTICS

In Figure 9-3 I see some suggestion of an upward drift, although weaker than what I noted in Figure 9-2.

Your interocular impression agrees with mine. Even though the reservation is minor, it is still a relevant reservation about the model.

But we aren't given much warning of this in the runs count or in the autocorrelation coefficients.

The RUNS count gives a slight hint, but AUTO is not particularly sensitive to the gentle kind of drift seen in Figure 9-3. And there is no complete substitute for plotting the data.

As we proceed, we'll try to keep the upward drift of residuals in mind, but the next task is to learn what the multiple regression model is about and what it might contribute to our analysis of these data.

Before we move on, I'd like to check my understanding. Autoregression is a possible model for analysis of data. The autocorrelation coefficient, however, has value as a diagnostic check for randomness, either of original observations or of residuals from regression.

That's right. The AUTO command plays a role entirely analogous to that of RUNS.

THE AUTOCORRELATION COEFFICIENT MAY NOT BE SENSITIVE TO GENTLE DRIFTS

CHAPTER TEN: MULTIPLE REGRESSION

FURTHER EXAMINATION OF THE RESIDUALS FROM THE REGRESSION OF C IRON ON M IRON

In the regression of C IRON on M IRON in Chapter 8, what did the autocorrelations of the residuals look like?

The command AUTO yields the following:

```
> AUTO
* VARIABLE : RESIDU

* MAX ORDER : 13

  ORDER   AUTOCORR.
    1      +0.198
    2      +0.188
    3      +0.053
    4      -0.110
    5      +0.162
    6      +0.132
    7      +0.272
    8      +0.101
    9      -0.043
   10      +0.050
   11      +0.009
   12      -0.097
   13      +0.106

S.E. OF EACH COEF. GIVEN RANDOM MODEL=0.137

BOX-PIERCE STATISTIC= 12.8322
EXPECTATION GIVEN RANDOM MODEL= 13
     (OBS.-EXP.)/(STD.DEV.) = 9.77022E-02
```

The overall picture of the autocorrelations is not bad, but there is some ground for a slight uneasiness: the first two coefficients are positive and substantially larger than the standard error. In conjunction with the RUNS result and the sequence plot from Chapter 8, the evidence puts the simple regression model under a shadow.

When IDA went through the step of "ANALYZING RESIDUALS..." as part of the execution of REGR in Chapter 8, it gave a warning about runs but not about autocorrelations. Did IDA check autocorrelations of residuals?

Yes, but these were not sufficiently alarming to include in the warning. Also, IDA checked for extreme residuals but found none.

Why is it that RUNS and AUTO point to different verdicts about the model?

RUNS does suggest a severe deficiency, while AUTO is only mildly accusatory. The two procedures reflect, to some extent, different aspects of the time sequence. If you look at the sequence plot in Chapter 8 you will see some persistence *and* the appearance of gentle upward drift through most of the plot, with a downward drift at the end.

Now, in my experience, gentle drifts of this type seem to create little systematic displacement of autocorrelation coefficients away from zero, but they are often reflected, as here, in the RUNS count as a deficiency in the actual runs by comparison with the expected number.

170 □ CONVERSATIONAL STATISTICS

Drifts, as you call them, tend to result in fewer runs than expected?

Yes, and in general, it is a good idea to look at *both* RUNS and AUTO when assessing model adequacy.

We've gone at the problem of predicting C IRON by two different approaches: simple regression on M IRON in Chapter 8 and simple autoregression on CIRO-1 in Chapter 9. How do these fit together?

Both analyses offered a certain degree of predictability, with the simple regression displaying a smaller standard deviation of residuals but also more deficiencies of model adequacy.

Could both analyses have suggested completely adequate, but different, models?

Yes, that's possible.

How can that be? If the residuals of one model behave randomly, how can anything else contribute to improvement of predictability?

Remember that randomness refers to predictability from the *past sequence* of the variable to be predicted. If there is a degree of predictability in the sequence of residuals from a regression model, then we have a clear sign that work on improvement of the model is in order. Our attempts at improvement might, of course, include a search for additional promising independent variables.

But even if the residuals appear to behave randomly, the search for additional independent variables may still be rewarded. If we are successful in this search, we may end up with a *multiple regression* model for which residuals still behave randomly, but which offers better predictability of the dependent variable, as evidenced, say, by a smaller standard deviation of residuals.

Can you generalize to say that it may be possible to obtain improved predictability of something behaving randomly already, if only we can find another variable that is related to it?

Yes. For example, the price changes of a particular common stock may behave randomly, but it is conceivable that we could find *another* variable, other than price, that could be used to attain improved predictability.

Maybe now we had better get back to the prediction of C IRON. As I anticipate you, you will try to draw on what we have already learned in order to exploit both M IRON and CIRO-1 in prediction of C IRON?

And that will lead us into an example of multiple regression.

Is multiple regression going to be a lot more complicated?

Not really; we already have most of the ideas. You may wish to review page 128, however, before going on.

APPLYING REGRESSION WITH MORE THAN ONE INDEPENDENT VARIABLE

MULTIPLE REGRESSION □ 171

The multiple regression model for two independent variables contemplates that there is some "true" but unknown linear function of, say, x_1 and x_2 such that

$$\beta_0 + \beta_1 x_1 + \beta_2 x_2 ;$$

for any x_1 and x_2, the variable $\tilde{y}$ is distributed randomly and normally, with the same standard deviation, about the value taken by this function.

Do you mean that this function describes a straight line?

It is a function of two variables, x_1 and x_2, so the geometrical interpretation is that of a plane rather than a straight line. Mathematical usage, however, describes such a function as linear.

According to this model, the $\tilde{y}$'s are generated according to the scheme:

$$\beta_0 + \beta_1 x_1 + \beta_2 x_2 + \tilde{\epsilon},$$

where the interpretation follows that given for the simpler model of Chapter 8. Note especially that $\tilde{\epsilon}$ is a random drawing from a normal distribution with mean 0 and the same standard deviation, σ_ϵ, regardless of x_1 and x_2; as before, $\tilde{\epsilon}$ is called a disturbance.

Can you draw a picture of this?

The full picture takes three dimensions. A condensation to two dimensions is often useful, and it can easily be extended to more than two independent variables.

A typical point is shown, for which the disturbance is positive.

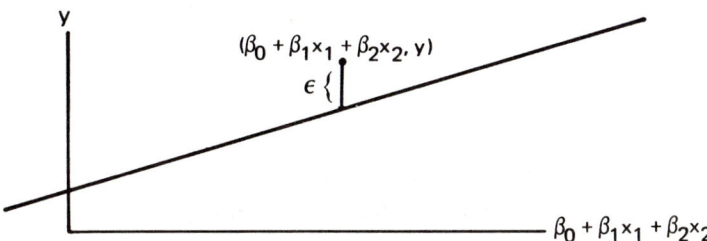

From the discussion in Chapter 8, I would understand that this figure is conceptual. We don't observe the line in applications, do we?

No, we don't. In fact, because of the condensation to two dimensions, we can't even plot a point on this chart; it would take a scatter plot in three dimensions to do that. In any event, the parameters β_0, β_1, β_2, and σ_ϵ are unknown or, in more accurate language, are not precisely known.

As before, we will make sample estimates. Before we get to that, let me give you brief interpretations of β_0, β_1, and β_2:

β_0 is the value of the regression function for $x_1 = x_2 = 0$. In a three-dimensional plotting of y against x_1 and x_2, it is the intercept; it is also called the *constant* term.

β_1 represents the change of the regression function for a unit increase of x_1, *at any given value of x_2*. (In a three-dimensional representation, β_1 is the slope of the regression plane in the direction of the x_1 axis.)

STANDARD MULTIPLE REGRESSION MODEL

INTERPRETATION OF REGRESSION COEFFICIENTS

β_2 represents the change of the regression function for a unit increase of x_2, at any given value of x_1.

Would it be correct to say that β_1, for example, is the expected change in $\tilde{y}$ for a unit increase in x_1 at any x_2?

Yes.

Then you can think of regression as a way of finding out the effect of, say, x_1 on $\tilde{y}$ while holding x_2 constant?

That gets at the right idea, but the wording is misleading in one important respect: it suggests that the statistician has *intervened* to keep x_2 from varying, which is typically untrue, except in strictly experimental applications (see Chapter 2). Both x_1 and x_2 will vary. It would be better to say that regression is a way of estimating the effect on $\tilde{y}$ of a unit increase in x_1 *after allowance for the variation of x_2*.

I suppose that you will fit a statistical relationship, $\hat{y} = b_0 + b_1 x_1 + b_2 x_2$, to sample data in order to estimate the regression.

Yes, this is the natural extension of what we have done in the simple regression applications already illustrated, and the fitting process will use the method of least squares by the same reasoning. Further, residuals will be defined by the same relationship:

$$\hat{\epsilon} = y - \hat{y}.$$

Are the same IDA commands used?

Yes. To illustrate, we shall let x_1 = CIRO-1 and x_2 = M IRON, and start with the command REGR:

```
> REGR
UPDATING CORR. MATRIX...
* DEP. VAR. = C IRON
HOW MANY INDEP. VAR. ?2
INDEP. VAR.  1 = ?CIRO-1
INDEP. VAR.  2 = ?M IRON
ANALYZING RESIDUALS ...
```

No warnings were given as a result of IDA's analyses of residuals. Could we look at AUTO and RUNS for residuals?

Here they are:

```
> AUTO
* VARIABLE : RESIDU

* MAX ORDER : 13
  ORDER  AUTOCORR.

    1    +0.007
    2    +0.136
    3    +0.034
    4    -0.151
    5    +0.155
    6    +0.072
    7    +0.226
    8    +0.077
    9    -0.108
   10    +0.092
   11    -0.002
   12    -0.097
   13    +0.146
 S.E. OF EACH COEF. GIVEN RANDOM MODEL=0.139
 BOX-PIERCE STATISTIC= 9.33191
 EXPECTATION GIVEN RANDOM MODEL= 13
       (OBS.-EXP.)/(STD.DEV.)=-.669418

> RUNS
* VARIABLE : RESIDU

  OBSERVED NUMBER OF RUNS= 23
   EXPECTED NUMBER OF RUNS= 26.8462
  STANDARD DEVIATION OF RUNS= 3.54848
       (OBS.-EXP.)/(STD.DEV.)=-1.08389
```

DIAGNOSTIC CHECKS FOR MULTIPLE REGRESSION

Certainly the runs picture is better than either of the earlier simple regressions. But shouldn't we look at the sequence plot to see if the apparent drift that I noted in the sequence plot of residuals from autoregression has been removed?

That's a good suggestion. The plot is shown as Figure 10-1. To my eye, Figure 10-1 represents a distinct improvement but there's still a hint of a systematic time drift. The visual impression, however, is not strongly traumatic.

The normal probability plot of the residuals is also encouraging, as you can see from Figure 10-2; I'm not inclined to be concerned about the negative kurtosis.

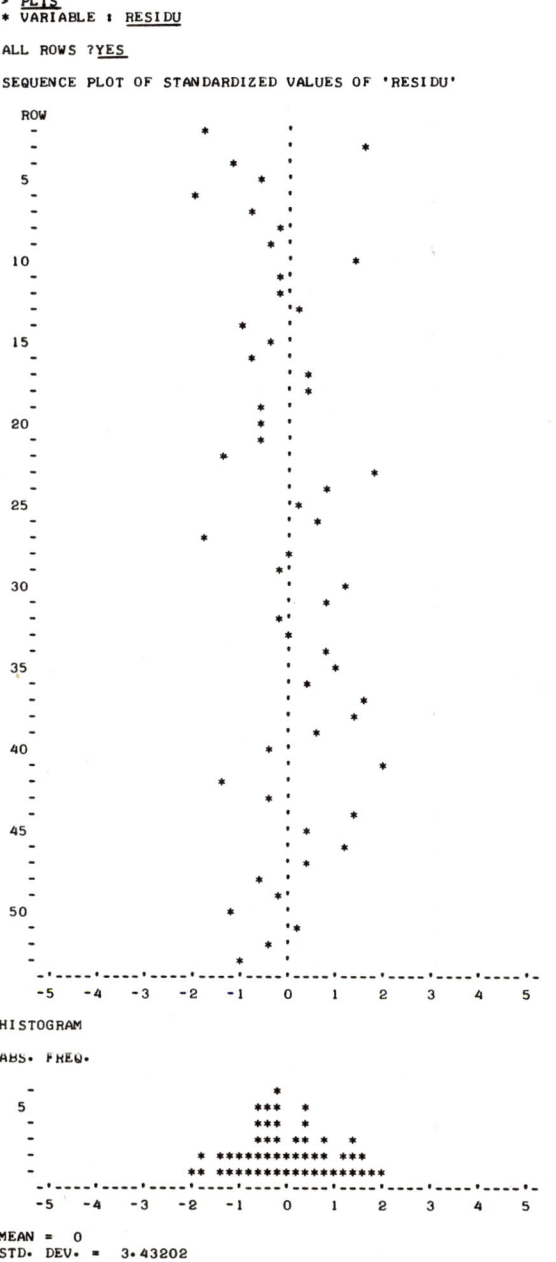

Figure 10-1. Sequence Plot and Histogram of Residual Values for Multiple Regression Model of C IRON on CIRO-1 and M IRON

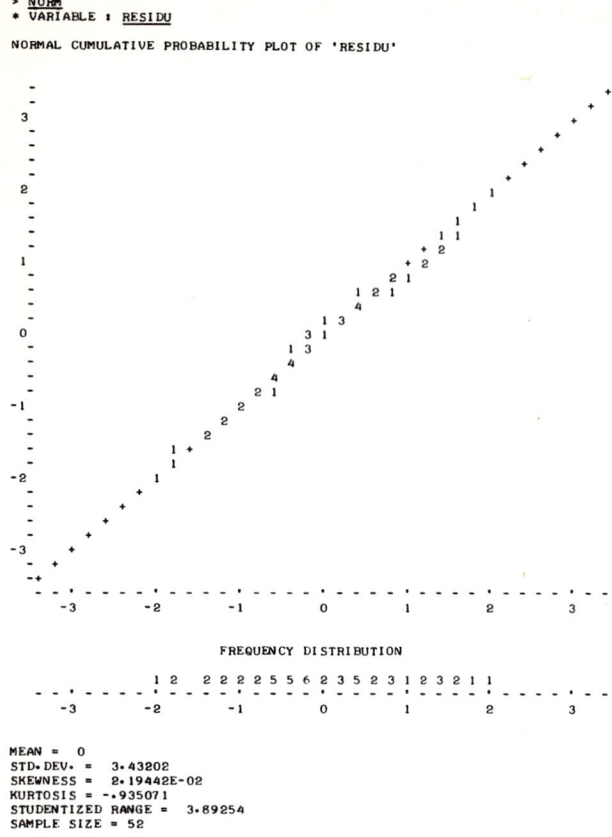

Figure 10-2. Normal Probability Plot of Residuals for Multiple Regression Model of C IRON on CIRO-1 and M IRON

How about the plot of residuals versus fitted?

Here it is:

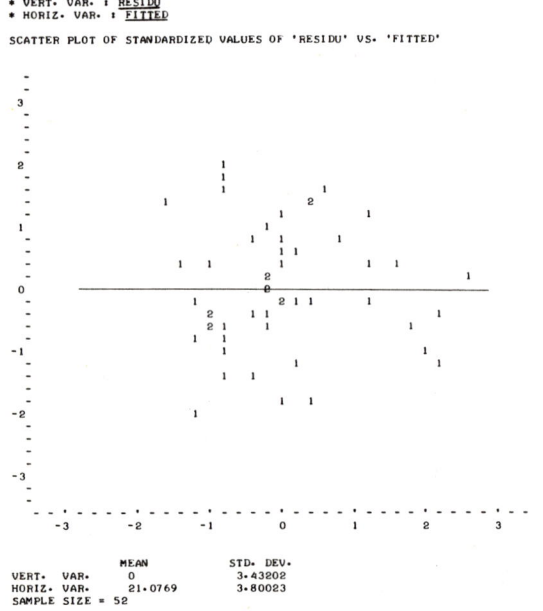

Figure 10-3. Scatter Plot of Standardized Values of 'RESIDU' vs. 'FITTED' for Multiple Regression Model of C IRON on CIRO-1 and M IRON

It seems to me that the hint of curvilinearity is now less pronounced than in the simple regression of C IRON on M IRON.

I agree.

This time, I notice, you talked about model adequacy before looking at the main regression results. Did you have any special reason for doing so?

As I said earlier, we're free to do things in either sequence. The only thing we should avoid is accepting as final the main results from a model for which the diagnostic checks of model adequacy have not been made.

Unfortunately, people often neglect residual analyses entirely. Although there's no best sequence in which to look at the analysis, the procedure followed above (of looking first at residuals) serves to remind you that the most important task is to find a model that corresponds reasonably well to what is going on in the data. If the model is grossly out of kilter, the analysis of the model is of interest only in suggesting how to approach an improved model.

But now we can look at the main regression analysis?

Here are results of two of the most important commands, COEF and SUMM:

OUTPUTS OF THE MAIN REGRESSION ANALYSIS

```
> COEF

VARIABLE    B(STD.V)      B           STD.ERROR(B)    T
CIRO-1      0.2123        2.1286E-01  1.2436E-01      1.712
M IRON      0.5939        4.7595E-01  9.9399E-02      4.788
CONSTANT                  6.7552E+00  2.0870E+00      3.237

> SUMM
                MULTIPLE R    R-SQUARE
UNADJUSTED      0.7488        0.5607
ADJUSTED        0.7367        0.5427
STD. DEV. OF RESIDUALS  =     3.43203
```

1. A point estimate of a single observation of C IRON is thus given by

   $\hat{y}$ = 6.7552 + 0.21286(CIRO-1) + 0.47595(M IRON).

2. Hence, for any given value of CIRO-1, the prediction goes up by 0.47595 units for each unit increase in M IRON. Similarly, for any given value of M IRON, the prediction goes up by 0.21286 units for each unit increase in CIRO-1.

It looks like we can be pretty sure that, if this model is appropriate, both $\tilde{\beta}_1$ and $\tilde{\beta}_2$ are positive.

Yes, since we see that b_1 is 1.712 standard errors from 0, while b_2 is 4.788 standard errors from 0. In either event, there is little probability on negative values of $\tilde{\beta}_1$ or $\tilde{\beta}_2$.

You can answer the question more precisely by the use of GAUS, right?

Yes, although recall that the normal distribution is only an approximation. I have obtained the answer for $\tilde{\beta}_1$ in two different ways below:

```
> GAUS

* MEAN     = ?.21286
  STD.DEV. = ?.12436

* LOWER LIMIT L = ?-100
  UPPER LIMIT U = ?0

  PR(L<X<U) =   0.04348

  MORE ?YES

* LOWER LIMIT L = ?0
  UPPER LIMIT U = ?100

  PR(L<X<U) =   0.95652

  MORE ?NO
```

Thus, the probability that we have the right sign, namely positive, for $\tilde{\beta}_1$ is 0.96?

Or, alternatively, our probability that $\tilde{\beta}_1$ is really negative is only about 0.04. Recall, too, that any such calculation presumes that the model used is adequate.

I notice also that b_1 and b_2 differ somewhat from the corresponding coefficients of the simple regressions.

That's right. In the simple regression of Chapter 8, for example, M IRON had a coefficient of 0.58664, as compared with 0.47595 above.

And the constant term is smaller.

Yes. Notice also that the standard deviation of the residuals (3.43203) is substantially smaller than for the autoregression (4.11636) and slightly smaller than for the M IRON simple regression (3.46359).

That lower standard deviation of residuals, along with the improved behavior of residuals, suggests that the multiple regression is a step forward?

Yes. So long as you are comparing alternative fits to the same basic data, and so long as the dependent variable is in the same units for all these alternatives, the standard deviation of residuals in combination with the behavior of the residuals in time sequence and on the normal probability plot provides a rough indication of the adequacy of the different regression models under consideration.

What do you mean by "dependent variable in the same units"?

For example, one model might use y as dependent variable, while another might use the transformation log y as dependent variable. Then it is not easy to compare standard deviations of residuals, because one is in units of y while the other is in units of log y.

What do you suggest in those circumstances?

CHOOSING BETWEEN ALTERNATIVE REGRESSION MODELS

My personal procedure is to make a comparison of the *adjusted multiple R's* for the models under consideration. Higher adjusted R, other things the same, is favorable for the model. (This criterion will not be satisfactory for all transformations, however.)

Why the adjusted multiple R?

The adjustment handicaps models according to the number of independent variables used. Without this handicapping, models with more independent variables would be at an artificial advantage because unadjusted R will increase with addition of independent variables even if these variables are simply random numbers!

When the dependent variable is in the same units, and you use the standard deviation of residuals as a guide, is there a similar handicapping for models with more independent variables?

Yes, since the sum of squared residuals is divided by n-k-1, where k is the number of independent variables. We'll see more of this point later in the chapter.

Are there other considerations in decisions about adequacy of competing regression models?

Yes. Some preference should be given to models in which the independent variables have some basis in theory and experience, rather than being introduced indiscriminantly because they happen to be available.

Further, there seems to be a good deal of experience to suggest that smaller models--models with fewer independent variables--will predict for new data better than larger models, other things nearly equal. This is an aspect of what is sometimes called the *principle of parsimony*.

What do you mean by "predicting for new data"?

For example, use of SEPR to make point estimates of $\tilde{y}$ given the x's of observations *not* used in fitting the regression.

The ultimate test of a model, then, is how well it predicts for new data?

That's right, and predictive tests may give disastrous results even when multiple R is high, when the other criteria discussed above are not met. By "disastrous", I mean much worse than suggested by the prediction error formulas of SEPR.

Now that you've mentioned it again, how about "MULTIPLE R"? How can I interpret it?

In this case, the unadjusted multiple R is 0.7488. R can be interpreted as follows:

1. Think of the scatter diagram of y versus $\hat{y}$; that is, of the dependent variable against the fitted values given by the multiple regression.

2. In this scatter diagram, let both y and $\hat{y}$ be expressed in standardized units, as we have usually done.

THE ULTIMATE TEST OF A MODEL

AN INTERPRETATION OF THE MULTIPLE CORRELATION COEFFICIENT

3. The sample regression line of y on $\hat{y}$ will pass through the point (0,0) and have slope equal to the multiple correlation coefficient R. This is illustrated in Figure 10-4, where the slope of the fitted line is 0.7488.

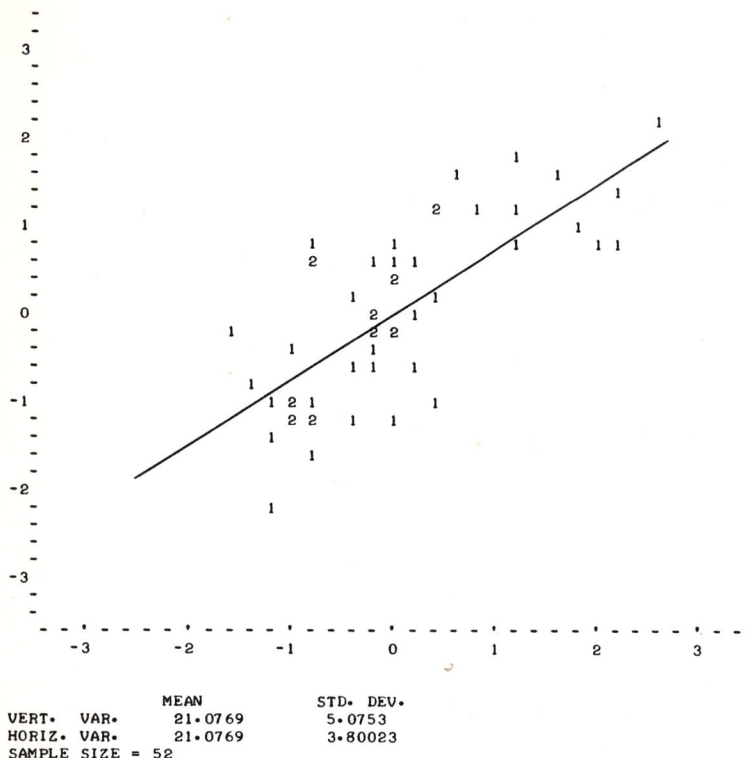

Figure 10-4. Scatter Plot of Standardized Values of 'C IRON' vs. 'FITTED' for Multiple Regression Model

Does this work regardless of the number of independent variables?

Yes. There must, of course, be at least one independent variable, and the problem won't make sense in the framework used here for more then (n-1) independent variables.

Does that mean that we are now satisfied with our model?

In a real sense, we are never completely satisfied with a statistical model, at least until we have been able to carry the search much further than has been possible here. At the same time, economic limitations, as well as the hazard of "overfitting" the particular set of data by giving undue attention to each little chance peculiarity of it, will dictate that we stop at some reasonable point.

I have carried this problem somewhat beyond the point that I have reported to you up until now. It might be useful just to outline the further investigation to give you some feeling for the possibilities and to suggest some of the hazards in the approach.

But what more could you do? No more variables are measured besides the two basic ones, C IRON and M IRON.

EXTENDING THE REGRESSION MODEL

I shall describe what I did, not so much to suggest a mechanical approach but to give you some idea of the opportunities for development of regression models.

The only basic independent variables at my disposal were M IRON and CIRO-1. The autoregression analysis of Chapter 9 suggests that it is not worthwhile to experiment with lags of C IRON beyond the first one.

There is, however, some hint that the relationship may not be linear to a satisfactory approximation if only M IRON and CIRO-1 are used. Now there is a simple and flexible multiple regression model that is worth exploring in these circumstances:

$$\beta_0 + \beta_1 x_1 + \beta_2 x_2 + \beta_3 x_1 x_2 + \beta_4 x_1^2 + \beta_5 x_2^2$$

If I remember my mathematics, this is no longer a linear equation in x_1 and x_2. Yet this is supposed to be linear regression?

Your recollection is right. But now define:

$$x_3 = x_1 x_2$$
$$x_4 = x_1^2$$
$$x_5 = x_2^2$$

Then the model is $\beta_0 + \beta_1 x_1 + \beta_2 x_2 + \beta_3 x_3 + \beta_4 x_4 + \beta_5 x_5$. *But is that fair?*

Fairness has nothing to do with it. It is a valid and often useful procedure for capturing certain nonlinearities by the standard model of regression. Whether it is advisable or not in a particular application is another question.

How do you use IDA to explore this approach?

An annotated printout is given on the next page. We started with C IRON, M IRON, and CIRO-1, respectively, in columns 1-3 of the data matrix.

Isn't it getting a bit hard to keep track of your variables?

Use the command NAME:

```
> NAME
THESE ARE THE VARIABLES IN THE DATA MATRIX :
COLUMN    NAME

   1      C IRON
   2      M IRON
   3      CIRO-1
   4      MX C-1
   5      MIROSQ
   6      C-1 SQ
```

180 □ CONVERSATIONAL STATISTICS

```
> DOTP
* COL# TO PLACE VAR. : 4
I,J = ?2,3
UPDATING MEAN, STD, ..., CORR, ...

GIVE NAME OF NEW VARIABLE
VAR.  4 ?MX C-1

> POWE

* COL# TO PLACE VAR. : 5
* COLUMN TO BE TRANSFORMED :2
* P = 2
UPDATING MEAN, STD, ..., CORR, ...
GIVE NAME OF NEW VARIABLE
VAR.  5 ?MIROSQ

> POWE

* COL# TO PLACE VAR. : 6
* COLUMN TO BE TRANSFORMED :3
* P = 2
UPDATING MEAN, STD, ..., CORR, ...

GIVE NAME OF NEW VARIABLE
VAR.  6 ?C-1 SQ
```

This command creates a new variable that is the product of two existing variables of the data matrix.

The name suggests by "X" that the new variable is the product of M IRON and CIRO-1.

This command raises a variable to any desired power. "2" is wanted for this application. If we had wanted the square root of the existing variable, the power would have been 0.5.

Here the desired power is entered.

Use of "SQ" suggests M IRON squared.

Another useful command at this stage is MEAN, which lists the variables plus the mean and standard deviation of each. For example,

```
> MEAN

VARIABLE      MEAN          STD. DEV.

  C IRON    2.10769E+01    5.07530E+00
  M IRON    2.06731E+01    6.33299E+00
  CIRO-1    2.10577E+01    5.06202E+00
  MX C-1    4.55635E+02    2.31800E+02
  MIROSQ    4.66712E+02    3.09113E+02
  C-1 SQ    4.68558E+02    2.16316E+02
```

Doesn't the simultaneous treatment of so many variables raise statistical problems?

It certainly does. I've avoided some of them by dealing only with a selected set of five independent variables that make sense *a priori*. Also, with the set of five variables used here, the linear assumption should be a good bet.

But there remains the problem of *multicollinearity* . . .

Multicollinearity? What is that?

It's an expression that refers to the degree of correlation between the independent variables in a regression model. To see the relevant correlation coefficients in our problem, use the command CORR:

```
> CORR
HOW MANY VARIABLES ?5
COL. #'S: ?2,3,4,5,6
* # DECIMALS = 3

          M IRON   CIRO-1   MX C-1   MIROSQ   C-1 SQ
  M IRON  1.000
  CIRO-1  0.646    1.000
  MX C-1  0.938    0.856    1.000
  MIROSQ  0.986    0.613    0.925    1.000
  C-1 SQ  0.665    0.992    0.874    0.635    1.000
```

THE PROBLEM OF MULTICOLLINEARITY

Some of those look pretty high. Especially the 0.992 between C-1 SQ and CIRO-1.

It often happens that when correlations between pairs of independent variables are extremely high, only one member of the pair will contribute to the model. We'll see what happens here in a moment.

But if the independent variables are really independent, shouldn't all the correlations be zero?

There's a play on words here. "Independent" in "independent variables" has nothing to do with the word "independent" in statistics and probability, which is closely related to our usage of "random". Statistical independence does imply zero correlation.

If, in an experimental study, one can control the values of all the independent variables, it is indeed desirable to arrange things so that the correlations are zero.

So multicollinearity isn't so bad as the name sounds?

A lot depends on the circumstances of the particular application, and on how the analysis is conducted. In general, it is likely to have more effect on inferences about individual regression coefficients than on the regression equation as a whole.

Let's see, then, how the analysis goes with all five independent variables:

```
> REGR
* DEP. VAR. = C IRON
HOW MANY INDEP. VAR. ?5
INDEP. VAR.  1 = ?M IRON
INDEP. VAR.  2 = ?CIRO-1
INDEP. VAR.  3 = ?MX C-1
INDEP. VAR.  4 = ?MIROSQ
INDEP. VAR.  5 = ?C-1 SQ
ANALYZING RESIDUALS ...


> COEF

VARIABLE   B(STD.V)      B            STD.ERROR(B)     T

M IRON     1.5283     1.2248E+00      5.8249E-01      2.103
CIRO-1     0.0989     9.9172E-02      7.8406E-01      0.126
MX C-1    -1.6593     -.3633E-01      3.2451E-02     -1.120
MIROSQ     0.1239     2.0341E-03      1.3147E-02      0.155
C-1 SQ     0.8611     2.0205E-02      2.7043E-02      0.747
CONSTANT              -.1936E+00      9.2812E+00     -0.021

> SUMM

              MULTIPLE R    R-SQUARE
UNADJUSTED      0.7606       0.5785
ADJUSTED        0.7299       0.5327
STD. DEV. OF RESIDUALS =      3.46932
```

THE MULTIPLE REGRESSION MODEL DOES NOT SPECIFY THAT INDEPENDENT VARIABLES MUST BE UNCORRELATED

The standard deviation of residuals is higher with five variables than it was with only two. How can that be?

In the computation, the sum of squared residuals is divided by n-k-1, where k is the number of independent variables. Addition of independent variables to a regression model will typically reduce the sum of squared residuals, and cannot increase it. But the denominator is reduced as k increases. In this case, the effect on the denominator more than offsets the effect on the numerator. Here we see an example of what I meant when I referred earlier to "handicapping" models that have more independent variables.

Should we discard a variable?

That's a sensible step. A rule-of-thumb is to discard the variable for which the ratio of coefficient to standard error is smallest in absolute value, in this case CIRO-1.

The command SWEEP enables us to eliminate CIRO-1 without respecifying the whole regression (and doing all the computations from scratch) by REGR:

```
> SWEEP
VAR. : CIRO-1
ANALYZING RESIDUALS ...

> COEF

VARIABLE   B(STD.V)     B           STD.ERROR(B)    T

 M IRON     1.5454    1.2385E+00    5.6625E-01     2.187
 C-1 SQ     0.9784    2.2955E-02    1.5904E-02     1.443
 MX C-1    -1.6980   -.3718E-01     3.1421E-02    -1.183
 MIROSQ     0.1290    2.1176E-03    1.2992E-02     0.163
 CONSTANT             6.6806E-01    6.2374E+00     0.107

> SUMM

              MULTIPLE R   R-SQUARE
 UNADJUSTED     0.7605      0.5784
 ADJUSTED       0.7366      0.5425
 STD. DEV. OF RESIDUALS =       3.43281
```

We've lost CIRO-1, which was one of the two variables in the original multiple regression earlier in this chapter. Also, the standard deviation of residuals is lower than with the original five independent variables, but it's still slightly larger than for our earlier regression.

Although CIRO-1 has dropped out directly, it's still reflected in two of the four remaining variables: MX C-1 and C-1 SQ.

The coefficient of MIROSQ still has a very small ratio to its standard deviation, so let's SWEEP out MIROSQ:

```
> SWEEP
VAR. : MIROSQ
ANALYZING RESIDUALS ...

> SUMM
              MULTIPLE R   R-SQUARE
 UNADJUSTED     0.7604      0.5782
 ADJUSTED       0.7428      0.5518
 STD. DEV. OF RESIDUALS =       3.39782

> COEF

VARIABLE   B(STD.V)     B           STD.ERROR(B)    T

 M IRON     1.5743    1.2616E+00    5.4259E-01     2.325
 C-1 SQ     0.9021    2.1165E-02    1.1381E-02     1.860
 MX C-1    -1.5390   -.3370E-01     2.2811E-02    -1.477
 CONSTANT             4.3113E-01    6.0038E+00     0.072
```

Well, the standard deviation of residuals is the lowest yet--3.39782. And there have been no serious warnings in the screening of residuals.

As you see below, the remaining three independent variables still have moderately high correlations. Since they have been built up as functions of M IRON and CIRO-1, this is inevitable. Notice, however, that the remaining T ratios are all substantially greater than one in absolute value. This suggests that we'd better stop sweeping at this point.

```
> CORR

 HOW MANY VARIABLES ? 3
 COL. #'S: ? 2,4,6
*   # DECIMALS = 3

            M IRON   MX C-1   C-1 SQ
 M IRON     1.000
 MX C-1     0.938    1.000
 C-1 SQ     0.665    0.874    1.000
```

Is this a good time to look at the detailed diagnostics on this model?

Yes, since we have tentatively singled out a promising combination of variables. The picture is encouraging.

EXAMINING DIAGNOSTICS ON THE MODEL AFTER SWEEPING PROCEDURE

```
> RUNS
* VARIABLE : RESIDU

    OBSERVED NUMBER OF RUNS= 23
    EXPECTED NUMBER OF RUNS= 26.8462
  STANDARD DEVIATION OF RUNS= 3.54848
       (OBS.-EXP.)/(STD.DEV.)=-1.08389

> AUTO
* VARIABLE : RESIDU

* MAX ORDER : 13

  ORDER   AUTOCORR.
    1      -0.071
    2      +0.079
    3      -0.042
    4      -0.184
    5      +0.134
    6      +0.033
    7      +0.242
    8      +0.022
    9      -0.129
   10      +0.072
   11      +0.007
   12      -0.104
   13      +0.138

 S.E. OF EACH COEF. GIVEN RANDOM MODEL=0.139

 BOX-PIERCE STATISTIC= 9.16992
 EXPECTATION GIVEN RANDOM MODEL= 13
       (OBS.-EXP.)/(STD.DEV.) =-.709277
```

The time sequence plot certainly looks better (*Figure 10-5*) but the histogram has a curious spike at -0.3, in standardized units.

If I saw a spike like that in original data, as opposed to residuals, I would look for some reporting anomaly, as in the Rockwell Hardness data. Here, I'm inclined to ascribe it to a sampling fluke. Note that the spike doesn't have much effect on the normal probability plot, which looks good.

The scatter plot (*Figure 10-6*) *of RESIDU versus FITTED seems to have ironed out any hint of curvilinearity.*

That is a gratifying result of the analysis.

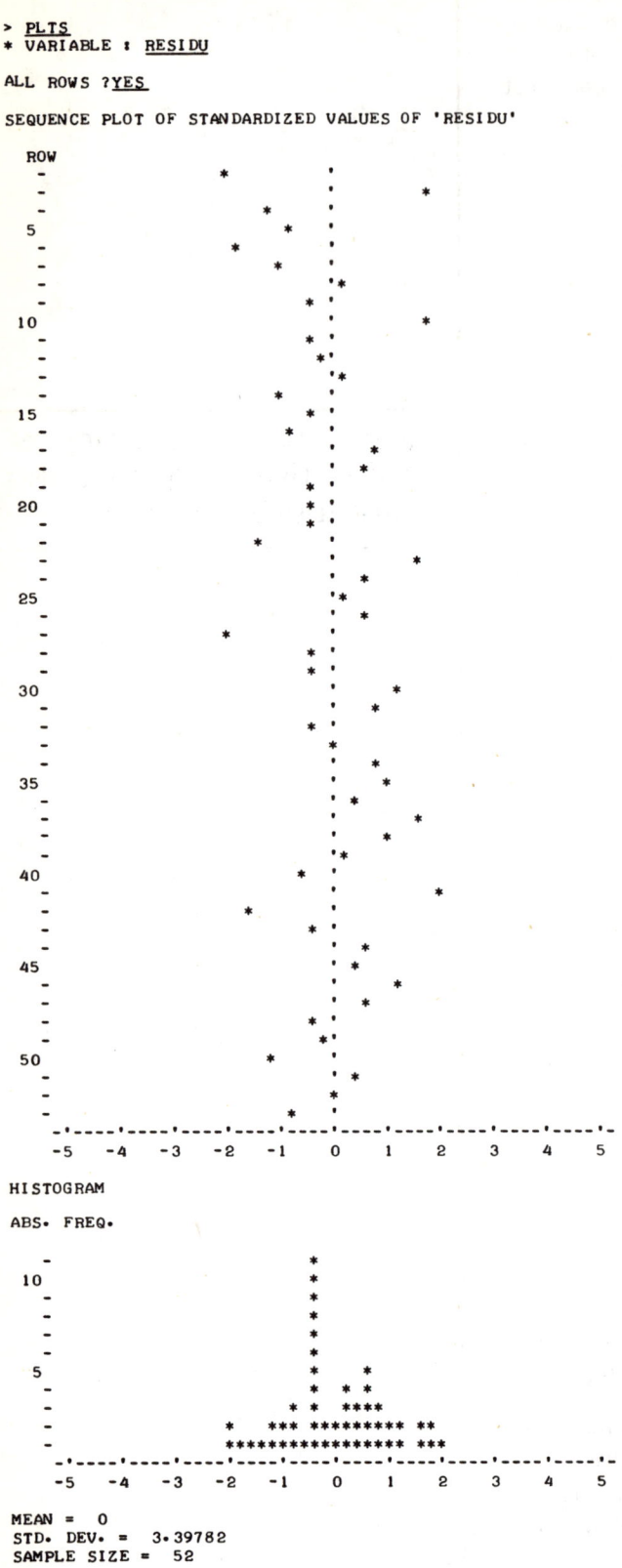

Figure 10-5. Sequence Plot of Standardized Values of 'RESIDU' for the Model After Sweeping

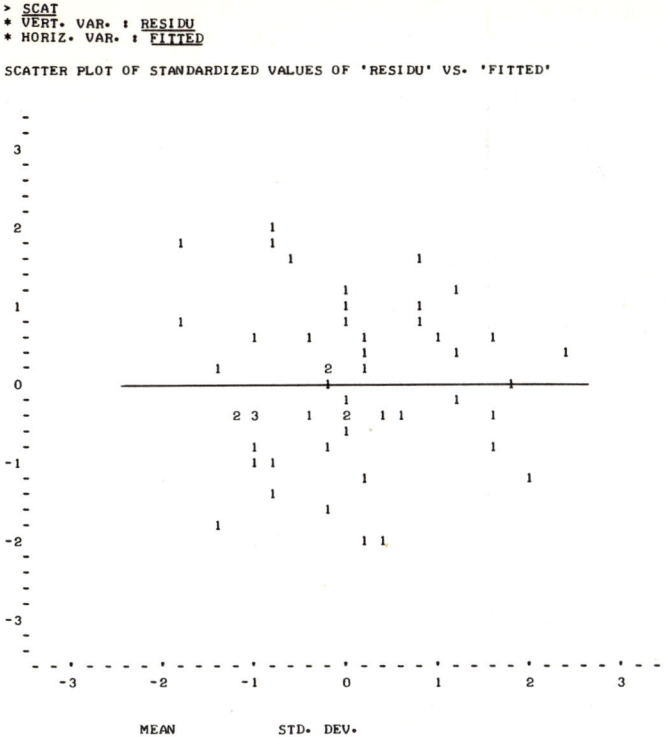

Figure 10-6. Scatter Plot of Standardized Values of 'RESIDU' vs. 'FITTED' for the Model After Sweeping

How would this rather more complicated model be interpreted in terms of its possible subject matter implications?

This is primarily a task for an expert in the subject matter. He might find it useful to consider the mathematical implications, remembering, of course, that the multicollinearity makes comparisons of the individual regression coefficients rather treacherous.

He may also have other information that would be suggestive in conjunction with the statistical analysis. For example, the data were not gathered at equal intervals over the approximately eight-month duration of the study. There was a gap of several weeks between the 21st and 22nd observation of the lagged analysis. Conceivably, the model reached here is a fortuitous outcome of noncontrolled variables that acted intermittently during the period.

If I wanted to know the correlations between variables in the posterior probability distribution of the "true" regression coefficients, $\tilde{\beta}_i$, how could I get them?

The command is BCOR:

```
> BCOR
* DECIMALS = ?3

M IRON    1.000
C-1 SQ    0.917   1.000
MX C-1   -0.983  -0.966   1.000
CONST.   -0.982  -0.930   0.963
```

JOINT POSTERIOR DISTRIBUTION OF THE REGRESSION COEFFICIENTS

186 ☐ **CONVERSATIONAL STATISTICS**

The fact that many of these correlations are so high means that our inference about any one $\tilde{\beta}_i$ is closely tied up with our inference about other $\tilde{\beta}_i$'s. For example, the correlation between $\tilde{\beta}_2$ (for the coefficient of C-1 SQ) and $\tilde{\beta}_3$ (for the coefficient of MX C-1) is -0.966. If you were somehow given the added information that $\tilde{\beta}_2$ for C-1 SQ was much higher than the sample value of 0.021165, then you would have to move your assessment of $\tilde{\beta}_3$ downward, though it is currently centered at -0.03370 (see previous COEF).

I notice that in this model the constant or intercept is estimated at 0.43113. This is much smaller than earlier models yielded. What am I to make of that?

Not too much:

1. The standard error is large: 6.0038.

2. As you see from BCOR, the distribution of this coefficient is strongly correlated with that of the slope coefficients.

3. Remember that the intercept is far removed from the data we are analyzing.

4. It is, however, encouraging that the point estimate is near the value 0 suggested by *a priori* reasoning.

```
> NORM
* VARIABLE : RESIDU
NORMAL CUMULATIVE PROBABILITY PLOT OF 'RESIDU'
```

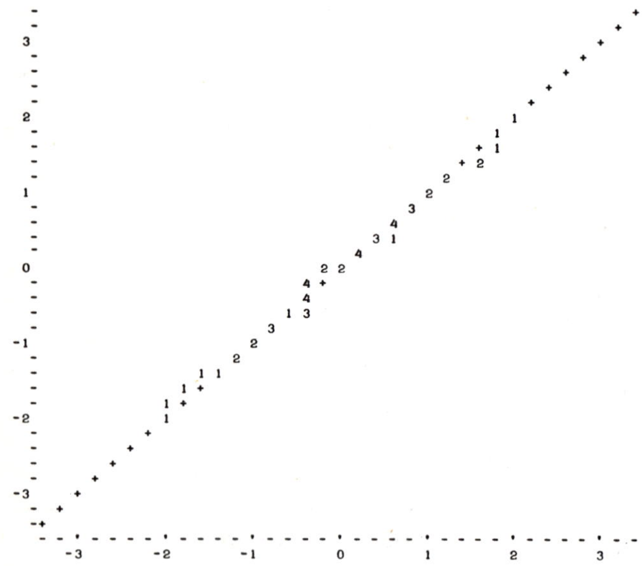

```
                                                          +
                                                       + +
 3                                                  + +
                                                +++
                                              ++
                                            ++
                                         + 1
 2                                      1
                                      + 1
                                     + 2
                                   2
                                  3
                               4
                             3 1
                            4
 0                       2 2
                        4 +
                        4
                       1 3
                     3
-1                  2
                   2
                 1 1
                 1 +
                1 +
-2             1
              +
            ++
          ++
        ++
-3    + +
     -+
   -----'----'----'----'----'----'---
      -3   -2   -1    0    1    2    3
                FREQUENCY DISTRIBUTION
         2 1 1 1 2 2 3 1 * 2 2 4 3 5 3 2 2   2 2 1
   -----'----'----'----'----'----'----'----'---
      -3   -2   -1    0    1    2    3

NOTE: FREQUENCIES OVER 9 INDICATED BY '*'

MEAN = 0
STD.DEV. = 3.39782
SKEWNESS = 1.94378E-02
KURTOSIS = -.857217
STUDENTIZED RANGE = 4.02213
SAMPLE SIZE = 52
```

Figure 10-7. Normal Cumulative Probability Plot of 'RESIDU' for Model After Sweeping

Is it possible that some other grouping of three of the five independent variables would lead to a smaller standard deviation of residuals?

Logically, it is possible. Practically, the method of sweeping used here (sometimes called *backwards stepwise regression*) will usually locate a model with a low standard deviation of residuals, if not the lowest possible.

IDA has an interesting command called SUBS (subsets of variables of a given size in regression) that gives all the possible regressions for a specified number of independent variables. I tried SUBS and found that the model reached above does have the smallest standard deviation of residuals among all combinations of three independent variables.

A still more ruthless command is ALLS, which does *all possible* combinations of independent variables.

Why not just put all the variables you can think of into the data matrix and then run ALLS?

This is an example of what I like to call the sledgehammer approach to regression analysis. It conflicts with the principle of parsimony that I discussed earlier, and often leads to absurd results because some of the independent variables created by transformations or otherwise introduced are senseless. For example, once you start manufacturing variables, you may well inadvertently make the *dependent variable*, or some simple function thereof, into an *independent variable*. Some of my more enthusiastic and energetic students have done this and been stunned by the absurd results of the analysis.

One statistician-friend of mine argues, in fact, that by clever transformation, one can often boil down several key independent variables into one or two that will do a good job. This approach has its hazards too, since a good deal of trial-and-error can be involved in the boiling-down process, with consequent danger of overfitting, but I much prefer this approach to the sledgehammer approach.

Has your friend tried this problem?

Yes, and he decided to regress C IRON on a single variable that was the product of $\sqrt{M\ IRON}$ and the natural logarithm (LOGE) of CIRO-1. You might enjoy looking at that model.

Aren't you yourself open to the charge of overfitting? After all, you took a good while to reach your preferred model.

I tried to guide my search by clear reasoning at each step. For example, my transformations were based on a commonplace of applied mathematics that many functions can be well approximated by a quadratic.

But why not just pick a single model based on a priori information, and then stick to it through thick and thin?

BEWARE THE SLEDGEHAMMER APPROACH

THE PROBLEM OF OVERFITTING

It would be delightful if we were smart enough to do this well, but experience strongly suggests that we must do a good deal of disciplined model-formulation and diagnostic checking, in the iterative fashion suggested by these last three chapters.

Is there any good precaution against the danger of overfitting?

One approach, when the data are reasonably abundant, is to fit the model on a random subsample of the data, then see how well the resulting model predicts the remaining data. It is planned to make this a special feature of the command called ENTS in a future version of IDA.

CHAPTER ELEVEN: TIME SERIES

Is it possible to extend the simple autoregressive analysis of time series treated in Chapter 9 by application of the multiple regression methods of Chapter 10?

Yes, it is. A number of the concepts and techniques treated in earlier chapters can be drawn together for a systematic attack on the problem of time series analysis.

What is the underlying strategy?

In terms of regression models, the idea is to find a model—possibly more complex in mathematical structure than any we have considered—in which the disturbances, $\tilde{\epsilon}$, conform to the simple model of random drawings from a normal distribution with mean zero.

The rationale is to capture *systematic* time effects in a model so that what is left over—the disturbances—behave randomly.

Will you be judging alternative models, as in earlier examples, by behavior of residuals?

Insofar as the judgment is to be based upon the evidence of the data, yes. We shall be computing residuals, $\hat{\epsilon}$, and studying their behavior with the same tools we have used before.

Remember, too, that non-sample information and knowledge may also enter into the choice of a model.

Are you going to give a complete coverage of the available tools for time series analysis?

That will not be possible within the scope of this elementary treatment. I shall, however, provide you with a kit of tricks that you can use to obtain some degree of success in finding a tolerable statistical model for many of the time series that arise in business and economics.

Will these tricks apply to very short time series—small sample sizes?

When the sample size is small, the sample itself gives very little evidence about the appropriate statistical model for its analysis, although certain models may be ruled out as outrageous. Then you must rely on non-sample information even more heavily than you ordinarily would when sample information is abundant.

For our main illustration, however, we shall use an example in which the sample size is large enough to shed a good deal of light upon the appropriateness of alternative models.

What illustration are you going to use?

I shall consider the Gross National Product (GNP) of the United States from the first quarter of 1947 through the fourth quarter of

STRATEGY OF TIME SERIES ANALYSIS

190 CONVERSATIONAL STATISTICS

1970. This GNP series is *not* seasonally adjusted. The data are in billions of dollars and are expressed in quarterly rates rather than the usual convention of annual rates, in which the quarterly rate is multiplied by 4.

Are these data in a file for convenient use in computer analysis?

Yes, they are in Column 1 of a file named $ECDAT1, which is in the HP library.

How did you access this file?

Using IDA, I simply entered the data from $ECDAT1 with the command ENTE, just as for any other file. Since $ECDAT1 has a total of 7 columns, each with 96 observations, I used SAVF to save only column 1 in a file in my personal library, which I then brought into my working space in IDA by another use of the command ENTE.

An alternative command for entering that you may wish to try is ENTS; it will work faster than ENTE since it permits you to designate the variables of $ECDAT1 to be entered in the IDA data matrix.

Why did you go to the trouble of this last step?

It just saves execution time for many of the commands used subsequently, since IDA updates summary statistics for all the variables in the current data matrix. It also makes it easier to keep track of the transformed variables created during the analysis.

In case I wanted to practice on other series, could you tell me what is stored in the remaining columns of $ECDAT1?

Column 2. Money Supply (M1) without seasonal adjustments. The data are means of daily data in billions of dollars.

Column 3. Federal government purchases of goods and services without seasonal adjustment at quarterly rates in billions of dollars.

Column 4. Standard and Poor's *Composite Index of Stock Prices* (means of daily data).

Column 5. Man-days idle due to strikes and lockouts in millions (means of monthly data without seasonal adjustment).

Column 6. GNP Implicit Price Deflators (1957-59 = 1).

Column 7. Time period identification (e.g., 1950 second quarter is designated 502).

For reference, these are printed out in Table 11-1.

ENTRY OF DATA FROM FILES

Table 11-1. Contents of the File $ECDAT1

```
> FPRS                                    > FPRS
CONSECUTIVE COL. ?YES                     CONSECUTIVE COL. ?YES
I1,I2,J1,J2 = ?1,96,1,5                   I1,I2,J1,J2 = ?1,96,6,7
FMT = ?#,4D.2D                            FMT = ?#,4D.2D

**  1 **   53.40 110.40   2.90 15.40  1.22    **  1 **   0.73 471.00
**  2 **   55.90 109.90   3.30 14.60  6.41    **  2 **   0.74 472.00
**  3 **   57.50 112.20   3.10 15.40  2.82    **  3 **   0.75 473.00
**  4 **   64.50 114.70   3.30 15.30  1.07    **  4 **   0.77 474.00
**  5 **   59.40 113.70   3.00 14.40  2.80    **  5 **   0.78 481.00
**  6 **   62.10 110.50   3.80 16.10  4.57    **  6 **   0.79 482.00
**  7 **   65.30 111.80   4.20 16.00  2.44    **  7 **   0.81 483.00
**  8 **   70.70 113.30   5.50 15.60  1.56    **  8 **   0.80 484.00
**  9 **   61.80 111.90   4.90 15.00  1.62    **  9 **   0.80 491.00
** 10 **   62.20 109.80   4.70 14.50  3.26    ** 10 **   0.79 492.00
** 11 **   64.50 110.50   5.10 15.20  3.59    ** 11 **   0.79 493.00
** 12 **   68.00 112.50   5.50 16.20  8.37    ** 12 **   0.79 494.00
** 13 **   64.00 112.70   4.70 17.10  5.06    ** 13 **   0.78 501.00
** 14 **   66.80 112.10   3.70 18.30  3.06    ** 14 **   0.79 502.00
** 15 **   73.50 114.30   4.30 18.30  2.97    ** 15 **   0.81 503.00
** 16 **   80.50 117.50   5.70 19.80  1.85    ** 16 **   0.82 504.00
** 17 **   76.60 118.10   6.70 21.60  1.64    ** 17 **   0.85 511.00
** 18 **   79.90 116.70   8.40 21.80  1.84    ** 18 **   0.85 512.00
** 19 **   83.20 118.60  10.70 22.80  2.35    ** 19 **   0.86 513.00
** 20 **   88.70 123.50  11.90 23.20  1.81    ** 20 **   0.87 514.00
** 21 **   82.00 124.70  11.60 23.90  1.44    ** 21 **   0.87 521.00
** 22 **   83.30 123.10  12.50 23.90  9.46    ** 22 **   0.87 522.00
** 23 **   85.70 124.50  13.40 25.00  6.30    ** 23 **   0.88 523.00
** 24 **   94.40 128.70  14.30 25.10  2.47    ** 24 **   0.88 524.00
** 25 **   87.40 128.80  13.90 26.00  1.24    ** 25 **   0.88 531.00
** 26 **   91.30 127.00  14.50 24.50  3.86    ** 26 **   0.88 532.00
** 27 **   90.30 127.30  13.50 24.00  2.82    ** 27 **   0.88 533.00
** 28 **   95.60 130.30  15.20 24.40  1.70    ** 28 **   0.88 534.00
** 29 **   86.50 130.30  12.60 26.00  1.13    ** 29 **   0.89 541.00
** 30 **   89.70 128.10  12.00 28.40  1.87    ** 30 **   0.90 542.00
** 31 **   90.00 129.30  10.90 30.80  3.32    ** 31 **   0.89 543.00
** 32 **   98.60 133.40  11.90 33.50  1.20    ** 32 **   0.90 544.00
** 33 **   92.60 134.70  10.60 36.30  0.89    ** 33 **   0.90 551.00
** 34 **   97.40 133.00  10.40 38.40  2.98    ** 34 **   0.91 552.00
** 35 **  100.40 133.50  11.10 43.20  3.05    ** 35 **   0.91 553.00
** 36 **  107.60 136.50  11.90 44.10  2.48    ** 36 **   0.92 554.00
** 37 **   98.60 136.80  10.60 45.40  2.15    ** 37 **   0.93 561.00
** 38 **  102.90 134.70  11.50 47.00  2.15    ** 38 **   0.93 562.00
** 39 **  104.10 134.60  10.80 48.00  5.70    ** 39 **   0.95 563.00
** 40 **  113.70 138.00  12.70 46.10  1.04    ** 40 **   0.95 564.00
** 41 **  104.40 137.90  12.10 44.30  0.78    ** 41 **   0.96 571.00
** 42 **  109.10 135.70  12.10 46.50  1.88    ** 42 **   0.97 572.00
** 43 **  110.60 135.80  12.10 46.10  1.97    ** 43 **   0.98 573.00
** 44 **  117.00 137.60  13.20 40.60  0.86    ** 44 **   0.99 574.00
** 45 **  103.90 136.90  12.40 41.50  0.75    ** 45 **   0.99 581.00
** 46 **  108.80 136.50  13.00 43.60  1.63    ** 46 **   1.00 582.00
** 47 **  111.70 137.90  13.30 47.50  2.24    ** 47 **   1.00 583.00
** 48 **  123.00 142.20  14.90 52.30  3.35    ** 48 **   1.01 584.00
** 49 **  113.10 143.00  13.10 55.50  1.48    ** 49 **   1.01 591.00
** 50 **  121.40 142.20  13.40 57.50  2.76    ** 50 **   1.02 592.00
** 51 **  119.30 143.00  13.10 58.70 12.14    ** 51 **   1.02 593.00
** 52 **  129.90 144.80  14.10 57.80  6.61    ** 52 **   1.02 594.00
** 53 **  120.50 142.70  12.70 56.30  1.31    ** 53 **   1.03 601.00
** 54 **  125.60 139.90  12.80 56.10  2.33    ** 54 **   1.03 602.00
** 55 **  124.60 140.40  13.80 55.70  1.83    ** 55 **   1.04 603.00
** 56 **  133.10 143.50  14.20 55.30  0.90    ** 56 **   1.04 604.00
** 57 **  120.60 143.10  13.50 62.00  0.61    ** 57 **   1.04 611.00
** 58 **  128.20 142.30  14.00 66.00  1.42    ** 58 **   1.05 612.00
** 59 **  129.10 142.90  14.10 66.80  1.79    ** 59 **   1.05 613.00
** 60 **  142.20 147.50  15.80 70.30  1.61    ** 60 **   1.05 614.00
** 61 **  131.30 147.20  15.20 69.90  0.90    ** 61 **   1.06 621.00
** 62 **  139.60 145.90  15.90 62.20  2.22    ** 62 **   1.06 622.00
** 63 **  138.10 145.40  15.30 57.80  1.85    ** 63 **   1.06 623.00
** 64 **  151.50 149.50  17.10 59.60  1.22    ** 64 **   1.06 624.00
** 65 **  137.80 150.00  15.70 65.60  1.41    ** 65 **   1.07 631.00
** 66 **  146.10 149.30  15.40 69.70  1.31    ** 66 **   1.07 632.00
** 67 **  146.50 150.40  16.00 71.00  1.38    ** 67 **   1.07 633.00
** 68 **  160.20 155.50  17.10 73.30  1.27    ** 68 **   1.08 634.00
** 69 **  148.50 155.40  15.90 77.50  0.92    ** 69 **   1.08 641.00
** 70 **  157.10 154.30  16.60 80.30  1.82    ** 70 **   1.09 642.00
** 71 **  156.30 156.70  15.80 82.90  1.78    ** 71 **   1.09 643.00
** 72 **  170.60 162.40  16.80 84.80  3.13    ** 72 **   1.10 644.00
** 73 **  158.20 162.30  15.40 86.60  1.65    ** 73 **   1.10 651.00
** 74 **  169.10 160.90  16.20 87.40  2.09    ** 74 **   1.11 652.00
** 75 **  168.90 162.60  16.20 86.90  2.67    ** 75 **   1.11 653.00
** 76 **  188.70 169.50  19.00 91.80  1.35    ** 76 **   1.12 654.00
** 77 **  176.20 170.60  17.60 91.60  1.14    ** 77 **   1.12 661.00
** 78 **  187.40 170.20  18.90 88.10  2.56    ** 78 **   1.13 662.00
** 79 **  186.30 169.50  20.10 81.40  2.75    ** 79 **   1.14 663.00
** 80 **  199.80 173.80  21.20 79.80  2.00    ** 80 **   1.15 664.00
** 81 **  186.50 173.80  21.50 87.10  1.34    ** 81 **   1.16 671.00
** 82 **  197.20 174.30  22.50 91.70  3.96    ** 82 **   1.17 672.00
** 83 **  198.40 178.10  22.30 94.40  4.45    ** 83 **   1.18 673.00
** 84 **  211.70 184.80  24.40 94.50  4.29    ** 84 **   1.19 674.00
** 85 **  199.90 185.60  23.00 91.60  3.48    ** 85 **   1.20 681.00
** 86 **  217.30 187.00  25.50 98.00  6.23    ** 86 **   1.22 682.00
** 87 **  215.80 190.70  24.30 99.90  3.91    ** 87 **   1.23 683.00
** 88 **  232.00 198.50  26.60 105.20 2.70    ** 88 **   1.24 684.00
** 89 **  217.50 200.10  24.50 100.90 2.72    ** 89 **   1.26 691.00
** 90 **  232.40 200.10  24.90 101.70 4.41    ** 90 **   1.27 692.00
** 91 **  234.80 200.80  26.00 94.50  3.38    ** 91 **   1.29 693.00
** 92 **  246.50 206.10  26.00 94.30  3.79    ** 92 **   1.31 694.00
** 93 **  229.30 206.30  24.50 88.70  2.76    ** 93 **   1.33 701.00
** 94 **  244.20 207.50  24.40 79.20  5.98    ** 94 **   1.34 702.00
** 95 **  242.60 209.70  23.40 78.70  5.88    ** 95 **   1.36 703.00
** 96 **  258.00 216.50  24.90 86.20  7.52    ** 96 **   1.37 704.00
```

192 ☐ **CONVERSATIONAL STATISTICS**

I'll be disappointed unless you attack the GNP with a sequence plot.

I won't disappoint you: PLTS is first! The result is shown in Figure 11-1.

I see a steady upward movement with only minor setbacks during the entire period of 24 years.

That's a good start. What else do you see?

Within each year, the fourth quarter is almost always higher than the first three quarters of the year, which in turn are relatively close to each other.

You've spotted evidence of a *seasonal pattern*. Go on.

Well, it looks as if the series gets a little more volatile as time goes on. By that I mean that GNP seems to vary more within short periods such as a year or two as we move from the late forties through the fifties and then through the sixties.

Good. I'm glad you noticed that, because it raises the question of the appropriate units of measurement. In analyzing a time series we want to transform the data, if necessary, so that the volatility will tend to be about the same everywhere in the series.

Does that mean that you will transform dollar GNP to the logarithm of GNP?

THE LOGARITHMIC TRANSFORMATION

That's a good suggestion. The logarithmic transformation promises several advantages:

1. If the volatility of the series *relative to its level* is about the same throughout, the logarithms will have the constant volatility we seek.

2. If the series tends to grow at a constant percentage rate, then the sequence plot will tend to follow a straight line.

3. As we have seen, the logarithmic transformation often tends to induce approximate normality in the residuals of any regression that we may later run.

So you will work with logarithms?

Well, no. I tried that, and it overcompensated; the logs show substantially more volatility in the *earlier* part of the series. That means that percentage variations of GNP have been milder in recent years.

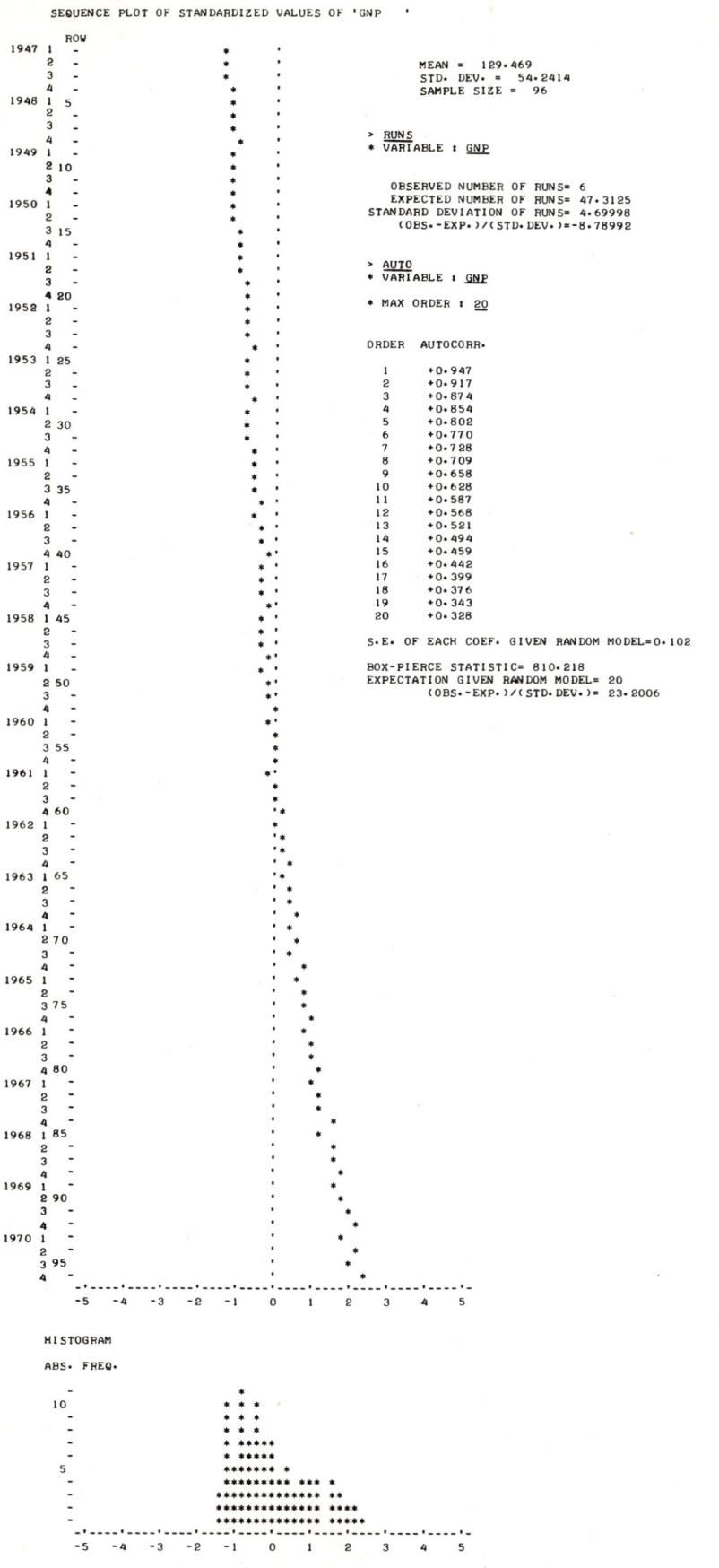

Figure 11-1. Sequence Plot and Histogram of GNP Data

194 □ CONVERSATIONAL STATISTICS

What can you do, then, to obtain constant volatility of the series?

A transformation known as the *square root transformation*, which is in a sense intermediate between the original units and the logarithms, works pretty well. The sequence plot of RTGNP is shown in Figure 11-2.

How do you get the square root transformation in $IDA?

As follows:

```
> POWE
* COL# TO PLACE VAR. : 1
* COLUMN TO BE TRANSFORMED :1
* P = .5
UPDATING MEAN, STD, ..., CORR, ...
GIVE NAME OF NEW VARIABLE
VAR. 1 ?RTGNP
```

Aside from the improvement with respect to "constant scatter", Figure 11-2 looks pretty much like Figure 11-1, and my earlier observations are applicable. What do we do next?

The key to the next step lies in the observation that you made about a steady upward movement. If this aspect of the sample record is adjudged a characteristic of the hypothetical process underlying the sample, then this process is *non-stationary*.

A *stationary process*, by contrast, generates data that vary about a constant level. Further, different samples from the process have a general statistical resemblance, whatever their detailed differences.

Hold it! What do you mean by those two conditions?

1. *Constant level.* Imagine an enormous sample from the process, plotted on a sequence plot of enormous length. Draw a horizontal line at height equal to the "true" mean. Then the series will tend always to keep returning to the horizontal line, even though it may from time to time wander pretty far away from it.

When you say "horizontal line", you mean that time is the horizontal axis?

Yes. In terms of our computer sequence plots, it would actually be a vertical line.

From the 96 observations in Figures 11-1 and 11-2, then, you're judging that there is no "true" horizontal line about which the series would tend to vary and to which it would tend to return, if only we could observe long enough.

That's the judgment I made.

In this application, my head swims to think of an enormous sample from the process, since that would take us far into the future: even another 96 observations would carry me so far that I wouldn't want to make any bets about the GNP—or even whether there would be a GNP!

The enormous sample is a conceptual device, one that we've talked about earlier in the context of the random model. The purpose of the scheme is to make inferences about events that are only a little way ahead of us. For inferences further in the future, we have the same skepticism we have about point estimates of regression intercepts far removed from the data.

A STATIONARY PROCESS

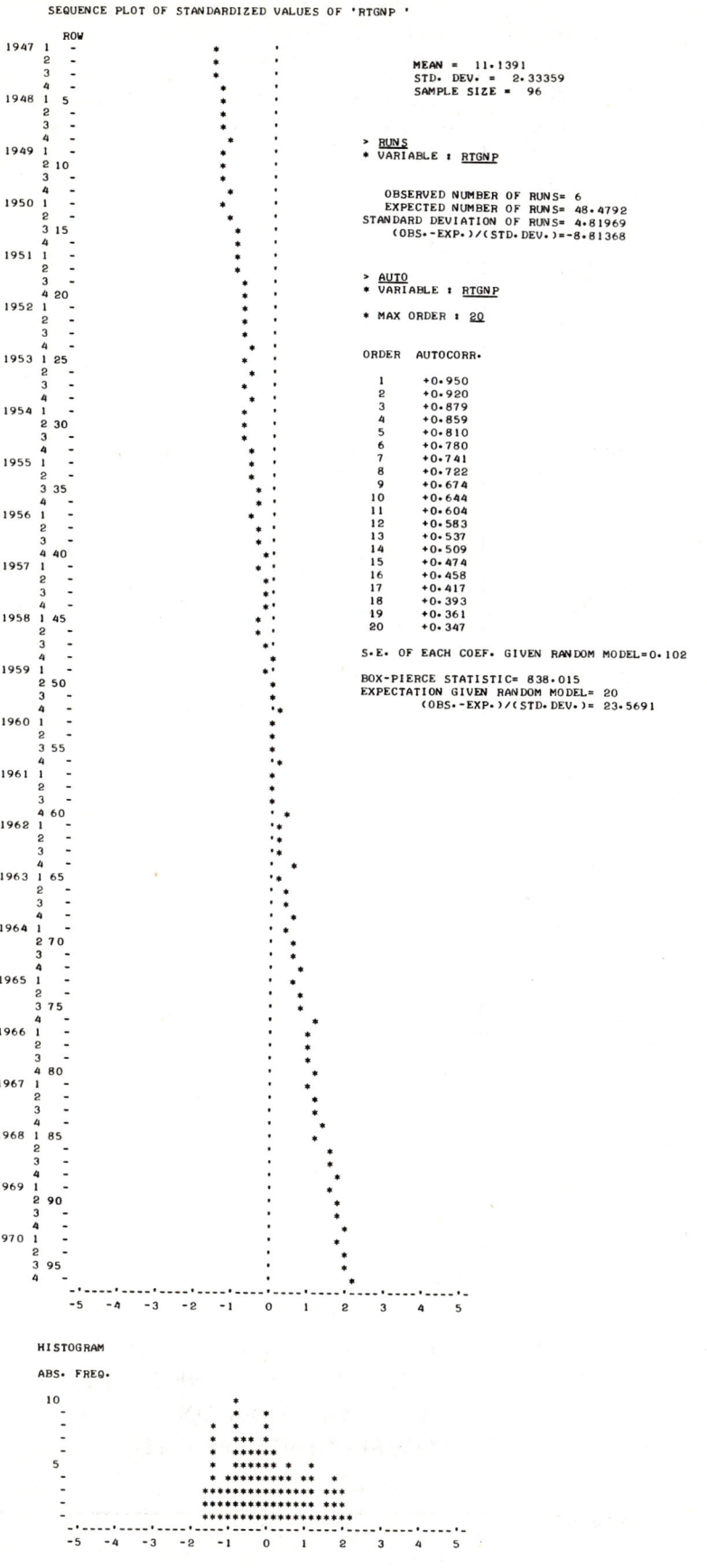

Figure 11-2. Square Root Transformation of GNP Data

What does your second condition for stationarity mean?

2. *Statistically similar appearance.* Let's illustrate this by the random, normal process. If we divide a long sequence plot into short snapshots, you have a pretty good idea of the statistical resemblance of the different snapshots. There can be, for example, no tendency to increasing scatter from one snapshot to the next; nor can there be any systematic pattern, such as a true "cycle", common to all. And this is true no matter whether each snapshot covers a short or a long period. The same idea is extended to stationary processes, except that stationary processes can show a statistical tendency for nearby observations to resemble each other more closely than do observations far removed. Autocorrelation may be present, for example.

Can the concept of stationarity be given a rigorous mathematical definition?

Yes, just as can a random, normal process.

Is a simple random process an example of a stationary process?

Yes, it is a special case. But stationary processes also include non-random (in the sense of autocorrelated, for example) processes.

What good is the concept of stationarity?

Stationary processes can be analyzed directly by a wide variety of statistical techniques.

It would appear that we are out of luck, then, with the GNP data.

Here comes one of our tricks. We will take *first differences* of the data and examine them.

The same trick that we used earlier with the Dow-Jones Industrial Average in Chapter 4?

Just so. But before we do that transformation, let me call your attention to the autocorrelation coefficients in Figures 11-1 and 11-2. Note that the first autocorrelation is about 0.95 and that the higher autocorrelations decline only very slowly. The 20th is still substantial. (And look at the enormous Box-Pierce statistics!)

Why is this important?

In case you'd like to have some numerical guidance as to whether a process is really non-stationary, here it is: the higher autocorrelation coefficients fail to die out rapidly.

Now look at the first differences in Figure 11-3.

That's the strangest sequence plot I've looked at yet! What's going on?

Look closer and tell me.

STATISTICALLY SIMILAR APPEARANCE IS SECOND CONDITION FOR STATIONARITY

FOR A NON-STATIONARY PROCESS, THE HIGHER AUTOCORRELATION COEFFICIENTS DIE OUT SLOWLY

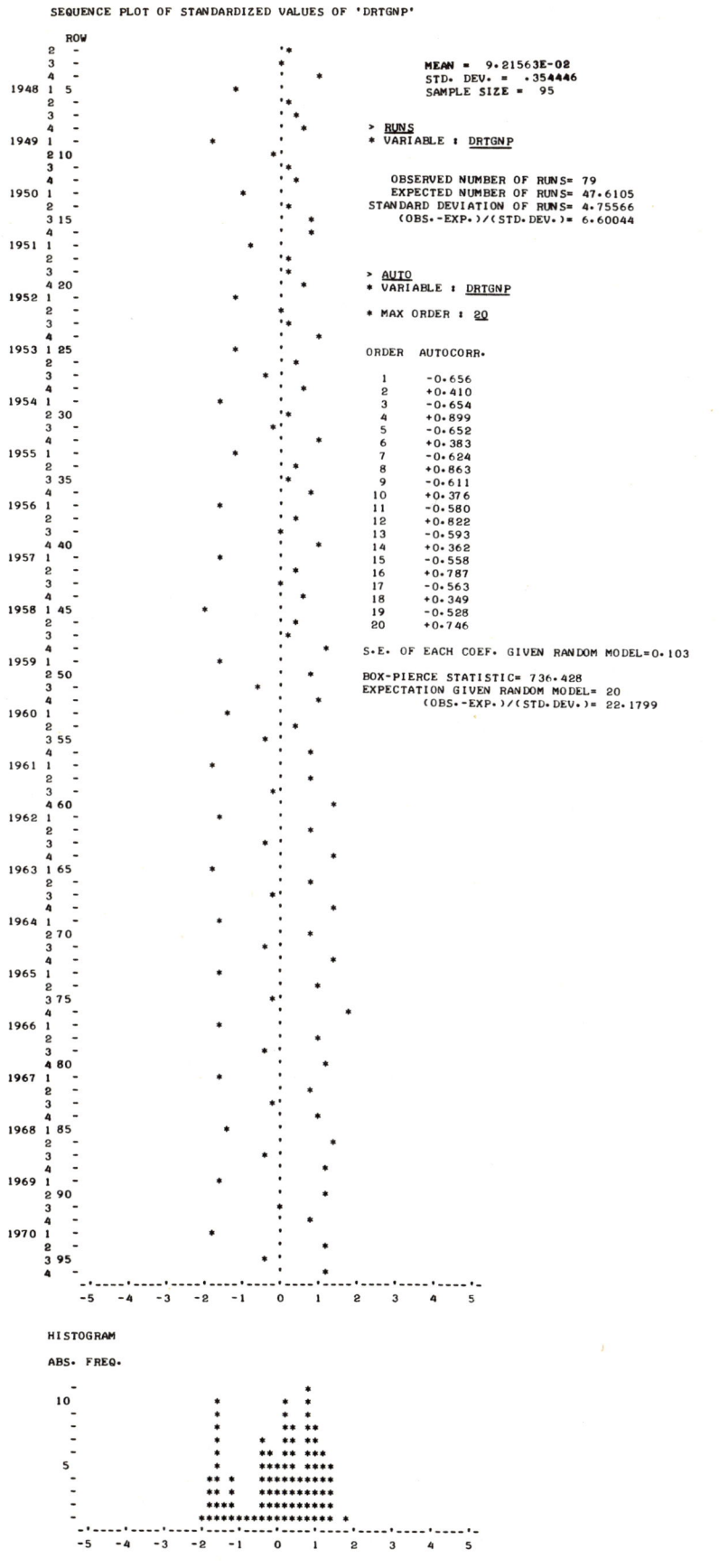

Figure 11-3. First Differences of GNP Data

Could it be a reflection of seasonality? I notice, for example, that there is a string of low values spaced four quarters apart.

Good. The string of low values comprises the differences from the fourth quarter of one year to the first quarter of the next.

But the data do seem to vary about a constant level. Is this an example of the behavior to be expected from a sample of a stationary process?

There's a catch because of the seasonality. The first quarter's change for each year is consistently low. If you look at the first quarter alone, it's reasonable to think of that subsample as displaying stationary behavior. Similarly, the fourth quarter is consistently high.

You can see this in vivid fashion in the histogram of the differences in Figure 11-3; see especially the spike at the left. Essentially, we may be observing four different stationary processes, but in a completely deterministic sequence: 1, 2, 3, 4.

I notice too that the autocorrelations alternate between - and + and do not show much tendency to die out.

That's a clear diagnostic sign, in case you were in doubt.

Do you have another trick to move the analysis ahead?

Since the behavior of the differences still suggests non-stationarity, a repetition of the differencing operation is in order. Since the source of non-stationarity has been traced to a seasonal pattern of the consecutive differences, the appropriate differencing now will involve a subtraction from the current difference of the difference a year ago, that is, the difference four periods ago. We can call this a fourth difference or seasonal difference. So that you can keep clear on the logistics, I show below the two differencing commands: first, the creation of DRTGNP that we have seen in Figure 11-3; second, the creation of the fourth difference of consecutive differences that yields D4DRTG, which is shown in Figure 11-4.

```
> DIFF
* COL# TO PLACE VAR. : 2
* COLUMN TO BE TRANSFORMED :1
GAP = ?1
UPDATING MEAN, STD, ...
GIVE NAME OF NEW VARIABLE
VAR.   2 ?DRTGNP


> DIFF
* COL# TO PLACE VAR. : 3
* COLUMN TO BE TRANSFORMED :2
GAP = ?4
UPDATING MEAN, STD, ...
GIVE NAME OF NEW VARIABLE
VAR.   3 ?D4DRTG
```

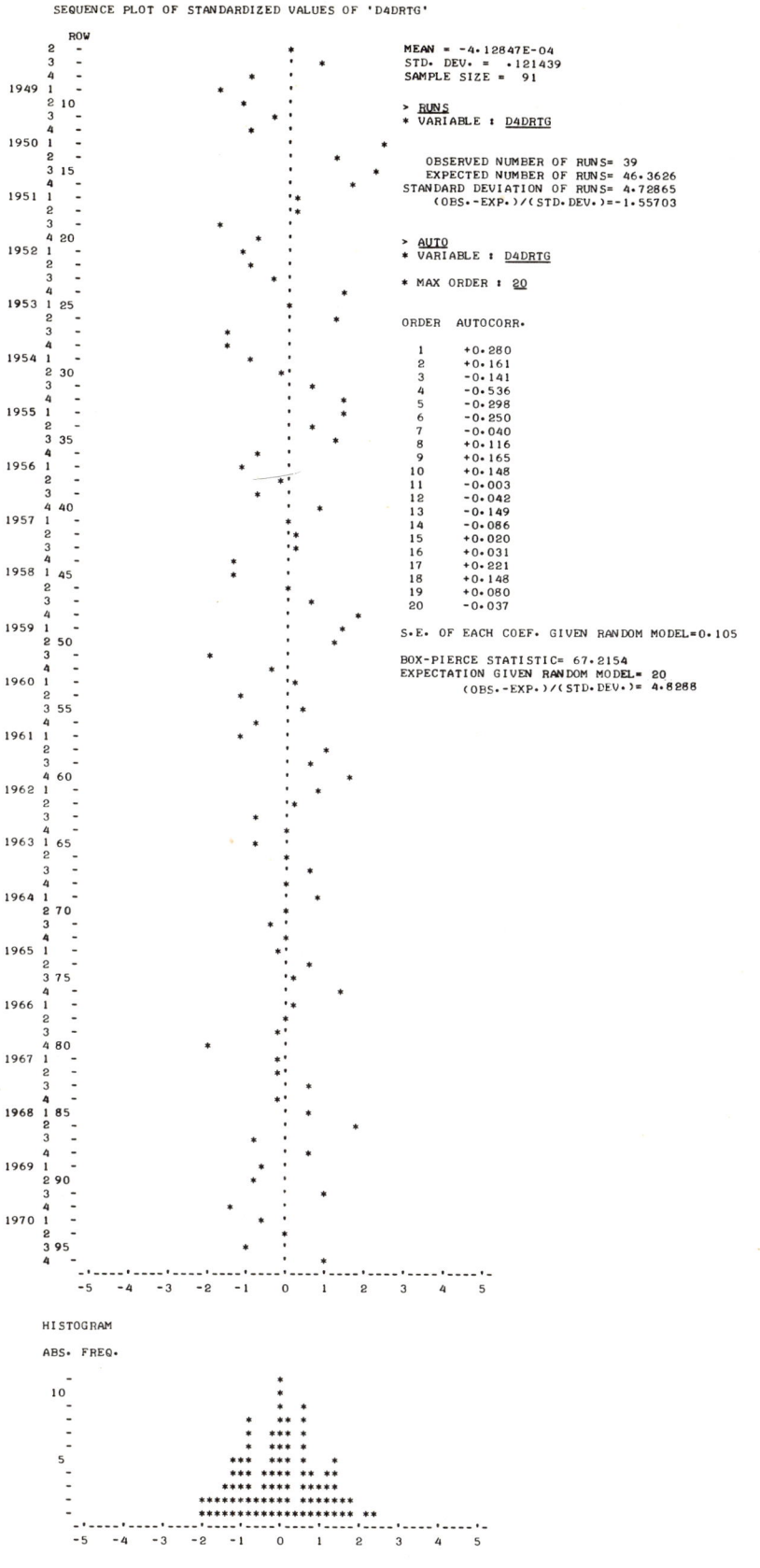

Figure 11-4. Fourth Differences of First Differences of GNP Data

I'm having trouble keeping track of all this. First you transformed GNP to its square root RTGNP in order to obtain constant scatter. Then you took consecutive differences to reach DRTGNP. Now you have taken the fourth differences of the consecutive differences to reach D4DRTG. Haven't we come a long way from GNP?

It takes a while to get used to this kind of procedure, but remember that we have not burned our bridges behind us. If you could predict the next value of D4DRTG, you could predict the next value of GNP. We'll come back to this point.

And, as you saw in Figure 11-4, it is reasonable to treat the behavior of D4DRTG as stationary.

But D4DRTG still does not behave randomly. I detect a tendency toward persistence or streakiness that at times looks almost cyclical.

Your interocular skill still serves you well. If you're up on your macroeconomics, you will notice some tendency for the apparently cyclical behavior to correspond with the so-called business cycles of this period. But the cycles that you seem to perceive here, and the so-called business cycles themselves, should not be confused with the deterministic or near-deterministic cycles often encountered in the natural sciences. These latter are often regular in both period and amplitude, except for measurement errors. In natural science, often the only "statistical" aspect of cyclical behavior is the measurement errors. In business and economics, the apparent cyclical behavior may be entirely statistical, aside from possible seasonal patterns.

You were right in saying that D4DRTG, though apparently stationary, is not random. You may confirm that by the runs count and, better, by the behavior of the autocorrelations. The autocorrelations are relatively small compared to those of the earlier charts, but some are still substantial. Further, although the autocorrelations die out, they swing in a systematic way from negative to positive. And the Box-Pierce statistic, though no longer astronomical, is still several standard deviations above the random expectation.

I notice that the largest autocorrelation in absolute value is the fourth, and that it is negative. Is that a hint of some remaining seasonality?

It is indeed. But it is a statistical tendency rather than the kind of seasonality we saw in Figure 11-3. If the observation four periods back is low, the current observation tends to be high, and vice versa. It isn't true that the fourth observation in each year tends to be consistently low or consistently high.

What is the appearance of the histogram?

It is also shown in Figure 11-4. It appears to conform reasonably well with the normal model, so far as I can judge without use of NORM. Here is a vivid example to counteract any remaining tendency to confuse normality with randomness.

But you wouldn't want to base predictions on this distribution, would you?

THE TRANSFORMED GNP NOW SUGGESTS STATIONARY BUT NON-RANDOM BEHAVIOR

No, because it loses all the information about the sequence of the observations. You have observed a persistence or streakiness and a statistical type of seasonality. We ought to be able to exploit this observation and use the recent behavior of the series as a guide to its prediction.

Are we now at the stage of autoregression?

Yes. We shall use lagged values of the series itself in a multiple regression in order to obtain a predictive relationship.

How do we decide which lagged values to use?

The behavior of the autocorrelation coefficients contains information that is useful in guiding us. Unfortunately, the full apparatus is more technical than we can handle in this course. A detailed treatment is given in the book by Box and Jenkins, *Time Series Analysis* (Holden-Day, 1970). A more elementary treatment is given by Charles R. Nelson in *Applied Time Series Analysis* (Holden-Day, 1973).

A few common sense hints, however, may lead the beginner a substantial part of the way toward a sensible choice.

1. Keep the number of lagged variables (independent variables) small.

2. Give preference to recent lags, especially the first two.

3. If there is a substantial autocorrelation at the seasonal lag, put that lag in.

4. Use the sweep command to eliminate variables with low ratios of slope coefficient to standard error (say, less than one in absolute value).

In this application, then, you might start with the first four lags?

Yes.

Just to refresh me, would you show how you set up the four lagged variables?

Starting with RTGNP, DRTGNP, and D4DRTG in Columns 1, 2, and 3, respectively, I proceeded as follows:

```
> LAGG

* COL# TO PLACE VAR. : 4
* COLUMN TO BE TRANSFORMED :3
GAP = ?1
UPDATING MEAN, STD, ...
GIVE NAME OF NEW VARIABLE
VAR.  4 ?L1D4DR

> LAGG

* COL# TO PLACE VAR. : 5
* COLUMN TO BE TRANSFORMED :3
GAP = ?2
UPDATING MEAN, STD, ...
GIVE NAME OF NEW VARIABLE
VAR.  5 ?L2D4DR

> LAGG

* COL# TO PLACE VAR. : 6
* COLUMN TO BE TRANSFORMED :3
GAP = ?3
UPDATING MEAN, STD, ...
GIVE NAME OF NEW VARIABLE
VAR.  6 ?L3D4DR

> LAGG

* COL# TO PLACE VAR. : 7
* COLUMN TO BE TRANSFORMED :3
GAP = ?4
UPDATING MEAN, STD, ...
GIVE NAME OF NEW VARIABLE
VAR.  7 ?L4D4DR
```

COMMON SENSE HINTS ON FORMULATING AUTOREGRESSION MODELS

202 ☐ CONVERSATIONAL STATISTICS

And show me now how the data matrix looks as we move to the regression analysis.

```
> MEAN

   VARIABLE      MEAN         STD. DEV.

   RTGNP      1.14830E+01   2.17463E+00
   DRTGNP     9.42652E-02   3.57975E-01
   D4DRTG     2.38948E-03   1.20732E-01
   L1D4DR     -.14333E-02   1.22550E-01
   L2D4DR     -.13302E-02   1.22444E-01
   L3D4DR     -.10303E-03   1.22865E-01
   L4D4DR     7.94970E-04   1.22604E-01

> NAME

   THESE ARE THE VARIABLES IN THE DATA MATRIX :
   COLUMN    NAME

      1      RTGNP
      2      DRTGNP
      3      D4DRTG
      4      L1D4DR
      5      L2D4DR
      6      L3D4DR
      7      L4D4DR
```

You started, then, with four independent variables: L1D4DR, L2D4DR, L3D4DR, and L4D4DR?

Yes. After looking at the results of COEF, I decided to sweep out L3D4DR:

```
> REGR
UPDATING CORR. MATRIX...
* DEP. VAR. = 3
HOW MANY INDEP. VAR. ?4
INDEP. VAR.   1 = ?4
INDEP. VAR.   2 = ?5
INDEP. VAR.   3 = ?6
INDEP. VAR.   4 = ?7
ANALYZING RESIDUALS...
WARNING: RESIDUAL IN ROW  51 IS -2.58 S.D. UNITS FROM 0

> SUMM

                MULTIPLE R   R-SQUARE
   UNADJUSTED     0.6399      0.4094
   ADJUSTED       0.6169      0.3806
   STD. DEV. OF RESIDUALS =    9.50181E-02

> COEF

   VARIABLE   B(STD.V)       B          STD.ERROR(B)     T

   L1D4DR      0.1375     1.3542E-01     9.0624E-02     1.494
   L2D4DR      0.2561     2.5256E-01     9.2057E-02     2.744
   L3D4DR     -0.0800     -.7857E-01     9.1499E-02    -0.859
   L4D4DR     -0.5573     -.5488E+00     9.0205E-02    -6.084
   CONSTANT               3.3478E-03     1.0188E-02     0.329

> SWEEP
VAR. : L3D4DR
ANALYZING RESIDUALS...
WARNING: RESIDUAL IN ROW  51 IS -2.71 S.S. UNITS FROM 0

> COEF

   VARIABLE   B(STD.V)       B          STD.ERROR(B)     T

   L1D4DR      0.1261     1.2423E-01     8.9541E-02     1.387
   L2D4DR      0.2395     2.3614E-01     8.9907E-02     2.627
   L4D4DR     -0.5800     -.5711E+00     8.6263E-02    -6.620
   CONSTANT               3.3357E-03     1.0172E-02     0.328

> SUMM

                MULTIPLE R   R-SQUARE
   UNADJUSTED     0.6357      0.4041
   ADJUSTED       0.6185      0.3826
   STD. DEV. OF RESIDUALS =    9.48677E-02

> CORR

HOW MANY VARIABLES ?3
COL. #'S: ?4,5,7
* # DECIMALS = 3

             L1D4DR    L2D4DR    L4D4DR
   L1D4DR    1.000
   L2D4DR    0.308     1.000
   L4D4DR   -0.139     0.160     1.000
```

The slope coefficients for the first two lagged variables are positive, but the slope coefficient for the seasonal lag is negative.

And the absolute value of the seasonal coefficient is larger than the sum of the coefficients of the first two lagged variables.

Note that the constant or intercept is 0.0033357, quite close to 0. When the dependent variable has been twice differenced as here, it is to be expected that the constant will be small.

Also, note that the correlation coefficients among independent variables are relatively small. The differencing done in creating the lagged variables has effected this, since the original levels (RTGNP) were highly correlated, as you can see from AUTO on Figure 11-2.

How about the model adequacy checks?

The sequence plot of residuals along with RUNS and AUTO are shown in Figure 11-5. The remaining plots are shown in Figures 11-6 and 11-7. As you can see, the picture is pretty good.

You didn't say anything about the warning given by IDA, as it executed REGR, concerning the 51st residual.

A good point. Let's see if you can locate it on the sequence plot.

That's the third quarter of 1959, but so what?

I didn't have any special reaction either, but a student from the steel industry has pointed out to me that there was a nationwide steel strike at that time. So the poorest performance of GNP, relative to what would be expected from its past performance, occurred in 1959 and is traceable to this event.

DIAGNOSTIC CHECKS FOR THE AUTOREGRESSION MODEL ON THE TRANSFORMED GNP APPEAR SATISFACTORY

204 ☐ CONVERSATIONAL STATISTICS

```
SEQUENCE PLOT OF STANDARDIZED VALUES OF 'RESIDU'

                            MEAN = 0
                            STD. DEV. = 9.48677E-02
                            SAMPLE SIZE = 87

                          > RUNS
                          * VARIABLE : RESIDU

                            OBSERVED NUMBER OF RUNS= 46
                            EXPECTED NUMBER OF RUNS= 44.4483
                            STANDARD DEVIATION OF RUNS= 4.63092
                               (OBS.-EXP.)/(STD.DEV.)= .335078

                          > AUTO
                          * VARIABLE : RESIDU

                          * MAX ORDER : 20

                            ORDER   AUTOCORR.

                              1     -0.041
                              2     +0.018
                              3     -0.144
                              4     -0.082
                              5     -0.011
                              6     +0.038
                              7     -0.052
                              8     -0.169
                              9     -0.001
                             10     -0.060
                             11     -0.066
                             12     -0.001
                             13     -0.003
                             14     +0.028
                             15     +0.088
                             16     -0.042
                             17     +0.116
                             18     +0.061
                             19     -0.060
                             20     -0.039

                            S.E. OF EACH COEF. GIVEN RANDOM MODEL=0.107

                            BOX-PIERCE STATISTIC= 8.98104
                            EXPECTATION GIVEN RANDOM MODEL= 20
                               (OBS.-EXP.)/(STD.DEV.)=-2.11668
```

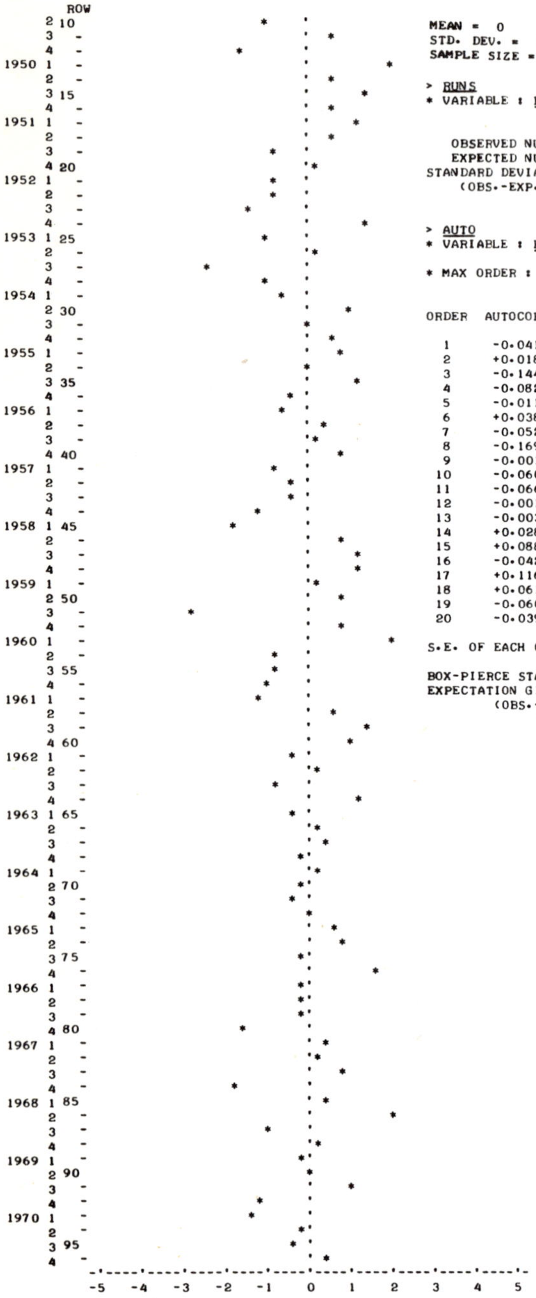

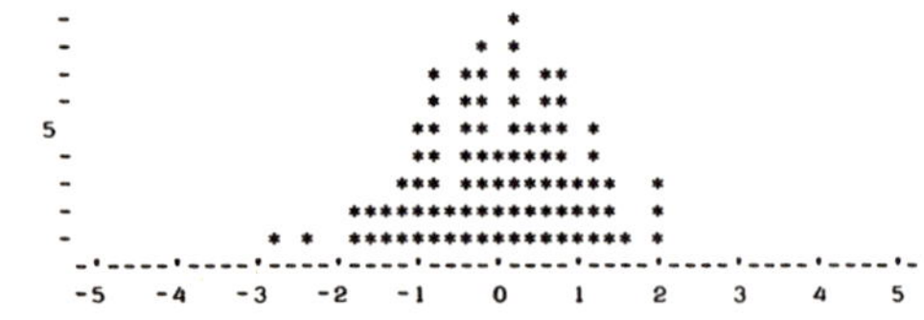

Figure 11-5. Sequence Plot and Histogram for Residuals for Autoregression of Transformed GNP Data

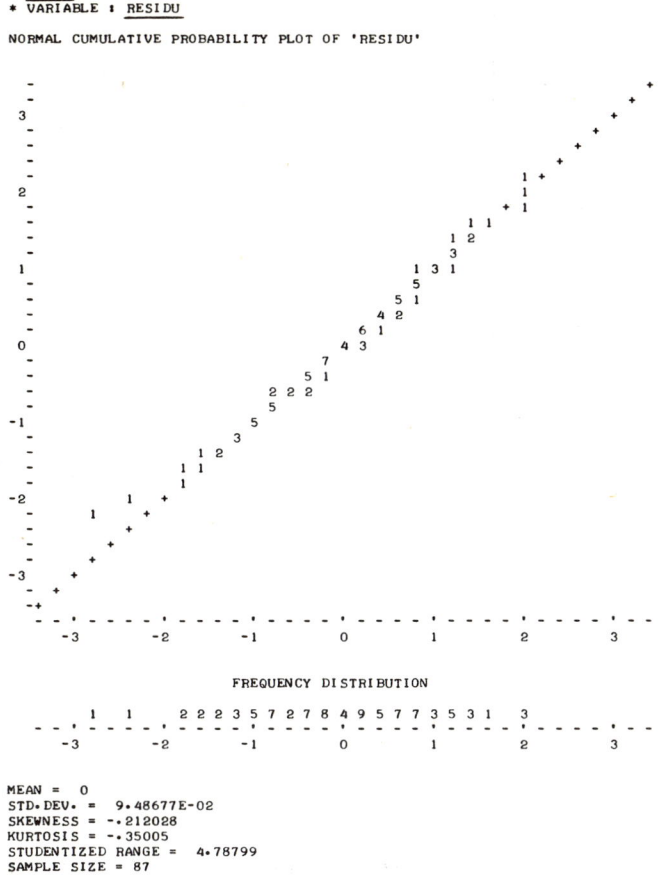

Figure 11-6. Normal Probability Plot of Residuals for Autoregression of Transformed GNP Data

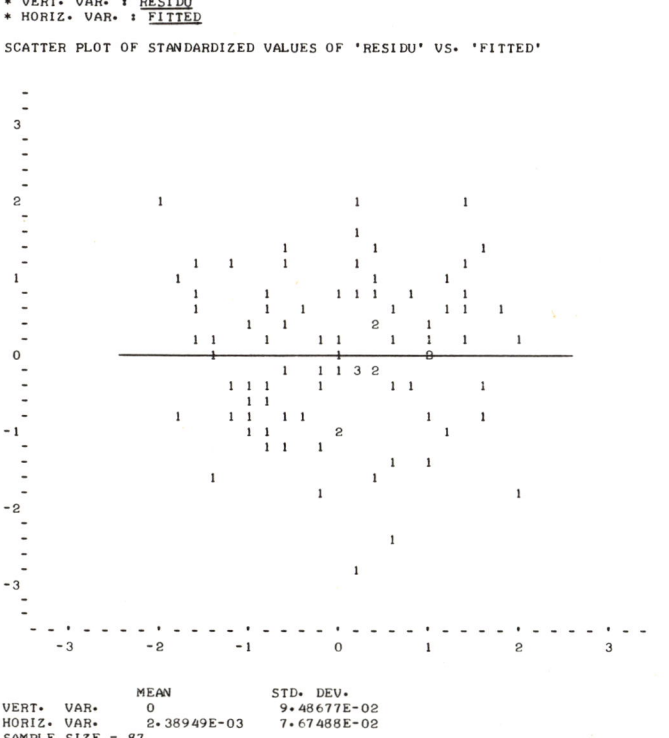

Figure 11-7. Scatter Plot of Standardized Values of 'RESIDU' vs. 'FITTED' for Autoregression of Transformed GNP Data

206 □ CONVERSATIONAL STATISTICS

Would you show me how SEPR can be used here?

First, I'll show the last few values of the dependent variable, D4DRTG, so that you can keep track; then I'll apply SEPR to the 97th observation.

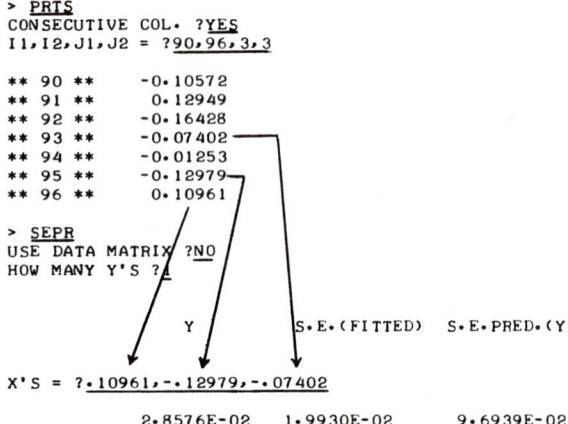

```
> PRTS
CONSECUTIVE COL. ?YES
I1,I2,J1,J2 = ?90,96,3,3

** 90 **      -0.10572
** 91 **       0.12949
** 92 **      -0.16428
** 93 **      -0.07402
** 94 **      -0.01253
** 95 **      -0.12979
** 96 **       0.10961

> SEPR
USE DATA MATRIX ?NO
HOW MANY Y'S ?1

              Y           S.E.(FITTED)    S.E.PRED.(Y)
X'S = ?.10961,-.12979,-.07402
           2.8576E-02      1.9930E-02      9.6939E-02
```

In other words, you went back one, two, and four periods in order to find the appropriate independent variables.

Yes.

The point forecast for period 97 is 2.8576E-02 = 0.028576. That's in square root units, isn't it?

Yes. Getting back to the original units requires a little care, since we have differenced twice, once consecutively and once seasonably in defining our variables. The variable from which we started differencing was RTGNP—

That's the square root of GNP—

Right.

Now a little algebra, in which y_t = RTGNP for time t and d_t = D4DRTG for time t. Note that:

$$\tilde{d}_{97} = (\tilde{y}_{97} - y_{96}) - (y_{93} - y_{92}),$$

where the tildes on $\tilde{d}_{97}$ and $\tilde{y}_{97}$ indicate that, standing at the end for t = 96, we are uncertain about these two variables.

But you don't have tildes on y_{96}, y_{93}, and y_{92} because they've already been observed?

Right. Now let's just rearrange the equation as follows:

$$\tilde{y}_{97} = \tilde{d}_{97} + y_{96} + y_{93} - y_{92}.$$

We have a point estimate of $\tilde{d}_{97}$, namely, 0.028576, from SEPR.

So now I just have to get y_{96}, y_{93}, and y_{92}.

TRANSLATING THE PREDICTION OF THE TRANSFORMED VARIABLE BACK TO THE ORIGINAL VARIABLE

Shown here is the printout of the last part of Column 1 (RTGNP) of the data matrix:

```
> PRTS
  CONSECUTIVE COL. ?YES
  I1,I2,J1,J2 = ?90,96,1,1

  ** 90 **      15.24468
  ** 91 **      15.32319
  ** 92 **      15.70033
  ** 93 **      15.14265
  ** 94 **      15.62691
  ** 95 **      15.57563
  ** 96 **      16.06237
```

So $y_{96} = 16.06237$, $y_{93} = 15.14265$, and $y_{92} = 15.70033$. Hence the point estimate for $t = 97$ is

$$\hat{y}_{97} = \hat{d}_{97} + y_{96} + y_{93} - y_{92}$$

$$= 0.028576 + 16.06237 + 15.14265 - 15.70033$$

$$= 15.533266.$$

Does S.E.PRED(Y) apply to this as well as to the forecast of the differenced variable?

Yes.

Then the standard error of 0.096939 is about 0.6% of the predicted value of 15.533266?

That's right.

We're still in square root units, aren't we?

We are, but for many purposes it will be satisfactory *for the purpose of point forecasting only* to simply square the point estimate of 15.533266; this gives a point estimate of 241.3 in billions of dollars. This is, of course, on a quarterly basis, not converted to an annual rate. Hence, the prediction for period 97 is 241.3 billions. Sometimes, however, a more elaborate procedure, not yet available in IDA, will be required.

Suppose that in a regression analysis of time series data, we want to forecast more than one period ahead. Can we do that?

POINT FORECASTS FOR MORE THAN ONE PERIOD AHEAD

In general, we have to have information about the independent variables. In this example which uses purely an autoregression model, in terms of doubly-differenced variables, we know d_{96}, d_{95}, and d_{93} as we contemplate $\tilde{d}_{97}$.

But, standing at $t = 96$, we are in a different position as we look forward to $\tilde{d}_{98}$, simply because $\tilde{d}_{97}$ is still uncertain. So, even though we know the independent variables d_{96} and d_{94}, we do not know what to enter in SEPR for $\tilde{d}_{97}$.

Why not just use the point estimate $\hat{d}_{97}$ that you've already obtained?

An excellent suggestion. We can then get our point prediction of $\tilde{d}_{98}$ from $\hat{d}_{97}$, d_{96}, and d_{94}, using SEPR. However, you must then ignore the standard errors printed out by SEPR.

Would you illustrate?

First, for easy reference, we print out the recent d's (D4DRTG):

```
> PRTS
CONSECUTIVE COL. ?Y
I1,I2,J1,J2 = ?90,96,3,3

** 90 **      -0.10572
** 91 **       0.12949
** 92 **      -0.16428
** 93 **      -0.07402
** 94 **      -0.01253
** 95 **      -0.12979
** 96 **       0.10961
```

and then recall that $\hat{d}_{97} = 0.028576$.

Then you use SEPR with $\hat{d}_{97} = 0.028576$, $d_{96} = 0.10961$, $d_{94} = -0.01253$?

As shown below:

```
> SEPR
USE DATA MATRIX ?NO
HOW MANY Y'S ?1

              Y         S.E.(FITTED)   S.E.PRED.(Y)

X'S = ?2.8576E-02,.10961,-.01253
            3.9925E-02    1.4038E-02     9.5901E-02
```

So $\hat{d}_{98} = 3.9925E-02 = 0.039925$?

Yes. And recall that the S.E.'s above should be ignored!

And for further periods into the future we can continue in the same way?

That's what's involved. For arriving at $\hat{d}_{99}$, we use $\hat{d}_{98}$ and $\hat{d}_{97}$ along with the known value d_{95}. That's the first of the two repeated applications of SEPR shown below. (I've removed the S.E.'s to remind you they're not relevant.)

```
> SEPR
USE DATA MATRIX ?NO
HOW MANY Y'S ?1

              Y

X'S = ?3.9925E-02,2.8576E-02,-.12979
            8.9167E-02

> SEPR
USE DATA MATRIX ?NO
HOW MANY Y'S ?1

              Y

X'S = ?8.9167E-02,3.9925E-02,.10961
           -.3876E-01
```

As I think on what we have done, I recall three major types of "tricks" used to reach the goal of randomly-behaving residuals: transformation, differencing, and autoregression.

TOOLS FOR TREATING TIME SERIES

That's right. In addition, there is the technique of "leading indicators" — that is, the introduction as an independent variable in regression of some variable other than the time-series variable itself that can be observed in advance.

Could the slag problem of Chapter 10 be viewed in that light?

Yes, since in that application, M IRON served as a leading indicator for C IRON.

So the approach to time series analysis fits within the regression framework that we have already studied?

Our approach has worked that way.

Can we handle all time series problems with the tools at our disposal?

We can do a reasonable job on many, but not all.

1. We have not been able to do more than hint at diagnostic techniques for interpretation of the information in the autocorrelation coefficients.

2. Although with enormous sample sizes, autoregression will always work for characterization of a stationary series, it may demand an excessive number of lagged variables to do so. It is possible to get around this difficulty by the introduction of *moving average* terms into the model. Broadly speaking, the model is still a regression model, but it can no longer be treated by ordinary regression.

3. We have left the introduction of leading indicators as an opportunistic procedure without exploring efficient approaches for doing this.

4. When sample sizes are very small, as noted earlier, there is little evidence in the data about the appropriate time series model, regardless of the sophistication of the tools of analysis.

But at least I don't have to give up in despair if the time series doesn't conform to the simple random model?

And that is no small consolation!

One final question about the example: would it be possible to improve the model further once we have reached the stage of randomly-appearing residuals?

Yes. I would look, however, in the direction of leading indicators more than in the elaboration of the model for the time series RTGNP itself.

Before I encountered this chapter, I'd read something about "time series analysis". It seems to me that there was a lot of discussion of "trend analysis", especially linear and quadratic. Are you going to tell me about this?

Now that you've asked, I will. But the main purpose is to illustrate that trend analysis is usually a catastrophic way to treat time series data arising in business, economics, and social science.

Really? How can it be all that bad?

As an example, let's take the Dow-Jones Industrial Index, August, 1968 through July, 1972. We studied this series in detail in Chapters 4-6. The main conclusion was that the first differences of monthly closing prices behaved nearly randomly with a standard deviation of 33.1968.

What's the idea of trend-fitting?

"Time" and simple transformations of "time" are used as independent variables in a regression model in which the time series of interest is the dependent variable.

What do you mean by "time"?

For example, a variable that takes value "1" for the first observation, "2" for the second, and so on.

How can you introduce such a variable in IDA?

By the command INDX, which is used in the following sequence of commands with the Dow-Jones data.

```
> ENTER
* MODE = F
NAME IS ?$DJ
* N,K GIVEN IN FILE? YES
DATA MATRIX HAS 48   OBSERVATIONS ON 1    VARIABLE(S)
COMPUTING...
            WANT TO SUPPLY NAMES ?YES
VAR.  1 = ?DJ

> CHGP
* LEVEL = 1

> INDX
COLUMN TO PLACE INDEX VARIABLE = ?2
VALUE OF 1ST INDEX = ?1
INCREMENT = ?1

GIVE NAME OF NEW VARIABLE
VAR.  2 ?TIME
UPDATING MEAN, STD, ...
```

At this point, then, a linear trend analysis would include TIME as independent variable and DJ as dependent variable?

Yes, but I'll leapfrog the linear trend analysis by setting up a variable TIMESQ, the square of TIME, which will be a second independent variable. This will give us a quadratic trend analysis, of which the linear is a special case.

TREND ANALYSIS DOES NOT USUALLY PROVIDE A GOOD MODEL FOR TIME SERIES

A QUADRATIC TREND FOR THE DOW-JONES LEVELS LEADS TO DISASTROUS DIAGNOSTIC CHECKS

```
        > CHGP
      * LEVEL = 3


        > DOTP

      * COL# TO PLACE VAR. : 3
        I,J = ?2,2
        UPDATING MEAN, STD, ..., CORR, ...

        GIVE NAME OF NEW VARIABLE
        VAR.  3 ?TIMESQ
```

I'd feel a little easier if I could see what the data matrix looks like before we move ahead.

See Table 11-2, below.

Table 11-2. Data Matrix for Dow-Jones Data with Trend Variables

```
        > FPRS
        CONSECUTIVE COL. ?YES
        I1,I2,J1,J2 = ?1,48,1,3
        FMT = ?#,6D.2D

        **   1 **    896.01    1.00      1.00
        **   2 **    935.79    2.00      4.00
        **   3 **    952.39    3.00      9.00
        **   4 **    985.08    4.00     16.00
        **   5 **    943.75    5.00     25.00
        **   6 **    946.05    6.00     36.00
        **   7 **    905.21    7.00     49.00
        **   8 **    935.48    8.00     64.00
        **   9 **    950.18    9.00     81.00
        **  10 **    937.56   10.00    100.00
        **  11 **    873.25   11.00    121.00
        **  12 **    815.47   12.00    144.00
        **  13 **    836.72   13.00    169.00
        **  14 **    813.09   14.00    196.00
        **  15 **    855.99   15.00    225.00
        **  16 **    812.30   16.00    256.00
        **  17 **    800.36   17.00    289.00
        **  18 **    744.06   18.00    324.00
        **  19 **    779.59   19.00    361.00
        **  20 **    785.57   20.00    400.00
        **  21 **    736.07   21.00    441.00
        **  22 **    700.44   22.00    484.00
        **  23 **    683.50   23.00    529.00
        **  24 **    734.12   24.00    576.00
        **  25 **    764.58   25.00    625.00
        **  26 **    760.68   26.00    676.00
        **  27 **    755.61   27.00    729.00
        **  28 **    794.09   28.00    784.00
        **  29 **    838.90   29.00    841.00
        **  30 **    868.50   30.00    900.00
        **  31 **    878.80   31.00    961.00
        **  32 **    904.37   32.00   1024.00
        **  33 **    941.75   33.00   1089.00
        **  34 **    907.80   34.00   1156.00
        **  35 **    891.14   35.00   1225.00
        **  36 **    858.40   36.00   1296.00
        **  37 **    898.07   37.00   1369.00
        **  38 **    887.19   38.00   1444.00
        **  39 **    839.00   39.00   1521.00
        **  40 **    831.75   40.00   1600.00
        **  41 **    890.20   41.00   1681.00
        **  42 **    902.17   42.00   1764.00
        **  43 **    928.13   43.00   1849.00
        **  44 **    940.70   44.00   1936.00
        **  45 **    954.17   45.00   2025.00
        **  46 **    960.72   46.00   2116.00
        **  47 **    929.03   47.00   2209.00
        **  48 **    924.74   48.00   2304.00
```

Now we look at REGR:

```
        > REGR
        UPDATING CORR. MATRIX...
      * DEP. VAR. = DJ
        HOW MANY INDEP. VAR. ?2
        INDEP. VAR.  1 = ?TIME
        INDEP. VAR.  2 = ?TIMESQ
        ANALYZING RESIDUALS ...
        WARNING: RESIDUALS MAY BE EXCESSIVELY CORRELATED :
```

```
         OBSERVED NUMBER OF RUNS= 9
         EXPECTED NUMBER OF RUNS= 25
       STANDARD DEVIATION OF RUNS= 3.42705
           (OBS.-EXP.)/(STD.DEV.)=-4.66874

    WARNING:  AUTO( 1) = +0.74
    WARNING:  AUTO( 2) = +0.49
    WARNING:  AUTO( 3) = +0.30
    WARNING:  AUTO( 8) = -0.34
    WARNING:  AUTO( 9) = -0.36
    WARNING:  AUTO(10) = -0.39
    WARNING:  AUTO(11) = -0.35
```

Wow! IDA really blew the whistle on that! How does the sequence plot look?

Traumatic. The worst we've seen yet for residuals from a regression model. Note its resemblance to the original sequence plot in Figure 4-3.

How's the full AUTO?

Pretty horrendous:

```
    > AUTO
    * VARIABLE : RESIDU

    * MAX ORDER : 12

      ORDER   AUTOCORR.
        1      +0.742
        2      +0.491
        3      +0.304
        4      +0.282
        5      +0.154
        6      -0.039
        7      -0.235
        8      -0.342
        9      -0.361
       10      -0.388
       11      -0.351
       12      -0.389

      S.E. OF EACH COEF. GIVEN RANDOM MODEL=0.144

      BOX-PIERCE STATISTIC= 82.3988
      EXPECTATION GIVEN RANDOM MODEL= 12
            (OBS.-EXP.)/(STD.DEV.)= 6.75483
```

But at least the histogram (Figure 11-8) doesn't look too bad. How is NORM?

In itself, not bad, as you can see in Figure 11-9.

I take it, however, that that's no real consolation.

You're right. Normality is no assurance of randomness!

How about RESIDU versus FITTED?

Look at Figure 11-10.

I get the impression of strong nonlinearity from Figure 11-10.

So do I. If, in earlier chapters, you had to strain your eyes on the interocular tests, your confidence in your vision should be restored by these plots.

```
>  PLTS
*  VARIABLE : RESIDU

ALL ROWS ?YES

SEQUENCE PLOT OF STANDARDIZED VALUES OF 'RESIDU'
```

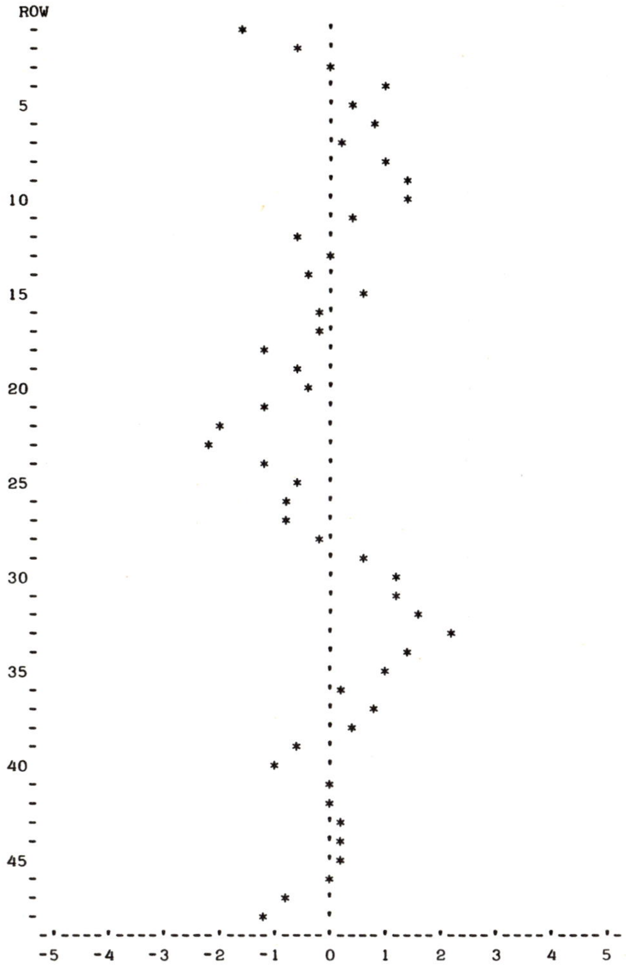

<pre>
    ROW
      -                       *   '
      -                            * '
      -                              *'
      -                               '   *
     5   ' *
      -                               ' *
      -                               '*
      -                               ' *
      -                               '   *
    10   '   *
      -                              *'
      -                           *   '
      -                              *'
     15   ' *
      -                             *'
      -                             *'
      -                    *   '
      -                          *   '
    20   *   '
      -                       *   '
      -           *   '
      -                           * '
    25   *   '
      -                        *   '
      -                       *   '
      -                         *   '
      -                               ' *
    30   '   *
      -                               ' *
      -                               '   *
      -                               '   *
    35   '   *
      -                              *'
      -                              '*
      -                               ' *
      -                          *   '
    40   *   '
      -                               '*
      -                               '*
      -                              '*
    45   '*
      -                               '*
      -                     *   '
      - *   '
   -'----'----'----'----'----'----'----'----'----'----'-
   -5   -4   -3   -2   -1   0   1   2   3   4   5
</pre>

HISTOGRAM

ABS. FREQ.

<pre>
    5   * **
    -                         * * ' **
    -                   * ** **** '   * *
    -                      * ********'********
    -              ** * ***********'* *
   -'----'----'----'----'----'----'----'----'----'----'-
   -5   -4   -3   -2   -1   0   1   2   3   4   5
</pre>

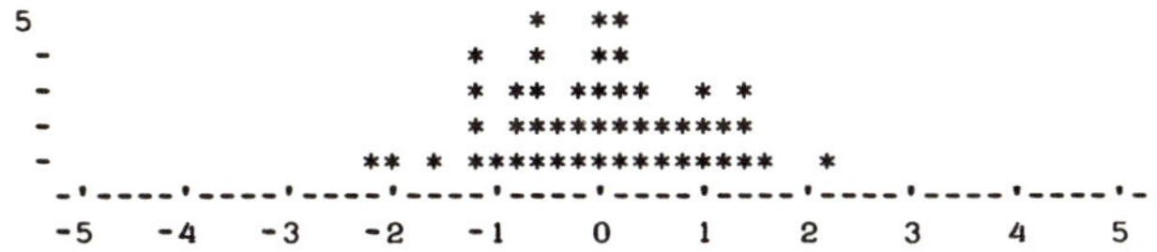

```
MEAN      =   0
STD. DEV. = 51.969
SAMPLE SIZE =  48
```

Figure 11-8. Sequence Plot and Histogram for Residuals from "Trend Analysis" of Dow-Jones Levels

214 ☐ **CONVERSATIONAL STATISTICS**

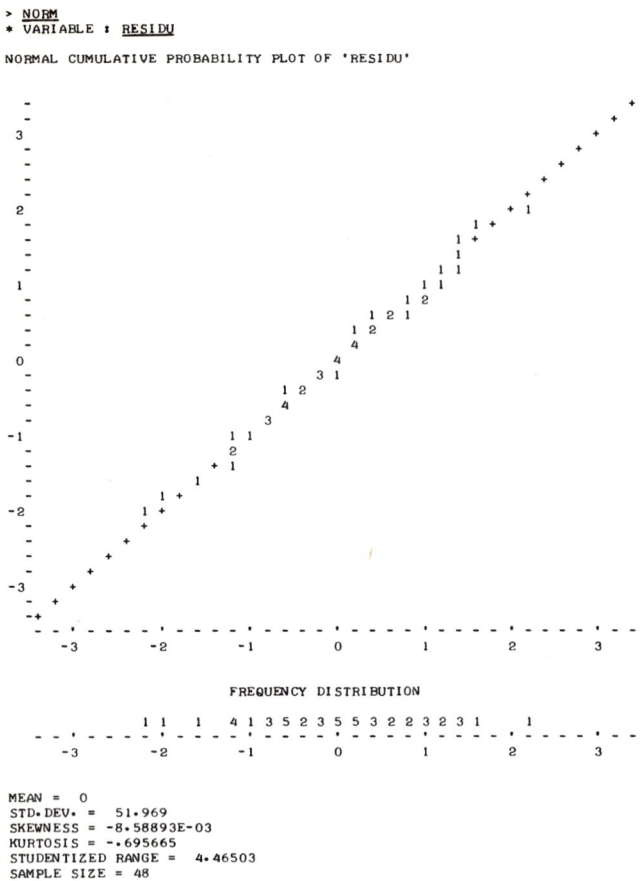

Figure 11-9. Normal Cumulative Probability Plot of 'RESIDU' for "Trend Analysis" of Dow-Jones Levels

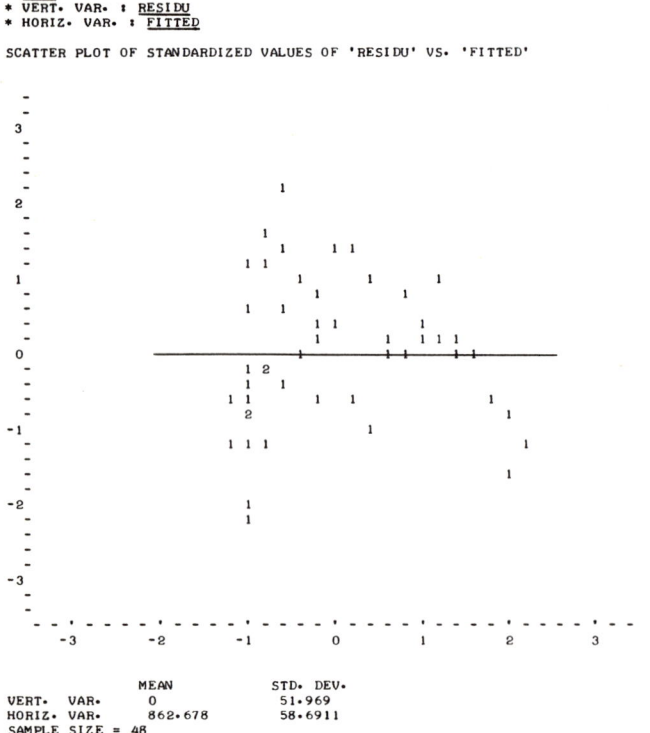

Figure 11-10. Scatter Plot of Standardized Values of 'RESIDU' vs. 'FITTED' for "Trend Analysis" of Dow-Jones Levels

Suppose we'd done none of the checks for model adequacy. What would we have obtained?

Here are COEF and SUMM:

```
> COEF

   VARIABLE   B(STD.V)      B         STD.ERROR(B)      T

   TIME       -2.9392    -.1630E+02    2.2100E+00    -7.377
   TIMESQ      3.0790     3.3792E-01   4.3727E-02     7.728
   CONSTANT              9.9441E+02    2.3474E+01    42.361

> SUMM

              MULTIPLE R   R-SQUARE
   UNADJUSTED   0.7558      0.5712
   ADJUSTED     0.7431      0.5521
   STD. DEV. OF RESIDUALS =       51.9686
```

I guess that if I'd looked at these first, I would have been impressed. After all, R = 0.7558, and look at the T-values for TIME and TIMESQ.

But look at the standard deviation of residuals: it's 51.9686!

Is it appropriate to compare that with the standard deviation of DIFDJ from Chapter 4?

Yes, and that was only 33.1968, as I reminded you a while ago. Further, the SEPR command below shows that for a prediction of month 49, the S.E.PRED.(Y) is substantially inflated over the STD.DEV. of RESIDUALS of 51.9686. It's up to 57.024:

```
> SEPR
USE DATA MATRIX ?NO
HOW MANY Y'S ?1

                    Y         S.E.(FITTED)   S.E.PRED.(Y)

X'S = ?49,2401
                1.0069E+03    2.3474E+01     5.7024E+01
```

Do you ever advocate this kind of trend-fitting for data arising in management problems?

There might be instances in research and development in which polynomial trends are a sensible statistical model, but ordinarily I think it is unwise to hope that TIME variables used in this way will be useful as a part of the final model, whether or not other independent variables are used.

The main use of TIME in regression is as a diagnostic check: if TIME *appears* to improve a model, look for the catch!

That remark sounds a little enigmatic, if not dogmatic. Would you amplify?

In the example just discussed, suppose that we had started out with the trend analysis, rather than as we did in Chapter 4. Then we are hit with the terrible diagnostics. This should then lead us to explore ways to improve the model. If we're alert, we'd then hit on the simple differencing of Chapter 4.

All right. Suppose, then, for the sake of argument, that the trend analysis were applied to the differences, and that the standard deviation of residuals was reduced and the diagnostics were good.

Then I'd say that the trend analysis had pointed to a flaw in the model of the random walk, though not necessarily a major one. The challenge would be to see whether the variable that is "really" causing a slow change — evidenced in the successful time trend — could be isolated.

Is it possible that my supposition would often be realized in practice — that is, that addition of TIME and TIMESQ as independent variables to an apparently successful regression model would reduce the standard deviation of residuals?

Yes, it's possible, and I've seen it happen very frequently, in a wide range of applications, even in engineering and natural science. Also with stock market data. You might try out the model of Chapter 10 from that point-of-view.

Why not use it for prediction? Is there a catch?

If you try out SEPR in these examples, you'll find that the STD. ERROR PREDICTED Y is surprisingly inflated by the fact that TIME is far removed from its mean when it is used as an independent variable for predicting to the future. Further, if you think that there is an omitted variable causing the TIME effect, you will think it unlikely that TIME represents this variable well enough to serve as a good proxy in prediction.

Why not?

Because then the omitted variable would itself be well fitted by a time trend, with good diagnostics. I know of few if any examples in which this is true. Look at the Dow-Jones example above for an illustration that is far from atypical, or try out a time trend on GNP.

PREDICTIONS BASED ON FITTING OF TIME TRENDS ARE LIKELY TO BE DISAPPOINTING.

APPENDIX A: APPLICATIONS

Instead of providing small problems or exercises with the individual chapters, I have collected a set of applications that invite various approaches to analysis based on the statistical ideas developed in the book, especially Chapters 4-11, but not always tied in an obvious way to a particular chapter. These applications can be developed by a series of "assignments" that direct the reader's attention to interesting aspects of the data, and that require the use of various statistical tools that become available as the study of Chapters 4-11 progresses. Suggested assignments are provided for the first few applications.

(1) MARKET MODEL

The following assignments are based on stock return data contained in a file called $CRSPR in the University of Chicago HP 2000F system library. The CRSP files were made by the Center For Research In Security Prices (sponsored by Merrill, Lynch, Pierce, Fenner and Smith of New York) of the Graduate School of Business of the University of Chicago. If you are interested in doing further work with other stocks in this file, you can access the $CRSPR file by the IDA command CRSP.

Using CRSP I set up a file $AT&T, in IDA format, that will serve as a basis for this series of assignments. Column 1 of this file contains returns by month for AT&T common stock for the 84 months from January, 1961, through December 1967. If the return in month t is denoted r_t, then

$$r_t = \frac{p_t + d_t - p'_{t-1}}{p'_{t-1}}$$

where
p_t = monthly closing price per share for month t,

d_t = dividend per share paid in month t,

p'_{t-1} = closing price per share of the previous month adjusted for splits, stock dividends or other capital changes.

Column 2 of $AT&T gives the corresponding return data for the same months for the New York Stock Exchange Arithmetic Index (NYSE).

Assignment 1

On the basis of the technical ideas of Chapter 4 and your common sense, discuss what contribution you can make to the question of predictability of AT&T returns, given only data on AT&T past returns.

Do the same for NYSE returns.

Assignment 2

To what extent would you be interested in a histogram of **AT&T** or some transformation thereof? What can you say about the model of normality? Express your inferences about the return for January, 1968, based on the past **AT&T** returns. Do the same for the long run mean return of **AT&T** assuming a continuation of the basic underlying conditions of 1961-1968.

Assignment 3

A simple form of the so-called market model envisages that the **AT&T** returns (or some transformation) can be expressed as a dependent variable in a simple linear regression in which NYSE return (or some transformation) is the independent variable.

Explore the adequacy of this statistical model for the data in the $AT&T file. Using the results of your exploration, carry out an analysis of the statistical relationship between **AT&T** and NYSE. Discuss concisely the interpretation of your key results, including the economic meaning of the various parameters that you estimate.

Assignment 4

Apply the simple autoregression model to **AT&T** returns and discuss your results, including a consideration of model adequacy.

Do the same for NYSE returns.

Apply the AUTO command to the residuals from the market model regression of Assignment 3. What, if anything, does this add to the conclusions of Assignment 3?

Assignment 5

One common technique for the analysis of time series data is to create a variable called TIME, which takes values 1, 2, . . . , n for the n observations of the time series, and use it as an independent variable in regression analysis, possibly in combination with one or more other independent variables. Sometimes a second variable TIMESQ, the square of TIME, is also used.

Does this technique add anything to the analyses that you have already done in the earlier assignments? Discuss carefully, and show relevant evidence. NOTE: The TIME variable can be created in IDA by the command INDX.

Table A-1. Data File: $AT&T

	AT&T	NYSE		AT&T	NYSE
** 1 **	0.07351	0.08232	** 43 **	-0.02381	0.02767
** 2 **	-0.00217	0.05936	** 44 **	-0.01742	-0.00896
** 3 **	0.07430	0.05037	** 45 **	-0.01607	0.03699
** 4 **	0.01655	0.00841	** 46 **	-0.01089	0.01700
** 5 **	-0.00712	0.04179	** 47 **	-0.03119	0.00072
** 6 **	-0.03463	-0.04242	** 48 **	0.04198	-0.00691
** 7 **	0.06417	0.01095	** 49 **	-0.00733	0.05871
** 8 **	-0.01588	0.02104	** 50 **	0.00554	0.02784
** 9 **	-0.01029	-0.02971	** 51 **	-0.01848	0.00528
** 10 **	0.00728	0.02133	** 52 **	0.02637	0.03592
** 11 **	0.09618	0.04641	** 53 **	0.03303	-0.00790
** 12 **	0.03223	-0.00351	** 54 **	-0.03578	-0.07433
** 13 **	-0.05969	-0.00793	** 55 **	-0.00928	0.02910
** 14 **	0.04512	0.01554	** 56 **	0.00936	0.04509
** 15 **	-0.02916	-0.00594	** 57 **	0.00935	0.03083
** 16 **	-0.03876	-0.06830	** 58 **	-0.02407	0.04739
** 17 **	-0.07944	-0.09800	** 59 **	-0.05806	0.03005
** 18 **	-0.08389	-0.08475	** 60 **	-0.01220	0.03267
** 19 **	0.07952	0.06369	** 61 **	-0.01852	0.04353
** 20 **	0.01473	0.02796	** 62 **	0.01132	0.01094
** 21 **	-0.05876	-0.06050	** 63 **	-0.05858	-0.02187
** 22 **	0.02827	-0.02192	** 64 **	0.00667	0.03374
** 23 **	0.05865	0.13730	** 65 **	-0.02781	-0.07238
** 24 **	0.01854	-0.00906	** 66 **	0.00917	-0.00483
** 25 **	0.03640	0.07816	** 67 **	-0.01364	-0.01266
** 26 **	-0.01116	-0.01508	** 68 **	-0.04055	-0.09305
** 27 **	0.02000	0.02095	** 69 **	-0.00485	-0.01432
** 28 **	0.03096	0.03902	** 70 **	0.10000	0.01269
** 29 **	-0.00781	0.03242	** 71 **	-0.03902	0.03825
** 30 **	-0.01931	-0.01583	** 72 **	0.02564	0.01623
** 31 **	-0.00622	-0.00952	** 73 **	0.05227	0.14285
** 32 **	0.04400	0.05056	** 74 **	0.01166	0.02094
** 33 **	0.02917	-0.01841	** 75 **	0.03448	0.05201
** 34 **	0.04399	0.01633	** 76 **	-0.01250	0.03648
** 35 **	0.04981	-0.00682	** 77 **	-0.06245	-0.01788
** 36 **	0.00000	0.00752	** 78 **	0.03636	0.05164
** 37 **	0.03950	0.02008	** 79 **	-0.08553	0.07092
** 38 **	-0.00726	0.02699	** 80 **	-0.01583	0.00280
** 39 **	-0.01244	0.03140	** 81 **	0.02217	0.03777
** 40 **	0.01170	-0.00305	** 82 **	-0.02410	-0.03594
** 41 **	-0.00890	0.01164	** 83 **	0.00444	0.00669
** 42 **	0.06329	0.01540	** 84 **	0.00249	0.05544

(2) FORECASTING DEGREE DAYS

A degree day is defined as the difference between 65 degrees F. and the actual daily mean temperature, if the latter is 65 degrees or less; otherwise, the value is 0. The number of degree days (actually, the sum of the degree days) per heating season is a basic measure of the severity of the season and its demands on fuel consumption, such as natural gas, oil, electricity, or coal. A heating season is defined as the 12-month period beginning on July 1 of one year and ending on June 30 of the following year; the year ending is its calendar designation. Thus the 1972 heating season ended on June 30, 1972.

Forecasting the number of degree days in the next heating season is important for planning purposes in a public utility serving a given area. The longer term outlook is also of interest: for example, is there a warming or cooling trend? One possible basis for a forecast is the past historical record itself. For Chicago, data collection has been standardized since the beginning of the 1932 heating season. The actual temperature readings are made by the National Weather Service at Midway Airport.

Assignment 1

On the basis both of the technical ideas of Chapter 4 and any other information you may have, discuss what contribution you can make to the forecasting problem.

Assignment 2

Assume for this assignment that the specification of randomness is tenable. (1) What can you say about the specification of normality? (2) Express your inference about a single future observation in the form of a probability distribution. (3) Same as (2), but for the long run mean $\tilde{\mu}$, given current underlying conditions.

Assignment 3

One common technique for the analysis of time series data is to fit a linear or quadratic "trend" to the series. This is done in IDA by using the command INDX to create a variable TIME that takes values 1, 2, ... , n for the n observations of the time series, and then creating a variable TIMESQ by DOTP. TIME and TIMESQ are then available as independent variables in regression. Fit a linear trend to the data of this application, and comment on the success of this model.

Assignment 4

Apply the simple autoregression model to the data, and comment critically on the adequacy of the statistical model, making any appropriate comparisons with the results of earlier assignments.

Assignment 5

Reconsider the data in the light of Chapter 10. Show your final preferred model and explain concisely why you prefer it.

Table A-2. Data File: $HALSEY

		DDAYS	YEAR			DDAYS	YEAR
**	1 **	5526	1932	**	21 **	6287	1952
**	2 **	6364	1933	**	22 **	5815	1953
**	3 **	6366	1934	**	23 **	5399	1954
**	4 **	6363	1935	**	24 **	5543	1955
**	5 **	7154	1936	**	25 **	6217	1956
**	6 **	6457	1937	**	26 **	5830	1957
**	7 **	6202	1938	**	27 **	6114	1958
**	8 **	5960	1939	**	28 **	6265	1959
**	9 **	6624	1940	**	29 **	6400	1960
**	10 **	6081	1941	**	30 **	5995	1961
**	11 **	5795	1942	**	31 **	6398	1962
**	12 **	6908	1943	**	32 **	6529	1963
**	13 **	6426	1944	**	33 **	5741	1964
**	14 **	6327	1945	**	34 **	6354	1965
**	15 **	6196	1946	**	35 **	6124	1966
**	16 **	6430	1947	**	36 **	6121	1967
**	17 **	6518	1948	**	37 **	6010	1968
**	18 **	5943	1949	**	38 **	6139	1969
**	19 **	6289	1950	**	39 **	6390	1970
**	20 **	6630	1951	**	40 **	6219	1971

(3) VELOCITY OF MONEY

The following application is based on a working paper by John P. Gould and Charles R. Nelson of the Graduate School of Business, University of Chicago. The following background is taken from their paper.

> Students of money and monetary policy have focused a substantial amount of attention on the behavior of the ratio of national income to the stock of money. In the most naive form of the quantity theory of money, this ratio--the velocity of money--is assumed to be a constant. More modern and sophisticated versions of the quantity theory treat velocity as a stable function of variables such as interest rates, changes in the price level and the ratio of current to permanent income. Indeed, much of the empirical money demand literature deals with estimation of money demand functions, which incorporate at least a subset of these variables. To the extent that these efforts have been successful, velocity may be said to be predictable, at least in the sense of *ex post* predictability given the determinants of velocity. It has also been maintained that velocity is predictable in an extrapolation sense; that is, movements in velocity may to a degree be related to preceding history, allowing not only *ex post* explanation of a sort, but also forecasting. According to that view, "trends" and episodic deviations from trend may be discerned in the time series of velocity. This paper studies the stochastic structure of velocity in order to determine whether there is a statistical basis for extrapolative prediction. It is important to emphasize that our statistical analysis focuses exclusively on the velocity series, since we are interested in knowing what a time series analyst could conclude about velocity without reference to other economic variables.

One of the most important velocity series is that of Friedman and Schwartz for the American economy from 1869 to 1960. The accuracy of the data is best from 1891 on, so for this assignment you will be given the velocity series from 1891 to 1960.

Assignments parallel those of FORECASTING DEGREE DAYS.

	VELMON	YEAR		VELMON	YEAR
** 1 **	2.94	1891	** 36 **	1.95	1926
** 2 **	2.81	1892	** 37 **	1.87	1927
** 3 **	2.87	1893	** 38 **	1.84	1928
** 4 **	2.55	1894	** 39 **	1.95	1929
** 5 **	2.71	1895	** 40 **	1.70	1930
** 6 **	2.67	1896	** 41 **	1.47	1931
** 7 **	2.81	1897	** 42 **	1.28	1932
** 8 **	2.55	1898	** 43 **	1.38	1933
** 9 **	2.48	1899	** 44 **	1.52	1934
** 10 **	2.53	1900	** 45 **	1.52	1935
** 11 **	2.47	1901	** 46 **	1.60	1936
** 12 **	2.35	1902	** 47 **	1.67	1937
** 13 **	2.34	1903	** 48 **	1.53	1938
** 14 **	2.21	1904	** 49 **	1.52	1939
** 15 **	2.18	1905	** 50 **	1.51	1940
** 16 **	2.32	1906	** 51 **	1.61	1941
** 17 **	2.30	1907	** 52 **	1.84	1942
** 18 **	2.08	1908	** 53 **	1.77	1943
** 19 **	2.23	1909	** 54 **	1.61	1944
** 20 **	2.20	1910	** 55 **	1.37	1945
** 21 **	2.09	1911	** 56 **	1.16	1946
** 22 **	2.15	1912	** 57 **	1.23	1947
** 23 **	2.17	1913	** 58 **	1.31	1948
** 24 **	1.91	1914	** 59 **	1.27	1949
** 25 **	1.90	1915	** 60 **	1.43	1950
** 26 **	2.12	1916	** 61 **	1.53	1951
** 27 **	2.18	1917	** 62 **	1.50	1952
** 28 **	2.51	1918	** 63 **	1.51	1953
** 29 **	2.28	1919	** 64 **	1.49	1954
** 30 **	2.20	1920	** 65 **	1.58	1955
** 31 **	1.90	1921	** 66 **	1.61	1956
** 32 **	1.88	1922	** 67 **	1.63	1957
** 33 **	2.04	1923	** 68 **	1.56	1958
** 34 **	1.97	1924	** 69 **	1.63	1959
** 35 **	1.88	1925	** 70 **	1.69	1960

Table A-3. Data File: $VELMON

(4) HARDNESS OF STEEL COILS

These data come from the same steel mill that provided the Rockwell Hardness Data introduced in Chapter 4. The same background is appropriate, except that these data refer to a later period of time. Assignments parallel those of FORECASTING DEGREE DAYS.

Table A-4. Data File: $ROCKY2

	HARDNS		HARDNS		HARDNS
** 1 **	64	** 35 **	63	** 68 **	60
** 2 **	63	** 36 **	53	** 69 **	63
** 3 **	57	** 37 **	60	** 70 **	64
** 4 **	63	** 38 **	61	** 71 **	60
** 5 **	57	** 39 **	58	** 72 **	62
** 6 **	58	** 40 **	57	** 73 **	58
** 7 **	65	** 41 **	55	** 74 **	60
** 8 **	74	** 42 **	54	** 75 **	60
** 9 **	61	** 43 **	58	** 76 **	64
** 10 **	50	** 44 **	58	** 77 **	58
** 11 **	60	** 45 **	58	** 78 **	59
** 12 **	58	** 46 **	59	** 79 **	58
** 13 **	64	** 47 **	55	** 80 **	60
** 14 **	59	** 48 **	55	** 81 **	62
** 15 **	55	** 49 **	62	** 82 **	56
** 16 **	64	** 50 **	54	** 83 **	52
** 17 **	54	** 51 **	55	** 84 **	63
** 18 **	53	** 52 **	65	** 85 **	57
** 19 **	60	** 53 **	67	** 86 **	59
** 20 **	57	** 54 **	66	** 87 **	57
** 21 **	63	** 55 **	59	** 88 **	63
** 22 **	63	** 56 **	55	** 89 **	59
** 23 **	60	** 57 **	50	** 90 **	63
** 24 **	62	** 58 **	54	** 91 **	60
** 25 **	61	** 59 **	57	** 92 **	54
** 26 **	56	** 60 **	58	** 93 **	55
** 27 **	63	** 61 **	61	** 94 **	55
** 28 **	65	** 62 **	62	** 95 **	53
** 29 **	60	** 63 **	66	** 96 **	62
** 30 **	62	** 64 **	63	** 97 **	62
** 31 **	64	** 65 **	63	** 98 **	62
** 32 **	58	** 66 **	58	** 99 **	50
** 33 **	63	** 67 **	59	**100 **	58
** 34 **	53				

(5) FORECASTING LAKE LEVELS

This application is based on a time series of the mean monthly levels for July of each year from 1875 through 1972 of Lake Huron, which has a common level with Lake Michigan. The levels are measured at Harbor Beach, Michigan, and are expressed as elevations in feet above mean water level at Father Point, Quebec, International Great Lakes Datum (1955). The following quotation from Louis D. Kirshner, "Forecasting Great Lakes Levels Aids Power and Navigation," *Civil Engineering*, 24, 1954, 98-99, gives some of the practical reasons for interest in these data:

> Levels of the water surfaces of the Great Lakes have varying effects on three major economic interests--shore property, lake shipping, and hydroelectric power. In general, high levels benefit shipping and power. Increased depths in harbors and channels, which permit vessels to load only an inch or two deeper, permit sizable increases in cargoes, particularly in the huge modern lake freighters. Production of hydroelectric power is obviously facilitated by an abundance of water. But high lake levels are extremely injurious to shore properties, particularly during storms....Periods of low lake levels likewise present problems. For example, maintenance of high flows for power would further decrease the drafts to which vessels on the Great Lakes could be loaded.

There is no way to solve these problems with the lakes in their present unregulated state, since the recurring highs and lows are natural and not man-made....

It has long been recognized that accurate forecasting of lake levels would enable each interest to gain some measure of protection against oncoming highs and lows which might be damaging. A number of studies to this end have been made in the past by leading hydraulic engineers, but until recently it was believed inadvisable to forecast lake levels more than one month in advance.

At the present time, as you may know, high water levels are of concern. Even some lakeshore apartments on the north side of Chicago have flooded basements during storms. Kirshner comments as follows about the forces determining lake levels:

Mean water-surface evaluations of the lakes are the result of all the factors which either add or subtract water. Water is added by precipitation on the lake surface, tributary stream runoff, diversions into the lakes, condensation on the surface, inseepage, and inflow from the lakes above. Water is subtracted by outflow to the lakes below, evaporation, diversions from the lakes, and outseepage.

Assignments parallel those of FORECASTING DEGREE DAYS. In Table A-5, I subtracted 570 from each original observation before entry into the file $LMICH. This was done to achieve tolerable numerical accuracy with the HP 2000F; all computations should be made on the data thus coded. (This shift of units changes computational results in no essential way. The original means can be recovered by adding 570 to each mean computed by IDA; standard deviations, regression coefficients (except the intercept), and correlation coefficients are unaffected.)

Table A-5. Data File: $LMICH

#	LEVEL	YEAR	#	LEVEL	YEAR
1	10.38	1875	50	7.79	1924
2	11.86	1876	51	6.75	1925
3	10.97	1877	52	6.75	1926
4	10.80	1878	53	7.82	1927
5	9.79	1879	54	8.64	1928
6	10.39	1880	55	10.58	1929
7	10.42	1881	56	9.48	1930
8	10.82	1882	57	7.38	1931
9	11.40	1883	58	6.90	1932
10	11.32	1884	59	6.94	1933
11	11.44	1885	60	6.24	1934
12	11.68	1886	61	6.84	1935
13	11.17	1887	62	6.85	1936
14	10.53	1888	63	6.90	1937
15	10.01	1889	64	7.79	1938
16	9.91	1890	65	8.18	1939
17	9.14	1891	66	7.51	1940
18	9.16	1892	67	7.23	1941
19	9.55	1893	68	8.42	1942
20	9.67	1894	69	9.61	1943
21	8.44	1895	70	9.05	1944
22	8.24	1896	71	9.26	1945
23	9.10	1897	72	9.22	1946
24	9.09	1898	73	9.38	1947
25	9.35	1899	74	9.10	1948
26	8.82	1900	75	7.95	1949
27	9.32	1901	76	8.12	1950
28	9.01	1902	77	9.75	1951
29	9.00	1903	78	10.85	1952
30	9.80	1904	79	10.41	1953
31	9.83	1905	80	9.96	1954
32	9.72	1906	81	9.61	1955
33	9.89	1907	82	8.76	1956
34	10.01	1908	83	8.18	1957
35	9.37	1909	84	7.21	1958
36	8.69	1910	85	7.13	1959
37	8.19	1911	86	9.10	1960
38	8.67	1912	87	8.25	1961
39	9.55	1913	88	7.91	1962
40	8.92	1914	89	6.89	1963
41	8.09	1915	90	5.96	1964
42	9.37	1916	91	6.80	1965
43	10.13	1917	92	7.68	1966
44	10.14	1918	93	8.38	1967
45	9.51	1919	94	8.52	1968
46	9.24	1920	95	9.74	1969
47	8.66	1921	96	9.31	1970
48	8.86	1922	97	9.89	1971
49	8.05	1923	98	9.96	1972

(6) SALES AND FORECASTS

In this application, monthly sales (in $1000) are available for a division of a corporation for a 24-month period. Also available are judgmental forecasts of each month's sales made by the sales manager for the division. The question arose as to whether the forecasts by the sales manager were of any value. The sample size is so small that many of the approximations underlying the approach to regression outlined in the main text become pretty crude, and it is necessary to keep this fact in mind in interpreting the results of an analysis.

This set of data does not lend itself well to a series of sequential assignments. It was used originally as a part of an in-class final examination. Students were given the data and the background of the problem in advance of the examination. The examination itself consisted of printout from one possible analysis based on IDA (a somewhat earlier version), with queries about the analysis typed on the printout itself, with space for answers thereon. The approach was based on the use of the judgmental forecast (F) as one independent variable, first and second lagged sales (S-1 and S-2) as two other independent variables, and two cross-product variables (F x S-1 and F x S-2) built up from these. An alternative and simpler model could be formulated based on judicious use of differencing. An assignment could be based on examination of these alternative approaches.

In Table A-6 below, note that entry 999 for the first forecast is a dummy; only the last 23 months were forecast:

Table A-6. Data File: $SALES1

	S	F
** 1 **	437	999
** 2 **	505	498
** 3 **	517	525
** 4 **	583	525
** 5 **	532	510
** 6 **	441	500
** 7 **	445	463
** 8 **	450	440
** 9 **	350	420
** 10 **	409	400
** 11 **	504	450
** 12 **	456	465
** 13 **	420	470
** 14 **	396	450
** 15 **	409	400
** 16 **	414	400
** 17 **	343	370
** 18 **	339	350
** 19 **	344	350
** 20 **	290	325
** 21 **	330	300
** 22 **	275	290
** 23 **	325	300
** 24 **	370	340

(7) MONEY SUPPLY

An interesting application is afforded by the treatment of the time series on the money supply (M1) from 1947-1970 along the broad lines sketched in Chapter 11. The data are means of daily data in billions of dollars, and they are not seasonally adjusted. Transformation and differencing appear to be indicated.

Table A-7. Data File: $ECDAT1, Column 2

	M1			M1			M1
** 1 **	110.4	** 33 **	134.7	** 65 **	150.0
** 2 **	109.9	** 34 **	133.0	** 66 **	149.3
** 3 **	112.2	** 35 **	133.5	** 67 **	150.4
** 4 **	114.7	** 36 **	136.5	** 68 **	155.5
** 5 **	113.7	** 37 **	136.8	** 69 **	155.4
** 6 **	110.5	** 38 **	134.7	** 70 **	154.3
** 7 **	111.8	** 39 **	134.5	** 71 **	156.7
** 8 **	113.3	** 40 **	138.0	** 72 **	162.4
** 9 **	111.9	** 41 **	137.9	** 73 **	162.3
** 10 **	109.8	** 42 **	135.7	** 74 **	160.9
** 11 **	110.5	** 43 **	135.8	** 75 **	162.6
** 12 **	112.5	** 44 **	137.6	** 76 **	169.5
** 13 **	112.7	** 45 **	136.9	** 77 **	170.6
** 14 **	112.1	** 46 **	136.5	** 78 **	170.2
** 15 **	114.3	** 47 **	137.9	** 79 **	169.5
** 16 **	117.5	** 48 **	142.2	** 80 **	173.8
** 17 **	118.1	** 49 **	143.0	** 81 **	173.8
** 18 **	116.7	** 50 **	142.2	** 82 **	174.3
** 19 **	118.6	** 51 **	143.0	** 83 **	178.1
** 20 **	123.5	** 52 **	144.8	** 84 **	184.8
** 21 **	124.7	** 53 **	142.7	** 85 **	185.6
** 22 **	123.1	** 54 **	139.9	** 86 **	187.0
** 23 **	124.5	** 55 **	140.4	** 87 **	190.7
** 24 **	128.7	** 56 **	143.5	** 88 **	198.5
** 25 **	128.8	** 57 **	143.1	** 89 **	200.1
** 26 **	127.0	** 58 **	142.3	** 90 **	200.1
** 27 **	127.3	** 59 **	142.9	** 91 **	200.8
** 28 **	130.3	** 60 **	147.5	** 92 **	206.1
** 29 **	130.3	** 61 **	147.2	** 93 **	206.3
** 30 **	128.1	** 62 **	145.9	** 94 **	207.5
** 31 **	129.3	** 63 **	145.4	** 95 **	209.7
** 32 **	133.4	** 64 **	149.5	** 96 **	216.5

(8) STRIKES AND LOCKOUTS

This application can serve as the basis for an assignment along the lines of MONEY SUPPLY. Unlike other time series considered, this series does not appear to invite differencing. The data give man-days idle due to strikes and lockouts in millions (means of monthly data without seasonal adjustment).

Table A-8. Data File: $ECDAT1, Column 5

	STRIKES			STRIKES			STRIKES
** 1 **	1.223	** 33 **	0.892	** 65 **	1.408
** 2 **	6.410	** 34 **	2.977	** 66 **	1.306
** 3 **	2.820	** 35 **	3.050	** 67 **	1.382
** 4 **	1.066	** 36 **	2.480	** 68 **	1.269
** 5 **	2.801	** 37 **	2.147	** 69 **	0.918
** 6 **	4.570	** 38 **	2.153	** 70 **	1.823
** 7 **	2.437	** 39 **	5.697	** 71 **	1.777
** 8 **	1.561	** 40 **	1.037	** 72 **	3.127
** 9 **	1.620	** 41 **	0.782	** 73 **	1.650
** 10 **	3.260	** 42 **	1.883	** 74 **	2.093
** 11 **	3.587	** 43 **	1.967	** 75 **	2.670
** 12 **	8.373	** 44 **	0.860	** 76 **	1.352
** 13 **	5.063	** 45 **	0.746	** 77 **	1.143
** 14 **	3.060	** 46 **	1.630	** 78 **	2.563
** 15 **	2.973	** 47 **	2.240	** 79 **	2.750
** 16 **	1.851	** 48 **	3.353	** 80 **	2.003
** 17 **	1.640	** 49 **	1.477	** 81 **	1.344
** 18 **	1.837	** 50 **	2.760	** 82 **	3.960
** 19 **	2.353	** 51 **	12.143	** 83 **	4.449
** 20 **	1.807	** 52 **	6.610	** 84 **	4.288
** 21 **	1.440	** 53 **	1.313	** 85 **	3.485
** 22 **	9.463	** 54 **	2.330	** 86 **	6.235
** 23 **	6.300	** 55 **	1.830	** 87 **	3.914
** 24 **	2.471	** 56 **	0.897	** 88 **	2.705
** 25 **	1.240	** 57 **	0.612	** 89 **	2.717
** 26 **	3.863	** 58 **	1.418	** 90 **	4.407
** 27 **	2.820	** 59 **	1.787	** 91 **	3.380
** 28 **	1.700	** 60 **	1.612	** 92 **	3.786
** 29 **	1.132	** 61 **	0.899	** 93 **	2.764
** 30 **	1.873	** 62 **	2.223	** 94 **	5.976
** 31 **	3.317	** 63 **	1.850	** 95 **	5.878
** 32 **	1.205	** 64 **	1.220	** 96 **	7.520

(9) PHONE CALLS AND SALES

This application is based on weekly data from the week ending 12 Jan 52 to the week ending 31 May 52, as printed out below. The background of the problem is as follows:

The data concern a regular commercial laundry employing about 125 people. In May and June, the plant handles a peak volume of about $12,000 a week. About 25% of this is work for commercial accounts such as hospitals and hotels. The remaining 75% is received from homes and consists of regular family laundry priced on a pound basis, bundle work or bachelor bundles priced on a piece basis and specialties such as curtains, blankets and pillows.

Although volume is fairly constant from week to week, there is some variation. For example, in April and May, 1952, family bundle work and specialty sales (total sales less commercial sales) varied from a low of $8,664 to a high of $9,479. The weather seems to be the principal reason for these ups and downs but there apparently are other less distinguishable causes.

The laundry, like all commercial laundries, operates on a regular pickup and delivery schedule. Work picked up on Monday must be ready for delivery Thursday morning and work picked up on Tuesday must be ready on Friday morning. An occasional high sales week will result in some difficulty in meeting these schedules.

For this reason, it is helpful to have some estimate early in the week of what the approximate sales for that week will be. It was thought that one possible basis for such an estimate might be the number of telephone calls, including orders and inquiries, received on Monday.

Table A-9. Data File: $ISAACS

	CALLS	SALES
** 1 **	189	9173
** 2 **	130	8589
** 3 **	123	8567
** 4 **	148	8551
** 5 **	137	8846
** 6 **	142	8395
** 7 **	132	8561
** 8 **	132	8588
** 9 **	143	8557
** 10 **	144	8572
** 11 **	155	8572
** 12 **	152	8879
** 13 **	131	8664
** 14 **	166	8883
** 15 **	126	8714
** 16 **	167	9479
** 17 **	168	9205
** 18 **	176	9171
** 19 **	140	8925
** 20 **	159	8787
** 21 **	149	8929

(10) EFFECTS OF EDA FUNDING ON UNEMPLOYMENT IN MICHIGAN

The data and background for this application have been supplied by Mr. James E. Peterson of the Department of Commerce.

Earlier governmental programs--the Area Redevelopment Administration (ARA) and the Public Works Acceleration Act (APW)--showed sufficient apparent promise to make possible the enactment by Congress in 1965 of the Public Works and Economic Development Act of 1965, which set up the Economic Development Administration (EDA). The EDA legislation states that the Agency's mission is "to provide grants for public works and development facilities, other financial assistance and the planning and coordination needed to alleviate conditions of substantial and persistent unemployment and underemployment in economically distressed areas and regions".

The basic tools furnished to EDA were these: Public works grants and loans were continued as a means of improving the social and economic overhead of communities and regions. Business loans were again provided to aid businesses in creating employment in designated areas. Technical assistance grants were retained as aids in identifying development opportunities in economically distressed counties and towns, and funds were provided to support economic research. In addition, the EDA legislation introduced planning and administrative grants to finance full-time planners at the multi-county level.

The Midwestern Regional Office began to compile employment and unemployment statistics for the counties in their six-state region in May of 1972 with the aim of evaluation of program effectiveness. They used the County Business Patterns, U. S. Department of Commerce, as source material for the employment data. These data are derived from the taxable wages paid for covered employment during any given quarter. In addition, employment of units of multiunit employers was obtained from three sources: (1) Treasury Form 941, as adapted by the Social Security Administration's Establishment Reporting Plan; (2) Survey of Multiunit Companies; and (3) the 1967 Economic Censuses.

Because of the availability of employment data, a decision was made to consider projects funded in counties between the years 1965 and 1969 only, and to consider employment figures for the years 1966 and 1970. In order to qualify for designation, the county unemployment rate had to be half again as large as the national unemployment rate.

In the entire Midwestern region, 53 counties had been designated and funded between 1965 and 1969. Another 94 counties had been designated but not funded. Initially, they took at random 47 of the counties designated but not funded and compared the increase in employment (arithmetic mean) in this group and in the 53 counties actually funded. As expected, the funded counties showed a higher mean percentage growth in employment than did the non-funded counties: 18.1% versus 13.8%.

The next step was to develop additional information for more detailed comparison. It was decided to study unemployment rates. Unfortunately, the U. S. Labor Department does not keep unemployment statistics by counties for previous years. It was necessary to go to the Employment Securities Division of each state, which are based upon the number of unemployed people registered with their local Unemployment Office. Those people not eligible for unemployment benefits generally do not register; in addition, those people whose benefits have expired are not technically registered. Hence the unemployment statistics used by the state tend to be understated.

The actual data given here shall be confined to the 83 counties of the state of Michigan, which are listed below in the same order that they appear in the data file $MICH1.

1 Alcona	2 Alger	3 Allegan	4 Alpena	5 Antrim
6 Arenac	7 Baraga	8 Barry	9 Bay	10 Benzie
11 Berrien	12 Branch	13 Calhoun	14 Cass	15 Charlevoix
16 Cheboygan	17 Chippewa	18 Clare	19 Clinton	20 Crawford
21 Delta	22 Dickinson	23 Eaton	24 Emmet	25 Genesee
26 Gladwin	27 Gogebic	28 Grand Trav.	29 Gratiot	30 Hillsdale
31 Houghton	32 Huron	33 Ingham	34 Ionia	35 Iosco
36 Iron	37 Isabella	38 Jackson	39 Kalamazoo	40 Kalkaska
41 Kent	42 Keweenaw	43 Lake	44 Lapeer	45 Leelanau
46 Lenawee	47 Livingston	48 Luce	49 Mackinac	50 Macomb
51 Manistee	52 Marquette	53 Mason	54 Mecosta	55 Menominee
56 Midland	57 Missaukee	58 Monroe	59 Montcalm	60 Montmorency
61 Muskegon	62 Newaygo	63 Oakland	64 Oceana	65 Ogemaw
66 Ontonagon	67 Osceola	68 Oscoda	69 Otsego	70 Ottawa
71 Presque Isle	72 Roscommon	73 Saginaw	74 St. Clair	75 St. Joseph
76 Sanilac	77 Schoolcraft	78 Shiawassee	79 Tuscola	80 Van Buren
81 Washtenau	82 Wayne	83 Wexford		

The file $MICH1 contains 83 observations on the following three variables: (1) a "dummy" variable that takes the number 1 if special federal funding was made for the county as a part of the program after 1966; otherwise it takes the number 0; (2) The percentage unemployment in the county in 1966; and (3) The percentage unemployment in the county in 1970.

Assignment 1

Apply the simple regression model to the regression of items (3) on (2) above. You may wish to try transformations of one or more of the variables. Include the appropriate diagnostic checks for model adequacy. What value, if any, are the sequence plot and AUTO and RUNS commands in a cross sectional application of regression like this one?

Assignment 2

Use the multiple regression approach to see what you can conclude about the effectiveness of the government program in Assignment 1.

Table A-10. Data File: $MICH1

	FUND	U66	U70		FUND	U66	U70
** 1 **	0.0	6.9	8.2	** 43 **	0.0	9.7	11.9
** 2 **	0.0	6.5	8.7	** 44 **	0.0	3.3	8.1
** 3 **	0.0	3.3	9.7	** 45 **	0.0	6.3	9.3
** 4 **	0.0	8.4	12.3	** 46 **	0.0	3.4	8.2
** 5 **	1.0	5.6	10.1	** 47 **	0.0	3.9	5.8
** 6 **	1.0	7.8	11.8	** 48 **	1.0	6.3	10.0
** 7 **	0.0	6.4	10.0	** 49 **	1.0	13.1	20.9
** 8 **	0.0	3.1	6.2	** 50 **	0.0	3.2	6.6
** 9 **	1.0	4.8	9.8	** 51 **	1.0	8.9	12.4
** 10 **	0.0	10.7	14.5	** 52 **	1.0	4.2	6.7
** 11 **	0.0	3.4	5.5	** 53 **	0.0	5.9	7.0
** 12 **	0.0	4.3	9.0	** 54 **	1.0	5.2	6.9
** 13 **	0.0	3.1	6.2	** 55 **	0.0	4.1	6.1
** 14 **	0.0	3.4	9.8	** 56 **	0.0	3.8	4.2
** 15 **	0.0	5.2	11.6	** 57 **	0.0	5.5	12.6
** 16 **	0.0	12.8	19.2	** 58 **	1.0	5.0	8.8
** 17 **	1.0	11.2	14.5	** 59 **	0.0	4.6	11.3
** 18 **	1.0	6.8	14.5	** 60 **	0.0	5.9	9.6
** 19 **	0.0	2.4	5.7	** 61 **	0.0	3.8	9.6
** 20 **	1.0	7.0	11.5	** 62 **	0.0	8.1	12.4
** 21 **	1.0	6.2	10.2	** 63 **	0.0	3.2	6.6
** 22 **	1.0	5.4	7.6	** 64 **	1.0	13.8	17.5
** 23 **	0.0	2.4	5.7	** 65 **	0.0	10.3	32.2
** 24 **	0.0	7.3	12.1	** 66 **	1.0	3.9	4.8
** 25 **	0.0	3.3	8.1	** 67 **	0.0	5.5	12.6
** 26 **	0.0	6.5	9.1	** 68 **	0.0	10.0	13.6
** 27 **	0.0	8.1	8.0	** 69 **	1.0	6.3	11.3
** 28 **	0.0	6.3	9.3	** 70 **	0.0	3.1	6.5
** 29 **	1.0	9.7	13.8	** 71 **	1.0	6.1	6.9
** 30 **	0.0	3.8	9.4	** 72 **	0.0	4.6	4.8
** 31 **	1.0	6.4	9.9	** 73 **	0.0	2.8	5.8
** 32 **	0.0	8.0	13.2	** 74 **	1.0	5.5	11.6
** 33 **	0.0	2.4	5.7	** 75 **	0.0	3.9	5.1
** 34 **	0.0	4.6	11.3	** 76 **	0.0	5.8	10.3
** 35 **	0.0	6.9	8.2	** 77 **	0.0	10.1	13.8
** 36 **	1.0	8.2	18.7	** 78 **	0.0	4.2	14.1
** 37 **	0.0	7.8	6.0	** 79 **	0.0	5.2	12.6
** 38 **	0.0	2.6	6.3	** 80 **	0.0	3.3	8.5
** 39 **	0.0	3.0	5.4	** 81 **	0.0	2.2	5.4
** 40 **	1.0	6.3	9.3	** 82 **	0.0	3.2	6.6
** 41 **	0.0	3.1	6.5	** 83 **	0.0	5.5	12.6
** 42 **	0.0	6.4	9.9				

Assignment 3

The final examination given as a sample, starting two pages below, involves in part an extension of the analysis of the problem. After as much further study as you wish, answer the queries on the second part of the examination.

(11) WINNEBAGO MOTORHOME SALES and MORE EFFECTS OF EDA FUNDING (continued)

Two applications are combined below in the format of a final examination given in the Spring of 1973. Students were told that they would be given printouts of two analyses, with queries to be answered. They were given background both for the nature of the problems and the data files to be used, so that they could conduct exploratory analyses of their own before the examination itself.

The students were told that the first application would be based on a time series of monthly sales of motorhomes for Winnebago Industries, Inc., for a period of 64 months starting with November, 1966, and concluding with February, 1972. Sales are in units. The name of the data file is $BLUME, and the data themselves are printed out on the second page of the examination itself.

The second application entailed a more extensive examination of the data on Michigan unemployment. An expanded data file, $MICH, (Table A-11), was made available, so that students could explore the problem to the depth permitted by the data then at hand. From this file, by use of various IDA commands, I built up the data matrix with variables defined in the examination and observations printed.

The definition of columns in $MICH is as follows:

(1) Designated for funding? (1 = yes, 0 = no)

(2) Funded? (1 = yes, 0 = no) (Column 1 of $MICH1)

(3) Unemployment percentage in 1966 (Column 2 of $MICH1)

(4) Unemployment percentage in 1970 (Column 3 of $MICH1)

(5) Labor Force 1966

(6) Labor Force 1970

(7) Dollar funding for public works

(8) Dollar funding for technical assistance

(9) Dollar funding for business loans

Since the examination was prepared in the Spring of 1973, the version of IDA used differs slightly from that used in the preparation of this book.

Table A-11. Data File: $MICH

** 1 **	1.0	0.0	6.9	8.2	790	834	0	0	0
** 2 **	1.0	0.0	6.5	8.7	1280	1757	0	0	0
** 3 **	0.0	0.0	3.3	9.7	8504	10535	0	0	0
** 4 **	1.0	0.0	8.4	12.3	5610	6827	0	0	0
** 5 **	1.0	1.0	5.6	10.1	1576	1997	132000	27185	0
** 6 **	1.0	1.0	7.8	11.8	1137	1368	284000	850	0
** 7 **	1.0	0.0	6.4	10.0	1281	1237	0	0	0
** 8 **	0.0	0.0	3.1	6.2	5150	4334	0	0	0
** 9 **	1.0	1.0	4.8	9.8	25220	26581	322000	2200	0
** 10 **	1.0	0.0	10.7	14.5	1291	1523	0	0	0
** 11 **	0.0	0.0	3.4	5.5	46680	54145	0	0	0
** 12 **	0.0	0.0	4.3	9.0	7337	6683	0	0	0
** 13 **	0.0	0.0	3.1	6.2	43747	44827	0	0	0
** 14 **	0.0	0.0	3.4	9.8	5460	6594	0	0	0
** 15 **	1.0	0.0	5.2	11.6	3826	3322	0	0	0
** 16 **	1.0	0.0	12.8	19.2	1916	2789	0	0	0
** 17 **	1.0	1.0	11.2	14.5	3745	3105	152000	34000	0
** 18 **	1.0	1.0	6.8	14.5	2245	2447	94000	0	0
** 19 **	0.0	0.0	2.4	5.7	3811	4286	0	0	0
** 20 **	1.0	1.0	7.0	11.5	1242	1490	575000	0	0
** 21 **	1.0	0.0	6.2	10.2	6062	6903	172000	2100	0
** 22 **	1.0	1.0	5.4	7.6	4558	5586	0	11250	0
** 23 **	0.0	0.0	2.4	5.7	6629	7873	0	0	0
** 24 **	1.0	0.0	7.3	12.1	4196	4763	0	0	0
** 25 **	0.0	0.0	3.3	8.1	140497	137201	0	0	0
** 26 **	0.0	0.0	6.5	9.1	914	916	0	0	0
** 27 **	1.0	1.0	8.1	8.0	2695	3477	0	3675	1892000
** 28 **	1.0	0.0	6.3	9.3	9301	10815	0	0	0
** 29 **	1.0	1.0	9.7	13.8	8415	8124	120000	0	0
** 30 **	0.0	0.0	3.8	9.4	6525	6509	0	0	0
** 31 **	1.0	1.0	6.4	9.9	5626	4628	223000	675	0
** 32 **	1.0	0.0	8.0	13.2	4416	5499	0	0	0
** 33 **	0.0	0.0	2.4	5.7	76739	83582	0	0	0
** 34 **	0.0	0.0	4.6	11.3	6726	7379	0	0	0
** 35 **	0.0	0.0	6.9	8.2	2185	3229	0	0	0
** 36 **	1.0	1.0	8.2	18.7	3792	1751	0	38675	0
** 37 **	0.0	0.0	7.8	6.0	4095	5606	0	0	0
** 38 **	0.0	0.0	2.6	6.3	41348	39439	0	0	0
** 39 **	0.0	0.0	3.0	5.4	57776	63564	0	0	0
** 40 **	1.0	1.0	6.3	9.3	380	438	482000	0	0
** 41 **	0.0	0.0	3.1	6.5	131444	141488	0	0	0
** 42 **	1.0	0.0	6.4	9.9	520	290	0	0	0
** 43 **	1.0	0.0	9.7	11.9	301	592	0	0	0
** 44 **	0.0	0.0	3.3	8.1	5135	5725	0	0	0
** 45 **	1.0	0.0	6.3	9.3	727	1050	0	0	0
** 46 **	0.0	0.0	3.4	8.2	22405	22824	0	0	0
** 47 **	0.0	0.0	3.9	5.8	5959	7449	0	0	0
** 48 **	1.0	0.0	6.3	10.0	622	584	0	252	0
** 49 **	1.0	1.0	13.1	20.9	840	1106	84000	2152	0
** 50 **	0.0	0.0	3.2	6.6	137060	151068	0	0	0
** 51 **	1.0	1.0	8.9	12.4	4629	4726	98000	0	0
** 52 **	1.0	0.0	4.2	6.7	11640	11588	429000	2023	0
** 53 **	0.0	0.0	5.9	7.0	4413	4734	0	0	0
** 54 **	1.0	0.0	5.2	6.9	3959	3761	0	550	0
** 55 **	0.0	0.0	4.1	6.1	3911	4960	0	0	0
** 56 **	0.0	0.0	3.8	4.2	20220	22570	0	0	0
** 57 **	0.0	0.0	5.5	12.6	412	432	0	0	0
** 58 **	1.0	1.0	5.0	8.8	15909	18461	421000	10865	0
** 59 **	0.0	0.0	4.6	11.3	9284	9210	0	0	0
** 60 **	0.0	0.0	5.9	9.6	733	633	0	0	0
** 61 **	0.0	0.0	3.8	9.6	45936	46506	0	0	0
** 62 **	1.0	0.0	8.1	12.4	3280	3717	0	0	0
** 63 **	0.0	0.0	3.2	6.6	192530	242675	0	0	0
** 64 **	1.0	1.0	13.8	17.5	1196	1407	405000	2060	0
** 65 **	1.0	0.0	10.3	32.2	1096	1412	0	0	0
** 66 **	1.0	0.0	3.9	4.8	2723	3651	2940000	804	0
** 67 **	0.0	0.0	5.5	12.6	2592	3060	0	0	0
** 68 **	1.0	0.0	10.0	13.6	279	461	0	0	0
** 69 **	1.0	1.0	6.3	11.3	1800	2350	0	725	0
** 70 **	0.0	0.0	3.1	6.5	26713	29960	0	0	0
** 71 **	1.0	1.0	6.1	6.9	1656	1845	0	685	0
** 72 **	0.0	0.0	4.6	4.8	928	1173	0	0	0
** 73 **	0.0	0.0	2.8	5.8	62240	64010	0	0	0
** 74 **	1.0	1.0	5.5	11.6	22582	24748	706000	38339	0
** 75 **	0.0	0.0	3.9	5.1	13333	14300	0	0	0
** 76 **	0.0	0.0	5.8	10.3	5200	6357	0	0	0
** 77 **	1.0	0.0	10.1	13.8	1335	1114	0	0	0
** 78 **	0.0	0.0	4.2	14.1	11198	11194	0	0	0
** 79 **	0.0	0.0	5.2	12.6	5696	5808	0	0	0
** 80 **	0.0	0.0	3.3	8.5	9500	9933	0	0	0
** 81 **	0.0	0.0	2.2	5.4	60341	66807	0	0	0
** 82 **	0.0	0.0	3.2	6.6	927593	922705	0	0	0
** 83 **	1.0	0.0	5.5	12.6	4870	5270	0	0	0

UNIVERSITY OF CHICAGO
Graduate School of Business

Business 320-81

Mr. Roberts
Spring, 1973

Final Examination

The examination consists of two problems. In each you are given an annotated printout of an analysis of a set of data for which the background and file reference have been previously furnished to you. Various questions are typed on the printout. You are to write your answers to these, at the point raised, upon the examination itself. If space is not adequate for your answer, write on the blank facing page, but make all answers as concise as possible consistent with the ideas that you wish to convey. Repetition of the same idea in different language will be counted slightly against you and will take time away from your consideration of other questions.

The examination is open-book: you may refer to any inanimate source. The previous work that you have done with these data may be referred to in your answers if you feel that this will be illuminating. Please give numerical details so that I can evaluate your contribution.

Do not be afraid to use your common sense!

PRINT YOUR NAME HERE: _____

Question 1. *Winnebago Motorhome Sales*

```
GET-$IDA
RUN
IDA

VERSION 17MAY73

GOOD MORNING.  DO YOU NEED HELP ?N

> CHGP
* LEVEL = 3

> ENTER

* MODE = F

NAME IS ?$BLUME

* N,K GIVEN IN FILE ?Y

DATA MATRIX HAS  64 OBSERVATIONS ON  1 VARIABLES
COMPUTING ...
WANT TO SUPPLY NAMES  ?Y
VAR.  1 = ?SALES
```

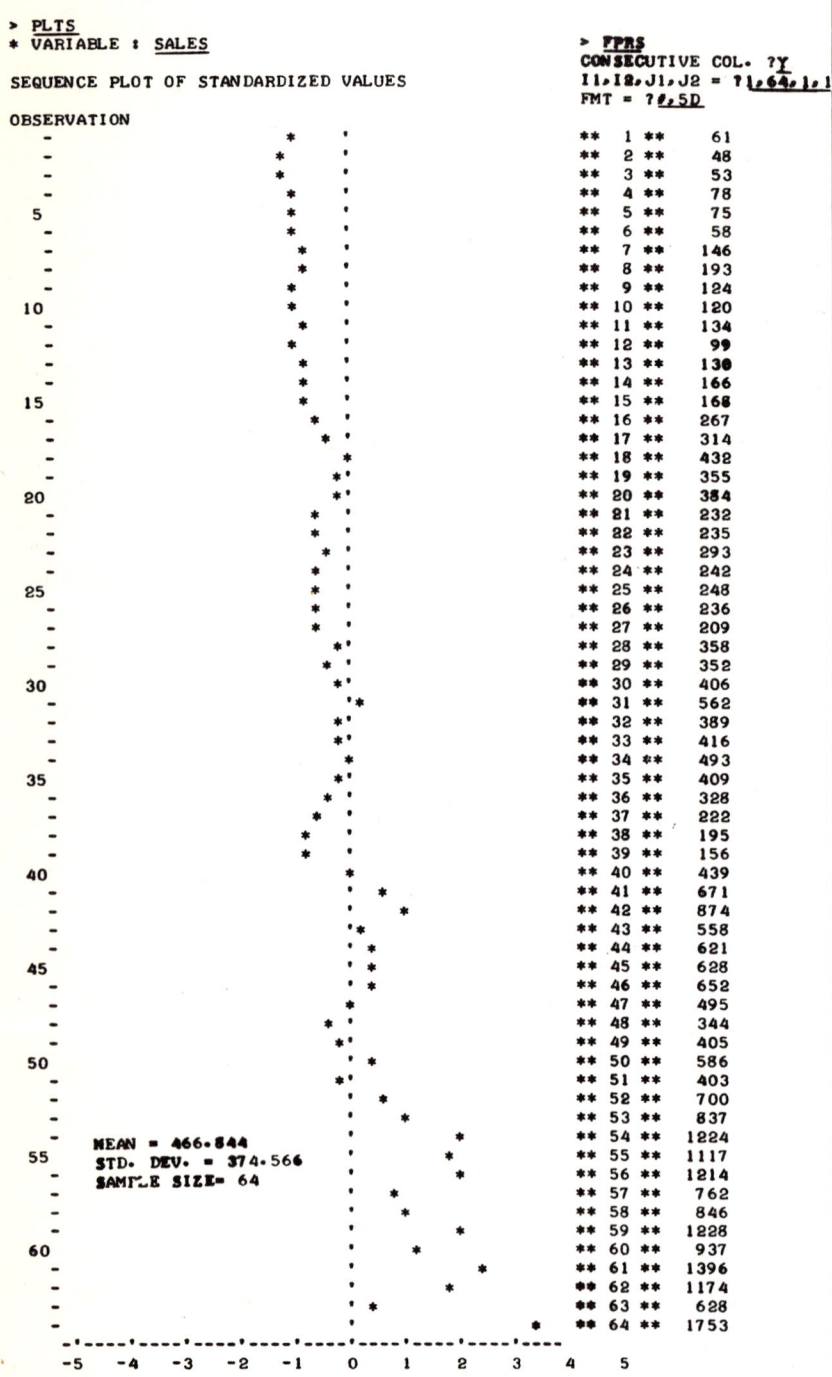

Figure A-1. Sequence Plot and Data Matrix for Motorhome Sales Data

```
> LOGE
* COL# TO PLACE VAR. : 1
* COLUMN TO BE TRANSFORMED : 1
UPDATING MEAN, STD, ..., CORR, ...

GIVE NAME OF NEW VARIABLE
VAR.  1 ?HGSALE\
LGSALE

> PLTS
* VARIABLE : LGSALE
```

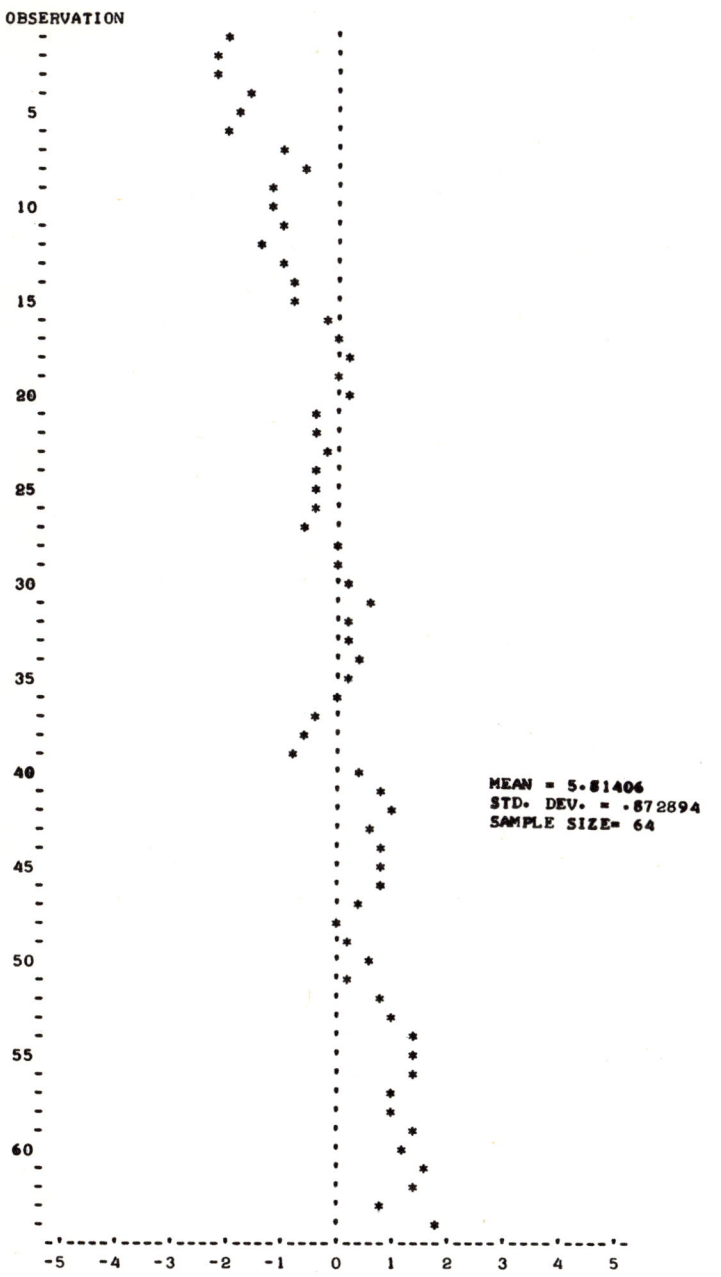

Figure A-2. Sequence Plot for Logarithmic Transformation of Motorhome Sales Data

What about the sequence plot in Figure A-1 suggested the logarithmic transformation? How successful has the transformation been? Discuss concisely.

```
HISTOGRAM

ABS. FREQ.
                        *
                        *
                    * ** *
 5              * ** * **  *
                * ** * **
              ****  ****** *
         **   ************  *
         ********************
    -'----'----'----'----'----'----'----'----'----'----'-
    -5   -4   -3   -2   -1    0    1    2    3    4    5
```

Figure A-3. Histogram of Logarithmic Transformation of Motorhome Sales Data

```
>  AUTO
*  VARIABLE = LGSALE
*  MAX ORDER : 16

  ORDER   AUTOCORR.     -1  -.75 -.50 -.25   0   .25  .50  .75  +1
                        '----'----'----'----'----'----'----'----'
    1      +0.859       -                    '                 *
    2      +0.765       -                    '               *
    3      +0.671       -                    '             *
    4      +0.574       -                    '           *
    5      +0.512       -                    '         *
    6      +0.445       -                    '       *
    7      +0.413       -                    '      *
    8      +0.419       -                    '      *
    9      +0.409       -                    '      *
   10      +0.388       -                    '     *
   11      +0.348       -                    '    *
   12      +0.307       -                    '   *
   13      +0.230       -                    '  *
   14      +0.192       -                    ' *
   15      +0.134       -                    '*
   16      +0.098       -                    '*

S.E. OF EACH COEF. GIVEN RANDOM MODEL=0.125

BOX-PIERCE STATISTIC= 227.911
EXPECTATION GIVEN RANDOM MODEL= 16
     (OBS.-EXP.)/(STD.DEV.)= 12.2016

>  RUNS
*  VARIABLE : LGSALE

   OBSERVED NUMBER OF RUNS= 6
   EXPECTED NUMBER OF RUNS= 32.5
   STANDARD DEVIATION OF RUNS= 3.90513
        (OBS.-EXP.)/(STD.DEV.)=-6.78595

>  DIFF

*  COL# TO PLACE VAR. : 2
*  COLUMN TO BE TRANSFORMED :1
GAP = ?1

UPDATING MEAN, STD.DEV....
OBSERVATIONS IN FIRST 1    ROWS NOW UNDEFINED.

GIVE NAME(S) OF NEW VARIABLE(S)
VAR. 2 ?DLGSAL
```

Based on the evidence presented here, do the logs of sales behave as a stationary process? As a random process? Explain concisely.

What made me think of differencing?

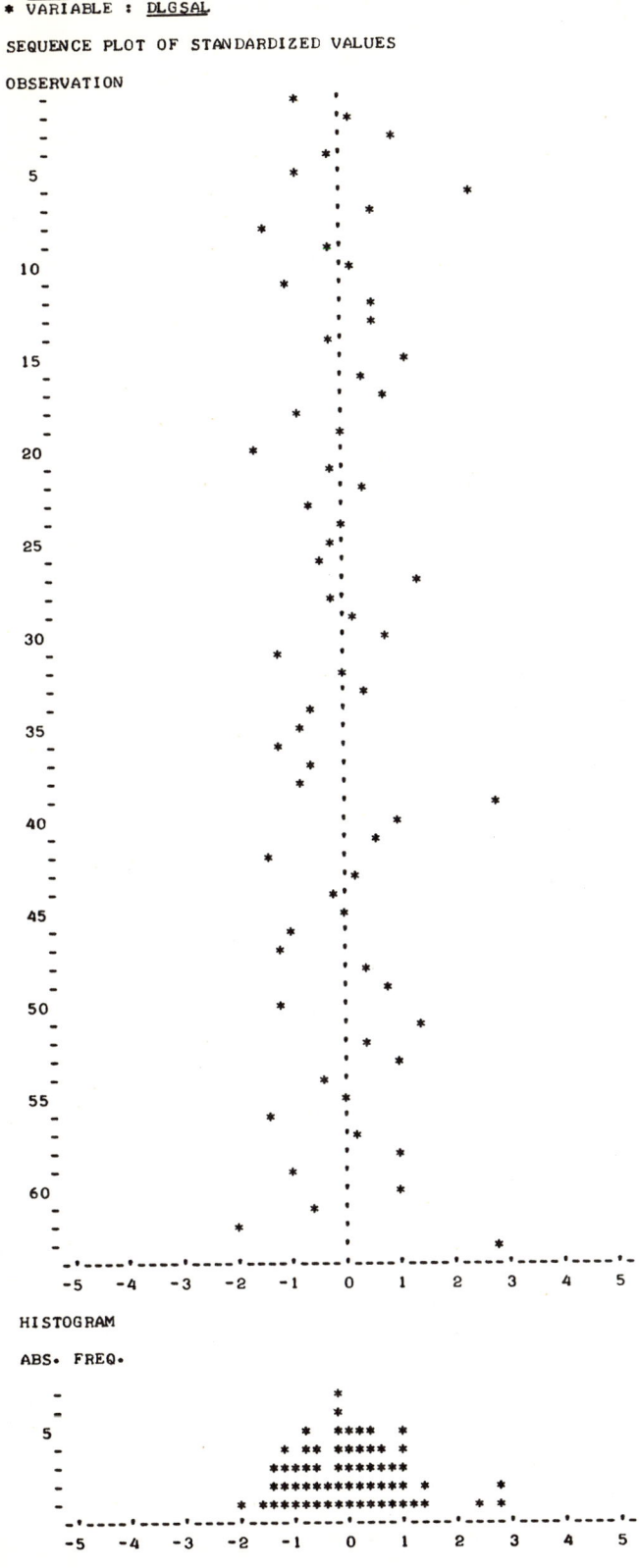

Figure A-5. Sequence Plot and Histogram for Logs of Motorhome Sales Data After Differencing

```
MEAN = 5.33049E-02
STD. DEV. = .352839
SAMPLE SIZE = 63

> AUTO
* VARIABLE = DLGSAL
* MAX ORDER : 16

ORDER   AUTOCORR.

  1     -0.145
  2     -0.132
  3     +0.045
  4     -0.156
  5     -0.109
  6     +0.051
  7     -0.167
  8     -0.110
  9     +0.052
 10     +0.110
 11     -0.001
 12     +0.304
 13     -0.075
 14     +0.028
 15     +0.039
 16     -0.083

S.E. OF EACH COEF. GIVEN RANDOM MODEL=0.126

BOX-PIERCE STATISTIC= 15.2168
EXPECTATION GIVEN RANDOM MODEL= 16
STANDARD NORMAL APPROXIMATION=-2.29262E-02

> RUNS
* VARIABLE = DLGSAL

EXPECTED NO. OF RUNS     =   32.4921
STANDARD DEVIATION (R)   =    3.93548
OBSERVED NO. OF RUNS     =   36
STANDARD NORMAL APPROX.  =    .891358

> MEAN

VARIABLE    MEAN         STD. DEV.

LGSALE    5.84109E+00   8.52469E-01
DLGSAL    5.33049E-02   3.52839E-01

> LAGG
* COL# TO PLACE VAR.    : 3
* COLUMN TO BE TRANSFORMED :2
GAP = ?1
UPDATING MEAN, STD, ...
GIVE NAME OF NEW VARIABLE
VAR.  3 ?L1DGLS


> LAGG
* COL# TO PLACE VAR.    : 4
* COLUMN TO BE TRANSFORMED :2
GAP = ?2
UPDATING MEAN, STD, ..., CORR, ...

GIVE NAME OF NEW VARIABLE
VAR.  4 ?L2DLGS

> LAGG
* COL# TO PLACE VAR.    : 5
* COLUMN TO BE TRANSFORMED :2
GAP = ?12
UPDATING MEAN, STD, ..., CORR, ...

GIVE NAME OF NEW VARIABLE
VAR.  5 ?L12DLG

> PRTS
CONSECUTIVE COL. ?Y
I1,I2,J1,J2 = ?1,15,1,5

** 14 **    5.11199    0.24445    0.27242   -0.30272   -0.23967
** 15 **    5.12396    0.01198    0.24445    0.27242    0.09909

> NAME
THESE ARE THE VARIABLES IN THE DATA MATRIX :
COLUMN    NAME

   1      LGSALE
   2      DLGSAL
   3      L1DLGS
   4      L2DLGS
   5      L12DLG

> REGR
* DEP. VAR. = 2
HOW MANY INDEP. VAR. ?3
INDEP. VAR.  1 = ?3
INDEP. VAR.  2 = ?4
INDEP. VAR.  3 = ?5
ANALYZING RESIDUALS ...
WARNING: AUTO( 3) = -0.18
WARNING: AUTO( 4) = -0.15
WARNING: AUTO( 7) = -0.24
WARNING: AUTO(10) = +0.14

> COEF

VARIABLE    B(STD.V)       B          STD.ERROR(B)      T

L1DLGS      -0.1499     -.1624E+00    1.4230E-01      -1.142
L2DLGS      -0.0991     -.1108E+00    1.4790E-01      -0.749
L12DLG       0.4102      4.3404E-01   1.3973E-01       3.106
CONSTANT                 4.0844E-02   4.6768E-02       0.873
```

The mean of the differences is 0.053. What does that fact tell you about the mean monthly percentage growth of sales over this period?

What is the single most striking diagnostic information given here by AUTO?

What made me think of the lag of 12? Does it make common sense?

Why did PRTS start with row 14 of the data matrix?

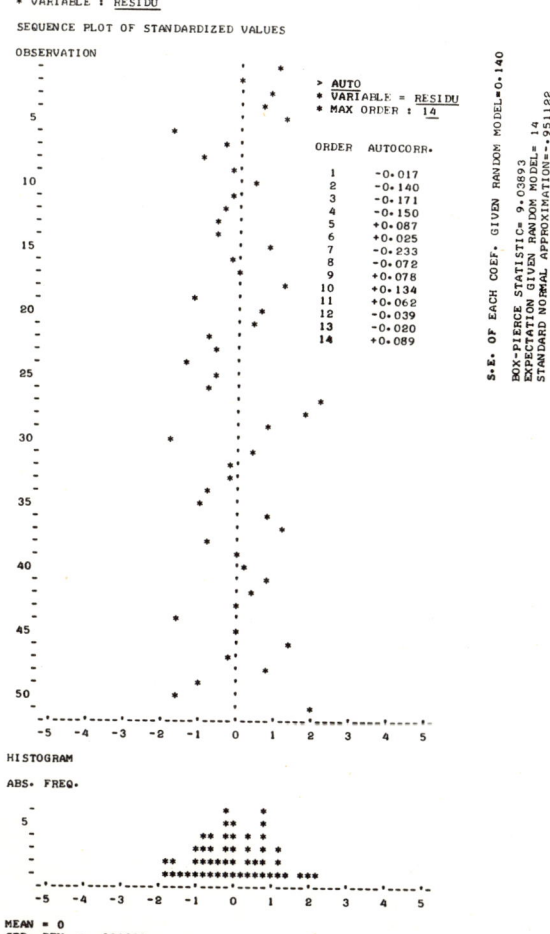

```
> SUMM
            MULTIPLE R   R-SQUARE
 UNADJUSTED    0.4384     0.1922
 ADJUSTED      0.3750     0.1406
 STD. ERROR OF RESIDUALS =     .3258
> SWEEP
 VAR. : L2DLGS
 ANALYZING RESIDUALS ...
 WARNING: AUTO( 3) = -0.17
 WARNING: AUTO( 4) = -0.15
 WARNING: AUTO( 7) = -0.23

> COEF

 VARIABLE  B(STD.V)       B        STD.ERROR(B)     T

 L1DLGS    -0.1443    -.1564E+00   1.4142E-01    -1.106
 L12DLG     0.3986     4.2182E-01  1.3814E-01     3.054
 CONSTANT              3.6492E-02  4.6194E-02     0.790

> CORR
 HOW MANY VARIABLES ?4
 VAR. #'S :?2,3,4,5
 # DECIMALS = ?3

 DLGSAL   1.000
 L1DLGS  -0.154   1.000
 L2GLGS  -0.042  -0.059   1.000
 L12DLG   0.402  -0.025   0.118   1.000

> SUMM
            MULTIPLE R   R-SQUARE
 UNADJUSTED    0.4273     0.1826
 ADJUSTED      0.3854     0.1485
 STD. ERROR OF RESIDUALS =    .324309
> RUNS
* VARIABLE = RESIDU

 EXPECTED NO. OF RUNS    =    26.0196
 STANDARD DEVIATION (R)  =     3.46688
 OBSERVED NO. OF RUNS    =    25
 STANDARD NORMAL APPROX. =    -.2941
```

Figure A-6. Sequence Plot and Histogram for Residuals for Motorhome Sales Data

Why did I give this command now for this particular variable?

Give a concise interpretation of: -.1564; -.14142; -1.106, which appear below COEF.

What does CORR tell me and what is it good for? Give an example of the interpretation of any one of the coefficients printed out.

Does it look as if the SWEEP at the top of the page was a good idea? Explain concisely.

Express in words the formula for predicting a given month's sales (point estimate) implied by the printout under COEF above.

Summarize concisely the evidence bearing on the specification of random disturbances for the regression model chosen.

238 □ CONVERSATIONAL STATISTICS

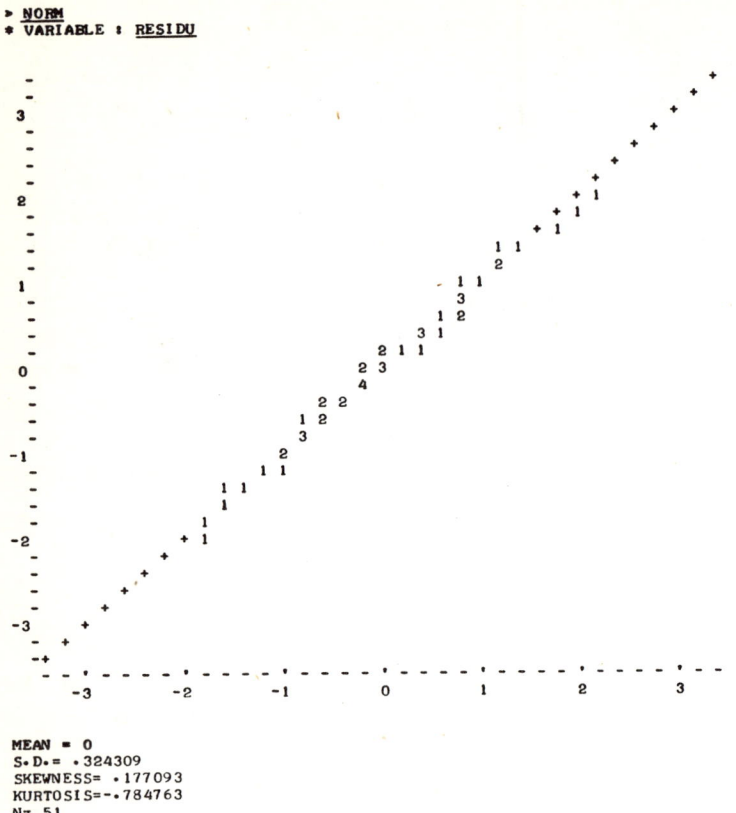

Figure A-7. Normal Probability Plot of Residuals for Motorhome Sales Data

What diagnostic information about the model is given by Figure A-7?

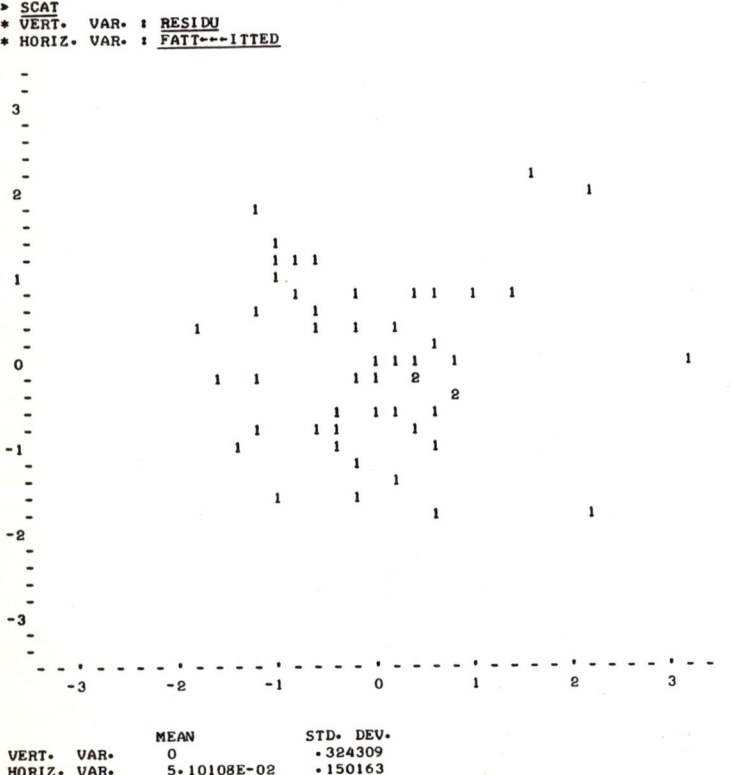

Figure A-8. Scatter Plot of Residuals of Motorhome Sales Data

Summarize the diagnostic information given by the above plot.

```
> PRTS
CONSECUTIVE COL.  ?Y
I1,I2,J1,J2 = ?46,64,2,5

** 46 **      0.03750      0.01121      0.10697      0.16982
** 47 **     -0.27549      0.03750      0.01121     -0.18679
** 48 **     -0.36392     -0.27549      0.03750     -0.22070
** 49 **      0.16325     -0.36392     -0.27549     -0.39034
** 50 **      0.36943      0.16325     -0.36392     -0.12968
** 51 **     -0.37438      0.36943      0.16325     -0.22314
** 52 **      0.55214     -0.37438      0.36943      1.03464
** 53 **      0.17874      0.55214     -0.37438      0.42427
** 54 **      0.38005      0.17874      0.55214      0.26431
** 55 **     -0.09148      0.38005      0.17874     -0.44872
** 56 **      0.08327     -0.09148      0.38005      0.10697
** 57 **     -0.46573      0.08327     -0.09148      0.01121
** 58 **      0.10457     -0.46573      0.08327      0.03750
** 59 **      0.37262      0.10457     -0.46573     -0.27549
** 60 **     -0.27046      0.37262      0.10457     -0.36392
** 61 **      0.39868     -0.27046      0.37262      0.16325
** 62 **     -0.17319      0.39868     -0.27046      0.36943
** 63 **     -0.62563     -0.17319      0.39868     -0.37438
** 64 **      1.02654     -0.62563     -0.17319      0.55214

> SEPR
USE DATA MATRIX ?N
HOW MANY Y'S ?1

                     Y         S.E.(Y)     S.E.PRED.(Y)

X'S = ?-.62563,.55214

              3.6722E-01   1.2392E-01   3.4718E-01

> SEPR
USE DATA MATRIX ?Y
HOW MANY Y'S ?1

                     Y         S.E.(Y)     S.E.PRED.(Y)

ROW : 64      3.6723E-01   1.2392E-01   3.4718E-01

> SEPR
USE DATA MATRIX ?N
HOW MANY Y'S ?1

                     Y         S.E.(Y)     S.E.PRED.(Y)

X'S = ?1.02654,.17874

             -.4863E-01   1.4875E-01   3.5680E-01
```

Why did this row come out the same as the one that was printed out for the previous use of SEPR?

What period of time does this application of SEPR refer to? What is the inference about sales for that period?

240 □ CONVERSATIONAL STATISTICS

Question 2. Effects of EDA Funding on Unemployment in Michigan

```
DATA MATRIX HAS  83 OBSERVATIONS ON  5 VARIABLES
COMPUTING ...

WANT TO SUPPLY NAMES   ?Y
VAR.  1 = ?DSIG
VAR.  2 = ?FUND
VAR.  3 = ?U66
VAR.  4 = ?U70
VAR.  5 = ?$/LF

> LOGE

* COL# TO PLACE VAR.  : 3
* COLUMN TO BE TRANSFORMED :3
UPDATING MEAN, STD, ..., CORR, ...

GIVE NAME OF NEW VARIABLE
VAR.  3 ?LU66

> LOGE

* COL# TO PLACE VAR.  : 4
* COLUMN TO BE TRANSFORMED :4
UPDATING MEAN, STD, ..., CORR, ...

GIVE NAME OF NEW VARIABLE
VAR.  4 ?LU70

> NAME
THESE ARE THE VARIABLES IN THE DATA MATRIX:
COLUMN    NAME

   1      DSIG
   2      FUND
   3      LU66
   4      LU70
   5      $/LF

> MEAN
VARIABLE      MEAN           STD. DEV.

DSIG       2.40964E-01       4.30268E-01
FUND       2.89157E-01       4.56127E-01
LU66       1.65907E+00       4.42536E-01
LU70       2.22000E+00       3.81714E-01
$/LF       5.57673E+01       2.04710E+02
```

(= 1 if designated but not funded; = 0 otherwise)
(= 1 if funded; = 0 otherwise)
(percentage unemployment in 1966)
(percentage unemployment in 1970)
(Total funding in dollars divided by size of labor force in 1966)

I'm going to log both U66 and U70. An alternative analysis would have logged U70 only.

Can you give a simple interpretation for the mean of FUND, .289157?

Table A-12. Data File: $MICH1

```
> FR-PRS
CONSECUTIVE COL. ?Y
I1,I2,J1,J2 = ?1,83,1,5
FMT = ?0,5D.2D
```

** 1 **	1.00	0.00	1.93	2.10	0.00
** 2 **	1.00	0.00	1.87	2.16	0.00
** 3 **	0.00	0.00	1.19	2.27	0.00
** 4 **	1.00	0.00	2.13	2.51	0.00
** 5 **	0.00	1.00	1.72	2.31	101.01
** 6 **	0.00	1.00	2.05	2.47	250.53
** 7 **	1.00	0.00	1.86	2.30	0.00
** 8 **	0.00	0.00	1.13	1.82	0.00
** 9 **	0.00	1.00	1.57	2.28	12.85
** 10 **	1.00	0.00	2.37	2.67	0.00
** 11 **	0.00	0.00	1.22	1.70	0.00
** 12 **	0.00	0.00	1.46	2.20	0.00
** 13 **	0.00	0.00	1.13	1.82	0.00
** 14 **	0.00	0.00	1.22	2.28	0.00
** 15 **	1.00	0.00	1.65	2.45	0.00
** 16 **	1.00	0.00	2.55	2.95	0.00
** 17 **	0.00	1.00	2.42	2.67	49.67
** 18 **	0.00	1.00	1.92	2.67	41.87
** 19 **	0.00	0.00	0.88	1.74	0.00
** 20 **	0.00	1.00	1.95	2.44	462.96
** 21 **	0.00	1.00	1.82	2.32	28.72
** 22 **	0.00	1.00	1.69	2.03	2.47
** 23 **	0.00	0.00	0.88	1.74	0.00
** 24 **	1.00	0.00	1.99	2.49	0.00
** 25 **	0.00	0.00	1.19	2.09	0.00
** 26 **	0.00	0.00	1.87	2.21	0.00
** 27 **	0.00	1.00	2.09	2.08	703.40
** 28 **	1.00	0.00	1.84	2.23	0.00
** 29 **	0.00	1.00	2.27	2.62	14.26
** 30 **	0.00	0.00	1.34	2.24	0.00
** 31 **	0.00	1.00	1.86	2.29	39.76
** 32 **	1.00	0.00	2.08	2.58	0.00
** 33 **	0.00	0.00	0.88	1.74	0.00
** 34 **	0.00	0.00	1.53	2.42	0.00
** 35 **	0.00	0.00	1.93	2.10	0.00
** 36 **	0.00	1.00	2.10	2.93	10.20
** 37 **	0.00	0.00	2.05	1.79	0.00
** 38 **	0.00	0.00	0.96	1.84	0.00
** 39 **	0.00	0.00	1.10	1.69	0.00
** 40 **	0.00	1.00	1.84	2.23	1268.42
** 41 **	0.00	0.00	1.13	1.87	0.00
** 42 **	1.00	0.00	1.86	2.29	0.00
** 43 **	1.00	0.00	2.27	2.48	0.00
** 44 **	0.00	0.00	1.19	2.09	0.00
** 45 **	1.00	0.00	1.84	2.23	0.00
** 46 **	0.00	0.00	1.22	2.10	0.00
** 47 **	0.00	0.00	1.36	1.76	0.00
** 48 **	0.00	1.00	1.84	2.30	0.41
** 49 **	0.00	1.00	2.57	3.04	102.56
** 50 **	0.00	0.00	1.16	1.89	0.00
** 51 **	0.00	1.00	2.19	2.52	21.17
** 52 **	0.00	1.00	1.44	1.90	37.03
** 53 **	0.00	0.00	1.77	1.95	0.00
** 54 **	0.00	1.00	1.65	1.93	0.14
** 55 **	0.00	0.00	1.41	1.81	0.00
** 56 **	0.00	0.00	1.34	1.44	0.00
** 57 **	0.00	0.00	1.70	2.53	0.00
** 58 **	0.00	1.00	1.61	2.17	27.15
** 59 **	0.00	0.00	1.53	2.42	0.00
** 60 **	1.00	0.00	1.77	2.26	0.00
** 61 **	0.00	0.00	1.34	2.26	0.00
** 62 **	1.00	0.00	2.09	2.52	0.00
** 63 **	0.00	0.00	1.16	1.89	0.00
** 64 **	0.00	1.00	2.62	2.86	340.35
** 65 **	1.00	0.00	2.33	3.47	0.00
** 66 **	0.00	1.00	1.36	1.57	1079.99
** 67 **	1.00	0.00	1.70	2.53	0.00
** 68 **	1.00	0.00	2.30	2.61	0.00
** 69 **	0.00	1.00	1.84	2.42	0.40
** 70 **	0.00	0.00	1.13	1.87	0.00
** 71 **	0.00	1.00	1.81	1.93	0.41
** 72 **	0.00	0.00	1.53	1.57	0.00
** 73 **	0.00	0.00	1.03	1.76	0.00
** 74 **	0.00	1.00	1.70	2.45	32.96
** 75 **	0.00	0.00	1.36	1.63	0.00
** 76 **	0.00	0.00	1.76	2.33	0.00
** 77 **	1.00	0.00	2.31	2.62	0.00
** 78 **	0.00	0.00	1.44	2.65	0.00
** 79 **	0.00	0.00	1.65	2.53	0.00
** 80 **	0.00	0.00	1.19	2.14	0.00
** 81 **	0.00	0.00	0.79	1.69	0.00
** 82 **	0.00	0.00	1.16	1.89	0.00
** 83 **	1.00	0.00	1.70	2.53	0.00

```
> REGR
* DEP. VAR. = 4
  HOW MANY INDEP. VAR. ?4
  INDEP. VAR.  1 = ?1
  INDEP. VAR.  2 = ?2
  INDEP. VAR.  3 = ?3
  INDEP. VAR.  4 = ?5
  ANALYZING RESIDUALS ...
  WARNING: 37-TH RESIDUAL IS -2.66 S.D. UNITS FROM 0
  WARNING: 65-TH RESIDUAL IS +3.12 S.D. UNITS FROM 0
  WARNING: AUTO( 2) = +0.23
  WARNING: AUTO( 3) = -0.15
  WARNING: AUTO( 4) = +0.13
  WARNING: AUTO( 7) = -0.34

> COEF

  VARIABLE    B(STD.V)       B         STD.ERROR(B)       T

  DSIG         0.0722    6.4040E-02     9.4953E-02      0.674
  FUND         0.0405    3.3890E-02     9.0909E-02      0.373
  LU66         0.7254    6.2573E-01     9.2774E-02      6.745
  $/LF        -0.1382   -.2572E-03      1.4932E-04     -1.725
  CONSTANT               1.1710E+00     1.2844E-01      9.117

> SUMM

              MULTIPLE R   R-SQUARE
  UNADJUSTED    0.7702      0.5932
   ADJUSTED     0.7566      0.5724
  STD. ERROR OF RESIDUALS =        .249611

> SWEEP
  VAR. : FUND
  ANALYZING RESIDUALS
  WARNING: 37-TH RESIDUAL IS -2.77 S.D. UNITS FROM 0
  WARNING: 65-TH RESIDUAL IS +3.11 S.D. UNITS FROM 0
  WARNING: AUTO( 2) = +0.22
  WARNING: AUTO( 3) = -0.14
  WARNING: AUTO( 4) = +0.13
  WARNING: AUTO( 7) = -0.33

> COEF

  VARIABLE    B(STD.V)       B         STD.ERROR(B)       T

  DSIG         0.0475    4.2168E-02     7.4251E-02      0.568
  $/LF        -0.1278   -.2383E-03      1.3923E-04     -1.711
  LU66         0.7506    6.4740E-01     7.1917E-02      9.002
  CONSTANT               1.1490E+00     1.1352E-01     10.122

> SUMM

              MULTIPLE R   R-SQUARE
  UNADJUSTED    0.7698      0.5925
   ADJUSTED     0.7596      0.5770
  STD. ERROR OF RESIDUALS =        .248247

> SWEEP
  VAR. : DSIG
  ANALYZING RESIDUALS ...
  WARNING: 37-TH RESIDUAL IS -2.86 S.D. UNITS FROM 0
  WARNING: 65-TH RESIDUAL IS +3.19 S.D. UNITS FROM 0
  WARNING: AUTO( 2) = +0.23
  WARNING: AUTO( 3) = -0.14
  WARNING: AUTO( 4) = +0.13
  WARNING: AUTO( 7) = -0.34

> COEF
  VARIABLE    B(STD.V)       B         STD.ERROR(B)       T

  LU66         0.7740    6.6764E-01     6.2196E-02     10.735
  $/LF        -0.1381   -.2575E-03      1.3445E-04     -1.915
  CONSTANT               1.1267E+00     1.0603E-01     10.626
```

Explain the statistical interpretation of the last three entries on the line for $/LF. If one were to take this association at face value from the point of view of causation, what would be your conclusion about the effectiveness of the EDA in reducing unemployment?

```
> SUMM
              MULTIPLE R   R-SQUARE
  UNADJUSTED    0.7687      0.5909
   ADJUSTED     0.7620      0.5806
  STD. ERROR OF RESIDUALS =        .247194
```

Why did I sweep FUND?

Was the SWEEP a good idea? Explain concisely.

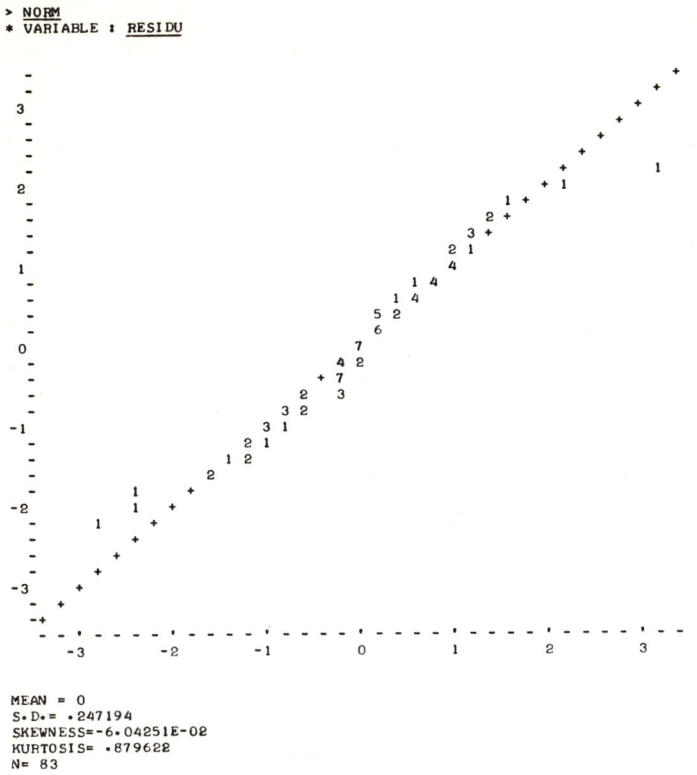

Figure A-9. Normal Probability Plot of Residuals for Unemployment Data

What does this plot tell about the adequacy of the regression model?

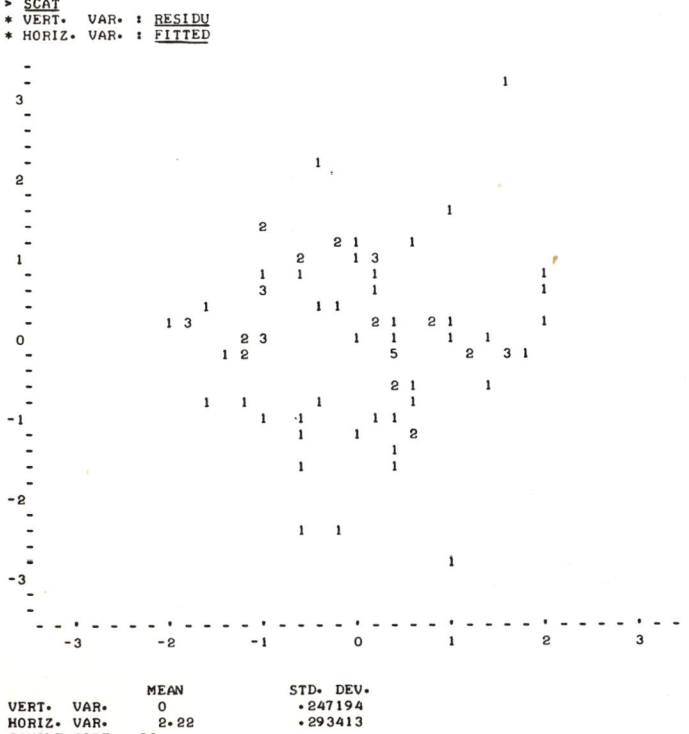

Figure A-10. Scatter Plot of 'RESIDU' vs. 'FITTED' for Unemployment Data

What does this plot tell about the adequacy of the regression model?

244 ☐ **CONVERSATIONAL STATISTICS**

```
> SCAT
* VERT.  VAR.  : RESIDU
* HORIZ. VAR.  : $/LF

   40-TH HOR. OBS. AN OUTLIER; PLOTTED AS 3.4
   66-TH HOR. OBS. AN OUTLIER; PLOTTED AS 3.4
 -
3                          1
 -
 -
 -
 -
2                          1
 -
 -                         1
 -                         2
 -                         3 1
1                          6
 -                         2  1
 -                         4        1           1
 -                         3
 -                         8  1  1  1
0                          9
 -                        *  2
 -                         4
 -                         2 1
-1                         3                 1
 -                         4
 -                         1
 -                         2
 -
-2
 -                         2
 -
 -                         1
-3
 -
   -.---.---.---.---.---.---.---.
     -3    -2    -1    0    1    2    3

                   MEAN        STD. DEV.
VERT.  VAR.         0            .247194
HORIZ. VAR.       55.7673       204.71
SAMPLE SIZE = 83
```

Figure A-11. Scatter Plot of 'RESIDU' vs. '$/LF' for Unemployment Data

What does this plot suggest about the nature of the statistical relationship between EDA funding and unemployment?

```
> CORR
HOW MANY VARIABLES ?5
VAR. #'S :?1,2,3,4,5
# DECIMALS = ?3

DSIG    1.000
FUND   -0.359   1.000
LU66    0.466   0.369   1.000
LU70    0.417   0.223   0.756   1.000
$/LF   -0.154   0.430   0.128  -0.039   1.000

> DELO
ROW = ?65
UPDATING MEAN, STD, ...
> REGR
UPDATING CORR MATRIX ...
* DEP. VAR. = 4
HOW MANY INDEP. VAR. ?2
INDEP. VAR.  1 = ?3
INDEP. VAR.  2 = ?5
ANALYZING RESIDUALS ...
WARNING: 37-TH RESIDUAL IS -2.94 S.D. UNITS FROM 0
WARNING: AUTO( 2) = +0.23
WARNING: AUTO( 3) = -0.16
WARNING: AUTO( 7) = -0.24
WARNING: AUTO( 8) = -0.16

> COEF

VARIABLE    B(STD.V)       B          STD.ERROR(B)     T

LU66         0.7752    6.3175E-01    5.9191E-02     10.673
$/LF        -0.1348   -.2342E-03     1.2617E-04     -1.857
CONSTANT               1.1750E+00    1.0033E-01     11.712

> SUMM

              MULTIPLE R    R-SQUARE
UNADJUSTED      0.7687       0.5909
  ADJUSTED      0.7619       0.5805
STD. ERROR OF RESIDUALS  =    .231648
```

What is the correlation coefficient between LU66 and LU70?

What was accomplished by this command and the analysis that follows? How does the conclusion modify, if at all, the previous analysis?

APPLICATIONS ☐ 245

The very rough analysis on the next two pages is intended to illuminate the possible causal interpretation of the regression model. What is the conclusion? Explain concisely.

```
> NAME
THESE ARE THE VARIABLES IN THE DATA MATRIX :
COLUMN     NAME

   1       DESIG.
   2       FUNDED
   3       UN66
   4       LGUN70
   5       FND/LF

> REGR
* DEP. VAR. = UN66
HOW MANY INDEP. VAR.   ?2
INDEP. VAR.  1 = ?1
INDEP. VAR.  2 = ?2
ANALYZING RESIDUALS ...
WARNING: 49-TH RESIDUAL IS +3.05 S.D. UNITS FROM 0
WARNING: 64-TH RESIDUAL IS +3.40 S.D. UNITS FROM 0
WARNING: AUTO( 3) = -0.18
WARNING: AUTO( 5) = -0.24
WARNING: AUTO( 8) = +0.11
WARNING: AUTO( 9) = -0.12

> COEF

VARIABLE       B(STD.V)        B         STD.ERROR(B)     T

DESIG.          0.6379      3.9073E+00    5.3656E-01     7.282
FUNDED          0.5626      3.2506E+00    5.0615E-01     6.422
CONSTANT                    3.9077E+00    3.1240E-01    12.509
```

The first two variables are the same as the first two before, though the names are slightly different. The third is 1966 percentage unemployment. The fourth and fifth are not used.

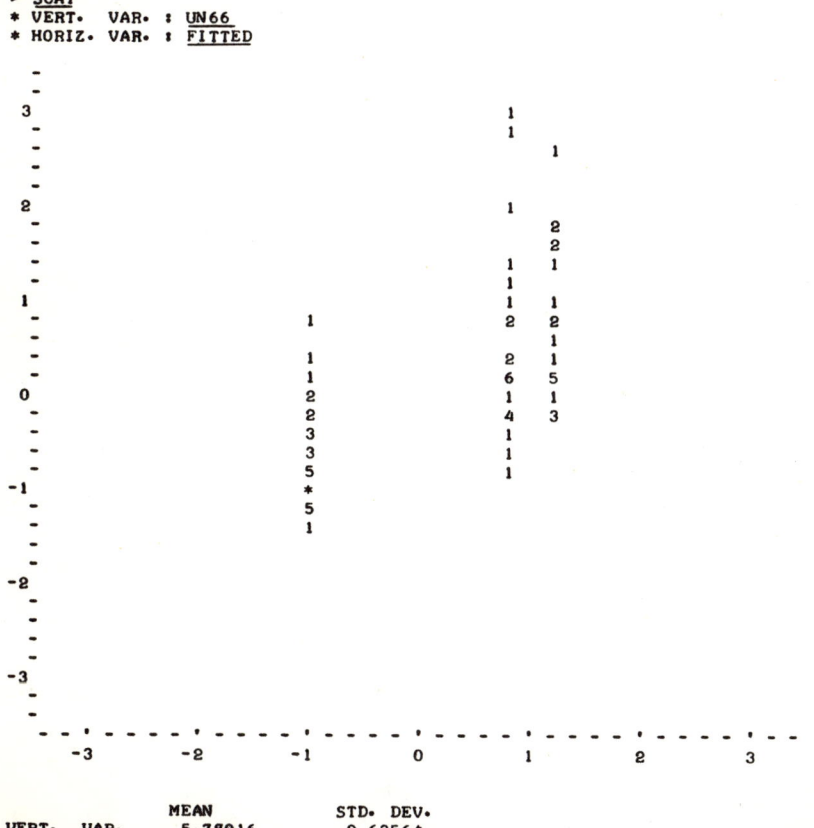

Figure A-12. Scatter Plot of 1966 Unemployment vs 'FITTED'

```
> NAME
THESE ARE THE VARIABLES IN THE DATA MATRIX :
COLUMN    NAME

   1      FXLU66  (product of 2 and 3)
   2      FUND
   3      LU66
   4      LU70
   5      $/LF

> CORR
HOW MANY VARIABLES ?5
VAR. #'S  ?1,2,3,4,5
# DECIMALS = ?3

FXLU66   1.000
FUND     0.981    1.000
LU66     0.439    0.369   1.000
LU70     0.296    0.223   0.756    1.000
$/LF     0.406    0.430   0.128   -0.039   1.000

> REGR
* DEP. VAR. = 4
HOW MANY INDEP. VAR. ?4
INDEP. VAR.  1 = ?1
INDEP. VAR.  2 = ?2
INDEP. VAR.  3 = ?3
INDEP. VAR.  4 = ?5
ANALYZING RESIDUALS ...
WARNING: 37-TH RESIDUAL IS -2.78 S.D. UNITS FROM 0
WARNING: 65-TH RESIDUAL IS +3.30 S.D. UNITS FROM 0
WARNING: AUTO( 2) = +0.22
WARNING: AUTO( 3) = -0.16
WARNING: AUTO( 4) = +0.14
WARNING: AUTO( 7) = -0.30

> COEF

VARIABLE   B(STD.V)       B           STD.ERROR(B)      T

FXLU66      0.5169     2.2167E-01     1.7446E-01      1.271
FUND       -0.4984    -.4171E+00      3.3274E-01     -1.253
LU66        0.7301     6.2972E-01     7.3396E-02      8.580
$/LF       -0.1282    -.2391E-03      1.4860E-04     -1.609
CONSTANT               1.1865E+00     1.1863E-01     10.001

> SUMM

              MULTIPLE R    R-SQUARE
UNADJUSTED      0.7741       0.5992
ADJUSTED        0.7607       0.5786
STD. ERROR OF RESIDUALS =     .247787

> BCOR
# DECIMALS = ?3

FXLU66    1.000
FUND     -0.977    1.000
LU66     -0.421    0.343    1.000
$/LF      0.079   -0.166   -0.000    1.000
CONST.    0.405   -0.357   -0.962    0.000
```

On this page I give summary output relative to a model alternative to that given at the beginning. On balance I preferred the earlier model. What do you see as the pros and cons of this choice? (NORM and SCAT for RESIDU versus FITTED were similar.) How might the policy conclusions of the two models be different?

APPLICATIONS 247

(12) IRON AND SULFUR

This application formed the basis of a final examination in the Fall of 1971. The examination was out-of-class.

The following information and data are real; only the source is not given, for reasons of company confidentiality. Your assignment on the exam, which is a take-home exam, is to draw what conclusions you can about the effects of coke moisture on quality at the blast furnace and basic oxygen furnace. You are free to use any references or computational facilities at your disposal. You are also free to ask questions of anyone, so long as you include acknowledgments and so long as the writeup is entirely your own.

Your writeup should be as concise as possible consistent with what you have to say. If you do any calculations or graphical work, include only what is necessary to support your written conclusions.

****

Iron produced in a blast furnace--referred to as hot metal--is the basic ingredient in the production of steel in the basic oxygen steelmaking furnace (BOF). Control of the sulfur content in the hot metal is important since it limits the range of steel specifications that the BOF can produce. By the nature of the BOF process, only a limited amount of sulfur can be removed. The BOF specification for hot metal calls for a maximum sulfur content of .035%

In the blast furnace operation, experience has shown that variation in the moisture of the charged coke causes variation in the sulfur content of the hot metal that is produced: higher moisture in blast furnace coke corresponds to higher sulfur in the hot metal. In turn, higher hot metal sulfur is associated with higher sulfur in basic oxygen steel. The blast furnace cannot correct for coke moisture variation because the process is continuous. However, the BOF operation is a batch process and some degree of control of sulfur in steel is possible based on the analysis of the hot metal prior to its use in the BOF.

The blast furnace department collected 39 daily observations of sulfur in basic oxygen steel, sulfur in hot metal, and coke moisture (Table A-13). The data were selected from days of normal operation and take into account the time lag between charging coke and casting hot metal.

If coke moisture affects sulfur in basic oxygen steel, it does so in two stages: (1) increasing coke moisture increases hot metal sulfur; and (2) increasing hot metal sulfur increases basic oxygen sulfur. Hot metal sulfur, however, does not account for all the basic oxygen steel sulfur. At very low hot metal sulfur, there may be more sulfur in the basic oxygen steel than there was in the hot metal from which it was made, although in theory there should not be more sulfur in the steel than in the hot metal. The most obvious explanation is poor operating practice that causes a mixture of blast furnace slag of high sulfur with the hot metal.

Table A-13. Observations of Sulfur and Other Variables in Daily Steel Production

OBSERVATION	BOF TURN DOWN SULFUR (PERCENT)	HOT METAL SULFUR (PERCENT)	COKE MOISTURE
1	0.021	0.037	0.148
2	0.025	0.039	0.114
3	0.026	0.045	0.165
4	0.028	0.039	0.184
5	0.025	0.033	0.134
6	0.026	0.036	0.155
7	0.028	0.024	0.116
8	0.021	0.024	0.151
9	0.024	0.033	0.155
10	0.027	0.028	0.129
11	0.022	0.022	0.122
12	0.032	0.036	0.153
13	0.037	0.043	0.166
14	0.035	0.044	0.178
15	0.035	0.042	0.169
16	0.035	0.050	0.174
17	0.028	0.040	0.162
18	0.025	0.027	0.150
19	0.021	0.018	0.147
20	0.025	0.024	0.158
21	0.029	0.031	0.147
22	0.031	0.030	0.150
23	0.027	0.025	0.142
24	0.041	0.040	0.144
25	0.029	0.034	0.142
26	0.025	0.023	0.114
27	0.030	0.036	0.151
28	0.038	0.063	0.154
29	0.034	0.038	0.129
30	0.034	0.056	0.127
31	0.027	0.034	0.123
32	0.029	0.028	0.099
33	0.028	0.032	0.169
34	0.025	0.027	0.122
35	0.037	0.041	0.160
36	0.027	0.029	0.149
37	0.026	0.027	0.135
38	0.024	0.020	0.161
39	0.029	0.026	0.110

(13) SALES AND QUOTES

This application formed the basis of an out-of-class examination in the Winter of 1972.

This problem is based on an Executive Program term paper written in 1953. The student who wrote the paper posed his problem as follows: "Study the sales and quotation figures of the XYZ Company for the purpose of ascertaining the relationship of the dollar volume of monthly quotations to the dollar volume of monthly sales. If relationships can be established, it is hoped that they can be used to predict sales for a few months ahead of current periods and rationalize sales budget figures."

The company manufactured several lines of engineered products: materials handling equipment, foundry equipment, machine tool equipment, railroad equipment, and chemical process equipment. The time interval extended from May 1, 1950 through April 30, 1953. The basic data are set forth below as I entered them in my data file. Month 1 is May, 1950, and month 36 is April, 1953. The first column of numbers after the code for month gives monthly sales quotations in thousands of dollars; the second column gives sales in the same units.

Table A-14. Data Matrix for XYZ Company

N= 36 K= 2

DATA MATRIX

1	6411	1451		19	8554	1314
2	7013	2095		20	7381	2621
3	6423	1380		21	8272	1764
4	7027	2811		22	8665	1357
5	4971	915		23	9151	1453
6	9339	2890		24	6406	1857
7	10846	2069		25	6678	1358
8	8736	2499		26	6624	1355
9	13660	2479		27	6901	2152
10	12134	2320		28	4648	1844
11	12198	3011		29	8407	1885
12	8748	2547		30	7219	2208
13	8632	2423		31	4641	916
14	7713	2434		32	5865	868
15	8697	2771		33	11499	851
16	6805	971		34	8568	1256
17	8880	1645		35	8891	1662
18	7973	3662		36	6855	2276

The student commented further: "For many years there has existed within the company a controversy between the sales and production departments over the reasons for periods of low sales volume. Production is critical of sales, saying that sales does not pursue an aggressive policy of developing new prospects. The sales people counter with the argument that the prices are too high. It is reasonable to expect that sales are dependent on the dollar volume of quotations. If there are no quotations, there will be no sales. An increase in quotations should bring an increase in sales if the pricing is correct. A plot of monthly quotations against monthly sales for a sufficient number of months might be expected to yield a curve of use in predicting sales, since there is usually a time lapse of varying magnitude between the date of a quotation and the date of the resulting sale."

Your main assignment is to develop the best predictive relationship you can, based solely on past data--sales and/or quotations that would be available at the time the prediction is to be made. For this purpose you may assume that sales and quotations for the current month are known at the time the next month is to be forecast. Actually, compute your forecast for the next month beyond the data--May, 1953.

Although you are asked only to develop a month-ahead forecasting scheme, it is possible to use it to predict a year ahead by making repeated monthly forecasts and treating these as actuals for the succeeding monthly forecast. I may have some time to discuss this in class. In any event, you may wish to try to apply the procedure to build up a forecast for the entire year from May, 1953, to April, 1954.

In writing your report, you should be careful to adhere to the standards of good business report writing. Make the exposition simple and clear, and as brief as possible consistent with the information you are conveying. Include summary tables or charts if they will help to focus and make vivid your main points. For readers who may wish to check the manner in which you arrived at your conclusions, prepare a statistical appendix (but do not clutter this up with every computer run that you made in the course of the study!) and label it carefully.

The main faults that I have noted in the past are these: rambling, disjointed exposition that is hard to follow and that loses the main point in a maze of details; padding the paper to make it look like you did a lot of work in preparing it; imitating the heavy, pedantic style and arcane jargon typical of much academic writing; including an undigested pile of computer output with little indication of the main conclusions. It is helpful, of course, to have thought through the analysis clearly, because no amount of literary skill can compensate for an inadequate statistical analysis.

(14) MARKET MODEL (IBM) AND M&M INDUSTRIES: SALARY AND JOB EVALUATION

Two applications are combined in the format of a final examination given in the Autumn of 1973. Instructions given in advance to students are shown below; the examination itself is reproduced on following pages.

The first application is based on an analysis of monthly return data on IBM common stock from January, 1961, through December, 1967. The background definitions are the same as for Application (1), Market Model, except that IBM takes the place of AT&T. The analysis will be an exploration of the adequacy of the market model for IBM and an examination of the inferences derived from the model. Queries about the analysis appear on the examination itself. The data are in a file $IBM, which is printed out below.

Table A-15. Contents of the File $IBM

ROW	IBM	NYSE	ROW	IBM	NYSE
1	0.07251	0.08232	43	-0.03145	0.02767
2	0.06250	0.05936	44	-0.04383	-0.00896
3	0.02963	0.05037	45	-0.00908	0.03699
4	0.02734	0.00841	46	-0.03780	0.01700
5	0.02752	0.04179	47	-0.01488	0.00072
6	-0.02661	-0.04242	48	-0.00727	-0.00691
7	0.02419	0.01095	49	0.09524	0.05871
8	0.06797	0.02104	50	0.01951	0.02784
9	0.03561	-0.02971	51	-0.00329	0.00528
10	0.09201	0.02133	52	0.06769	0.03592
11	-0.01174	0.04641	53	-0.01134	-0.00790
12	-0.00172	-0.00351	54	-0.04184	-0.07433
13	-0.06390	-0.00793	55	0.04585	0.02910
14	-0.00784	0.01554	56	0.04489	0.04509
15	-0.00791	-0.00594	57	0.02705	0.03083
16	-0.14782	-0.06830	58	0.04000	0.04739
17	-0.13381	-0.09800	59	-0.01220	0.03005
18	-0.13567	-0.08475	60	-0.04952	0.03267
19	0.14075	0.06369	61	-0.00601	0.04353
20	0.02584	0.02796	62	0.04133	0.01094
21	-0.10789	-0.06050	63	0.00194	-0.02187
22	-0.02122	-0.02192	64	0.08043	0.03374
23	0.15390	0.13730	65	-0.02202	-0.07238
24	-0.02133	-0.00906	66	-0.02964	-0.00483
25	0.08654	0.07816	67	-0.02782	-0.01266
26	-0.05428	-0.01508	68	-0.05620	-0.09305
27	0.05316	0.02095	69	-0.00936	-0.01432
28	0.10154	0.03902	70	0.04567	0.01269
29	0.03127	0.03242	71	0.13584	0.03825
30	-0.08591	-0.01583	72	-0.01197	0.01623
31	-0.00401	-0.00952	73	0.07537	0.14285
32	0.02589	0.05056	74	0.07910	0.02094
33	0.01630	-0.01841	75	0.04884	0.05201
34	0.09292	0.01633	76	0.10089	0.03648
35	-0.01518	-0.00682	77	-0.03523	-0.01788
36	0.04482	0.00752	78	0.06702	0.05164
37	0.06903	0.02008	79	0.02060	0.07092
38	0.05212	0.02699	80	-0.01359	0.00280
39	0.04438	0.03140	81	0.09705	0.03777
40	-0.04039	-0.00305	82	0.08254	-0.03594
41	0.05491	0.01164	83	0.03302	0.00669
42	-0.00625	0.01540	84	0.02451	0.05544

The second application presents an exploratory analysis of data on management ("exempt personnel") salaries and their job evaluations for M&M Industries. The data are given in the file $M&MIND, printed out in Table A-16.

The job evaluation works as follows. Each job is broken down into three components: (1) "Know-How" (KNOWHO); (2) "Problem-Solving" (PROBSO); and (3) "Accountability" (ACCTBL). Each of these components is evaluated by a consulting firm following detailed guidelines for the assignment of a numerical score. For example, "Know-How" is divided into three component dimensions: (a) Required managerial know-how, varying from limited performance in a narrow area to comprehensive integration and coordination; (b) Required technical know-how, varying from simple practical procedures to exceptional competence and unique mastery in scientific or other learned discipline; and (c) Required human relations skills, varying from ordinary courtesy in dealing with others to skill in selecting, developing, and motivating other people. The consulting firm rates the job on each of these three component dimensions and then, consulting a triple-entry table developed by the firm from past experience, gets a single numerical rating for "Know-How". Similar procedures are followed for the assignment of numerical ratings to "Problem-Solving" and "Accountability".

The point scores assigned to KNOWHO, PROBSO, and ACCTBL are added to obtain total points (TOTAL).

The resulting scores, together with the salaries for the positions, are evaluated in a number of ways, including job profiles showing the percentage of total points deriving from each of the three components and then scaled according to the degree to which action predominates or problem solving dominates. The most important output for any company, however, is a simple scatter plot showing salary versus total points for management personnel. Typically two or more straight lines are fitted, one for each range of total points (such as low and high) and possibly an overall line as well. This scatter plot is compared with plots from other local companies and possibly of composites of local companies or of national industry groups. The aim is both to draw recommendations about the general location of the company's plot by comparison with other companies, and to evaluate how salaries paid within a company relate to job content--essentially the idea of residuals from a regression, although formal statistical methods have not been extensively used. The exploratory analysis given on the examination is directed mainly toward this second objective, including such specific questions as these: (1) Can one single regression model serve for all ranges of total evaluation points? (2) How should the regression model or models be formulated? Are transformations of the variables desirable? Can the three components of total points (TOTAL) be exploited to improve the model? (3) How can the analysis be used to establish a range of guideline salaries for any given job--say, minimum, recommended, and maximum? (4) Can other useful applications be found within the company?

The data in Table A-16 constitute actual information for an industrial plant doing in excess of 18 million dollars in annual sales. The company, named here M&M Industries, is located in the Midwest more than fifty miles from any major metropolitan area. The positions are numbered here from 1 to 67, and I have generalized job titles that I am not free to disclose. However, the data do contain a variable Functional Classification (CLASS) which is coded as follows:

    0   Top Management
    2   Major Division 1
    3   Major Division 2
    4   Major Division 3
    5   Engineering
    7   Administration
   10   Sales

Partly on *a priori* grounds and partly from results of my first preliminary analysis, I eliminated the three jobs in Top Management. The analysis will therefore show only 64 positions.

Table A-16. Contents of the File $M&MIND

ROW	CLASS	KNOWHO	PROBSO	ACCTBL	TOTAL	ROW	SALARY
1	0	800	608	1056	2464	1	52000
2	2	528	304	460	1292	2	25740
3	3	460	264	460	1184	3	25740
4	5	528	304	304	1136	4	29172
5	4	460	264	400	1124	5	20000
6	0	460	264	400	1124	6	16536
7	0	528	304	264	1096	7	20000
8	7	460	230	264	954	8	18000
9	10	400	200	350	950	9	23140
10	7	400	175	230	805	10	16016
11	7	400	200	200	800	11	16016
12	5	400	175	200	775	12	21840
13	5	304	115	175	594	13	21580
14	2	264	100	175	539	14	15860
15	3	264	100	175	539	15	16432
16	10	230	100	132	462	16	14040
17	10	230	100	132	462	17	12610
18	7	230	87	132	449	18	15002
19	7	230	76	115	421	19	14001
20	5	230	76	115	421	20	16900
21	5	230	87	100	417	21	13000
22	5	230	87	100	417	22	13780
23	7	200	87	100	387	23	12000
24	7	200	76	100	376	24	11960
25	7	200	76	100	376	25	12012
26	7	200	76	87	363	26	12300
27	5	200	76	87	363	27	11960
28	7	200	66	87	353	28	11700
29	7	175	66	100	341	29	11440
30	2	175	57	100	332	30	12220
31	3	175	57	100	332	31	13260
32	7	175	57	100	332	32	9880
33	2	175	57	100	332	33	12480
34	3	175	57	100	332	34	13000
35	2	175	57	100	332	35	13260
36	3	175	57	100	332	36	12480
37	4	175	57	87	319	37	12480
38	7	175	57	87	319	38	11440
39	2	175	57	87	319	39	12064
40	3	175	57	87	319	40	11180
41	2	175	57	87	319	41	9100
42	3	175	57	87	319	42	9620
43	5	175	66	76	317	43	9880
44	5	175	66	76	317	44	10200
45	7	175	57	76	308	45	10140
46	7	175	57	76	308	46	11700
47	5	175	66	66	307	47	10000
48	7	152	50	87	289	48	10920
49	7	152	50	76	278	49	9100
50	3	152	50	76	278	50	11700
51	2	152	50	76	278	51	9880
52	3	152	50	76	278	52	11700
53	5	152	50	66	268	53	9360
54	5	152	43	66	261	54	10660
55	2	152	43	66	261	55	9984
56	2	152	43	66	261	56	10660
57	3	152	43	66	261	57	10920
58	3	152	43	66	261	58	10920
59	2	152	43	66	261	59	10920
60	3	152	43	66	261	60	10660
61	3	152	43	66	261	61	10660
62	7	152	43	66	261	62	8320
63	5	152	43	66	261	63	9360
64	2	152	43	66	261	64	10920
65	3	152	43	66	261	65	10920
66	4	152	43	66	261	66	10660
67	7	152	43	57	252	67	9880

My analysis is only exploratory, and you will see that the model I reached was not entirely satisfactory from a statistical point of view. Further, there are subtle questions, about which statisticians are not agreed, as to how cross sectional data of this kind can be given a probabilistic interpretation (note that these data do *not* comprise a *sample* of management personnel in the company; all are included). Nonetheless I have tried to point the analysis toward the objectives listed at the top of this page. The examination will pose questions about my analysis.

If your own analysis of either set of data prior to the examination reveals important points missed by me, you may report these on the examination itself for possible extra credit.

APPLICATIONS 253

UNIVERSITY OF CHICAGO
Graduate School of Business

Business 320

Mr. Roberts
Fall, 1973

Final Examination

The examination consists of two problems. In each you are given an annotated printout of an analysis of a set of data for which the background and file reference have been previously furnished to you. Various questions are typed on the printout. You are to write your answers to these, at the point raised, upon the examination itself. If space is not sufficient for your answer, write on the blank facing page, *but make all answers as concise as possible consistent with the ideas that you wish to convey*. Repetition of the same idea in different language will be counted slightly against you and will take time away from your consideration of other questions.

The examination is open-book: you may refer to any inanimate source. The previous work that you have done with these data may be referred to in your answers if you feel that this will be illuminating. Please give numerical details so that I can evaluate your contribution.

Do not be afraid to use your common sense!

PRINT YOUR NAME HERE: _____

Question 1. Market Model (IBM)

```
> NAME

DATA MATRIX HAS VARIABLE(S) IN FIRST  2    COLUMN(S)
THEY ARE :

COLUMN    NAME
   1      IBM
   2      NYSE
> MEAN

VARIABLE     MEAN          STD. DEV.

  IBM      1.59429E-02     5.96850E-02
  NYSE     1.33731E-02     4.26426E-02
```

What is IBM's mean return per month? What does this imply approximately for a yearly return?

If overall risk can be measured by the standard deviation of returns, which is riskier, IBM or NYSE? Explain briefly.

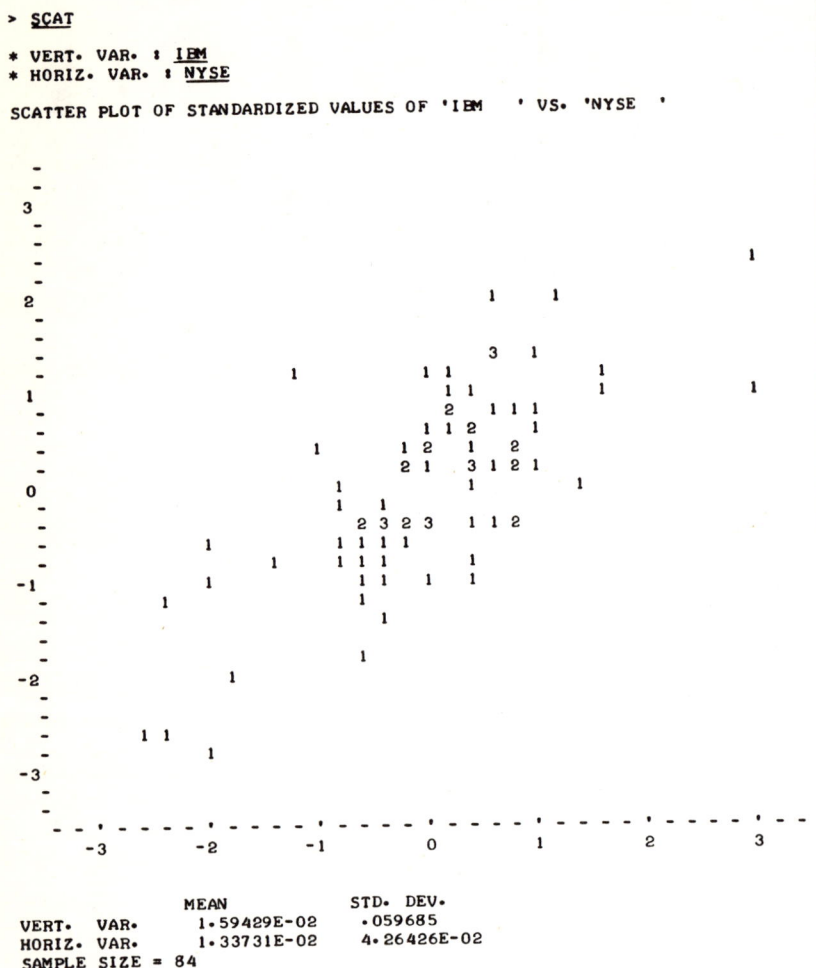

Figure A-13. Scatter Plot of Standardized Values of 'IBM' vs. 'NYSE'

Summarize briefly the extent to which, judging from information shown in the scatter plot above, the simple regression model looks promising. Place an "X" on the plot to show the intercept of the line (in standardized units) that would be fitted by least squares.

```
> REGR
UPDATING CORR. MATRIX...
* DEP. VAR. = IBM
HOW MANY INDEP. VAR. ?1
INDEP. VAR.  1 = ?NYSE
ANALYZING RESIDUALS ...
WARNING: RESIDUAL IN ROW  82 IS +2.75 S.D. UNITS FROM 0

> PLTS

* VARIABLE : RESIDU

ALL ROWS ?Y

SEQUENCE PLOT OF STANDARDIZED VALUES OF 'RESIDU'
```

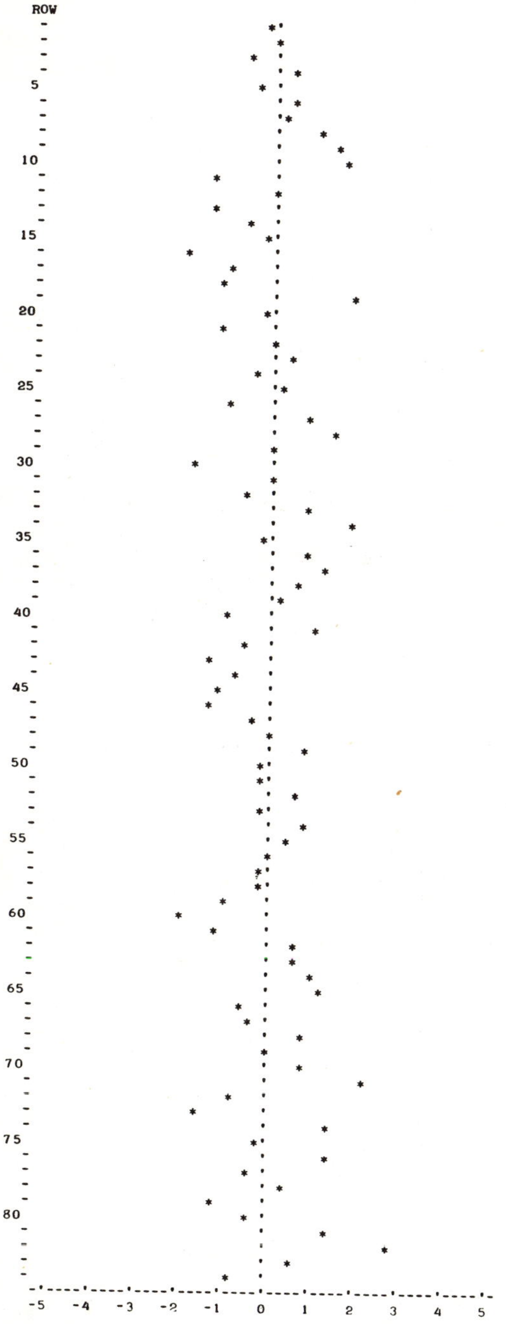

Figure A-14. Sequence Plot of Standardized Values of Residuals for IBM vs NYSE Data

What do you learn from the sequence plot in Figure A-14 about the adequacy of the regression model? What, if anything, is added by the histogram in Figure A-15? Explain what information is given by the STD. DEV. of residuals of 0.0421; in particular, what does this tell you about the appropriateness of the regression model?

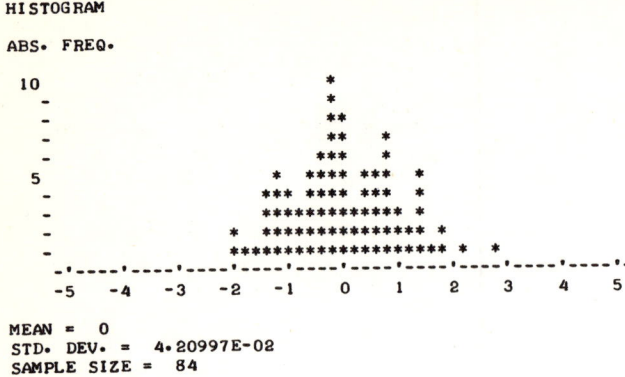

Figure A-15. Histogram for Residuals of IBM vs NYSE Data

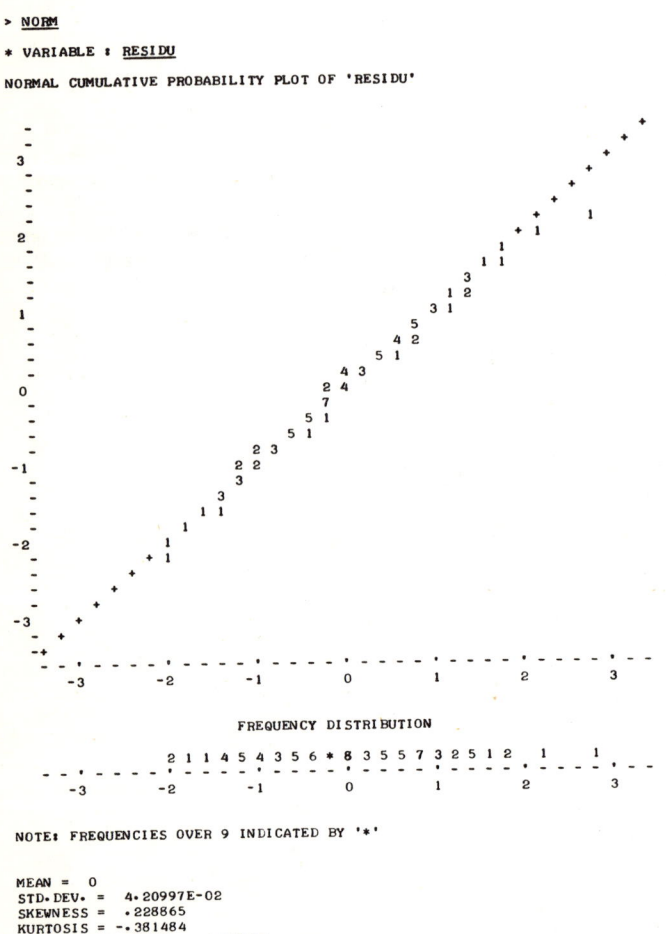

Figure A-16. Normal Probability Plot of Residuals for IBM vs. NYSE Data

Discuss the information on model adequacy provided by the normal probability plot and accompanying information (Figure A-16). After looking at this plot, how do you evaluate the warning about an extreme residual given by IDA at the top of Figure A-14?

```
> RUNS

* VARIABLE : RESIDU

    OBSERVED NUMBER OF RUNS= 41
    EXPECTED NUMBER OF RUNS= 42.619
   STANDARD DEVIATION OF RUNS= 4.51307
       (OBS.-EXP.)/(STD.DEV.)=-.358747

> AUTO

* VARIABLE : RESIDU

* MAX ORDER : 20

   ORDER   AUTOCORR.
     1      +0.194
     2      -0.034
     3      -0.064
     4      -0.083
     5      +0.011
     6      +0.019
     7      -0.088
     8      -0.131
     9      -0.151
    10      -0.063
    11      -0.043
    12      -0.021
    13      +0.132
    14      +0.119
    15      +0.035
    16      -0.008
    17      +0.031
    18      +0.047
    19      +0.080
    20      -0.082

   S.E. OF EACH COEF. GIVEN RANDOM MODEL=0.109

   BOX-PIERCE STATISTIC= 12.898
   EXPECTATION GIVEN RANDOM MODEL= 20
       (OBS.-EXP.)/(STD.DEV.)=-1.1851
```

What do the two diagnostic checks above tell you about the adequacy of the market model in this application?

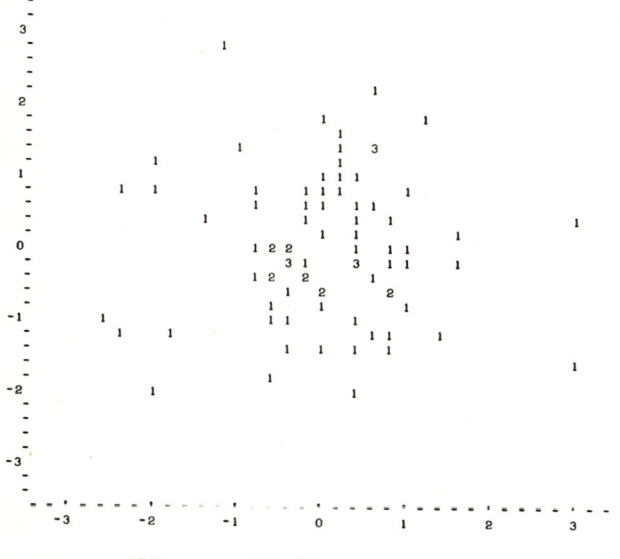

Figure A-17. Scatter Plot of Standardized Values of 'RESIDU' vs. 'FITTED' for IBM vs. NYSE Data

What do you learn about model adequacy from the scatter plot of RESIDUAL versus FITTED (Figure A-17)? Explain briefly the relationship of this scatter plot to the one in Figure A-13.

> RVSF

```
R !                        !
E !     1   1              !
S !  1  1   9 1            !
I !  2  1 * 6 3 1          !
D !  1    * * 1            !
U !  1  2  2 6 1   1       !
A !           1            !
L !                        !
       F I T T E D   Y
```

What, if anything, is added by RVSF above?

> SUMM

```
              MULTIPLE R   R-SQUARE
UNADJUSTED      0.7131      0.5085
ADJUSTED        0.7088      0.5025
STD. DEV. OF RESIDUALS =    4.20997E-02
```

What is the interpretation of the correlation coefficient 0.7131? Of the standard deviation of residuals 4.20997E-02? How do these numbers illuminate the "riskiness" of IBM?

> COEF

```
VARIABLE   B(STD.V)       B         STD.ERROR(B)     T

NYSE       0.7131     9.9804E-01    1.0837E-01    9.210
CONSTANT              2.5961E-03    4.8166E-03    0.539
```

Write the equation (with numerical coefficients) of the least-squares regression line $\hat{y}$.

What is the interpretation of the regression coefficient 0.99804? What additional information is provided by the standard error 0.10837 and the T value 9.210?

Same question as above except for the constant or intercept, 0.0025961, its standard error 0.0048166, and T value 0.539. Did IBM outperform NYSE during the period of the study? What, roughly, is the probability that $\tilde{\beta}_0$ was positive? (You may be aided in this latter by the GAUS execution below.)

> GAUS

```
* MEAN     = ? 2.5961E-03
  STD.DEV. = ? 4.8166E-03

* LOWER LIMIT L = ? -100
  UPPER LIMIT U = ?  0

PR(L<X<U) =   0.29495
```

> SEPR
```
USE DATA MATRIX ? N
HOW MANY Y'S ? 2

                   Y         S.E.(FITTED)    S.E.PRED.(Y)

X'S = ? .01
               1.2576E-02     4.6080E-03      4.2351E-02

X'S = ? -.01
              -.7384E-02      5.2455E-03      4.2425E-02
```

Given the information that the return on NYSE is +0.01, what is your point estimate of the return on IBM? What is the standard error?

Given that the return on NYSE is larger by 0.01 than its previous return, but no information about IBM's previous return, what is your point estimate of the increase in IBM's return associated with the increase in return on NYSE? (HINT: Don't look at SEPR for this.)

Same two questions as above, but for the number -0.01 instead of +0.01.

Summarize your evaluation of the adequacy of the market model in this application, including references to any work that you may have done that is not included above. What have you learned about the behavior of returns on IBM common stock during this period?

Assuming that the market model is adequate in this application, what is it good for in practice?

260 □ CONVERSATIONAL STATISTICS

PRINT YOUR NAME HERE: _____

Question 2. M&M Industries: Salary and Job Evaluation

```
> ENTER
* MODE = F
NAME IS ?$M&MIND
* N,K GIVEN IN FILE? Y
DATA MATRIX HAS 67  OBSERVATIONS ON 6   VARIABLE(S)
COMPUTING...
            WANT TO SUPPLY NAMES ?Y
VAR.  1 = ?CLASS
VAR.  2 = ?KNOWHO
VAR.  3 = ?PROBSO
VAR.  4 = ?ACCTBL
VAR.  5 = ?TOTAL
VAR.  6 = ?SALARY

> DELO
ROW = ?1
UPDATING MEAN, STD, ...
> DELB
I1, I2 = ?6,7
UPDATING MEAN, STD, ...

> CORR
UPDATING CORR. MATRIX...

HOW MANY VARIABLES ?4
COL. #'S: ?2,3,4,5
* # DECIMALS = 3

           KNOWHO  PROBSO  ACCTBL  TOTAL
KNOWHO     1.000
PROBSO     0.989   1.000
ACCTBL     0.935   0.951   1.000
TOTAL      0.989   0.993   0.977   1.000
```

What was accomplished by the two commands immediately to the left? Why?

What does the above execution of CORR tell you about the system of job evaluation?

```
    > POWE
    * COL# TO PLACE VAR. : 3
    * COLUMN TO BE TRANSFORMED :5
    * P = -1
    UPDATING MEAN, STD, ...
    GIVE NAME OF NEW VARIABLE
    VAR.  3 ?TOT↑-1

    > MULV
    * COL# TO PLACE VAR. : 2
    I,J = ?2,3
    UPDATING MEAN, STD, ...
    GIVE NAME OF NEW VARIABLE
    VAR.  2 ?RELKNO

    > MULV
    * COL# TO PLACE VAR. : 4
    I,J = ?3,4
    UPDATING MEAN, STD, ...
    GIVE NAME OF NEW VARIABLE
    VAR.  4 ?RELACC

    > CORR
    UPDATING CORR. MATRIX...

    HOW MANY VARIABLES ?3
    COL. #'S: ?2,4,5
    * # DECIMALS = 3

             RELKNO  RELACC  TOTAL
    RELKNO    1.000
    RELACC   -0.845   1.000
    TOTAL    -0.887   0.619   1.000

    > NAME
    DATA MATRIX HAS VARIABLE(S) IN FIRST  6   COLUMN(S)
    THEY ARE :

    COLUMN    NAME
       1      CLASS
       2      RELKNO
       3      TOT↑-1
       4      RELACC
       5      TOTAL
       6      SALARY
```

What is RELKNO in terms of original variables?

What is RELACC in terms of original variables?

What does the CORR command tell you about the new variables that have been created?

Have I somehow lost the information contained in PROBSO by these transformations? Explain. (Use common sense and/or elementary mathematics.)

In looking ahead to what might be required of an analysis, which would you prefer to work with, the present data matrix or the one with which we started? Explain your reasoning.

```
> SCAT
* VERT. VAR. : SALARY
* HORIZ. VAR. : TOTAL
SCATTER PLOT OF STANDARDIZED VALUES OF 'SALARY' VS. 'TOTAL'
  VERT. OBS. IN ROW   4 AN OUTLIER; PLOTTED AS 3.4
```

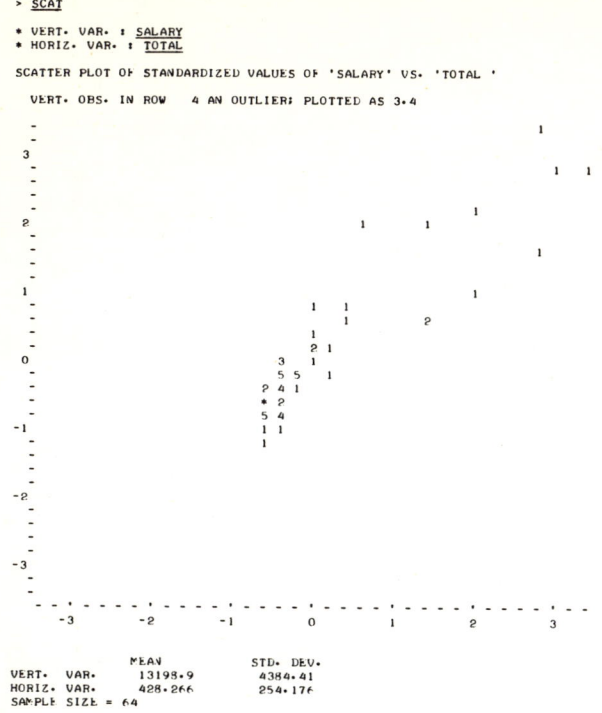

```
             MEAN          STD. DEV.
VERT. VAR.   13198.9       4384.41
HORIZ. VAR.  428.266       254.176
SAMPLE SIZE = 64
```

Figure A-18. Scatter Plot of Standardized Values for 'SALARY' vs. 'TOTAL' for M&M Data

Summarize the result of the interocular traumatic test as applied to the above plot.

```
> PRTS
1ST ROW, LAST ROW = ?4,4

* CONSECUTIVE COL(S). ? NO
HOW MANY COLUMNS ?1
COL. #'S :?6

   ROW      SALARY

*   4  *  29171.99967


> LOGE

* COL# TO PLACE VAR. : 5
* COLUMN TO BE TRANSFORMED :5
UPDATING MEAN, STD,...
GIVE NAME OF NEW VARIABLE
VAR.  5 ?LOGTOT
> LOGE

* COL# TO PLACE VAR. : 6
* COLUMN TO BE TRANSFORMED :6
UPDATING MEAN, STD,...
GIVE NAME OF NEW VARIABLE
VAR.  6 ?LOGSAL

> NAME

DATA MATRIX HAS VARIABLE(S) IN FIRST  6   COLUMN(S)
THEY ARE :

COLUMN    NAME
   1      CLASS
   2      RELKNO
   3      TOT!-1
   4      RELACC
   5      LOGTOT
   6      LOGSAL


> CORR
UPDATING CORR. MATRIX...

HOW MANY VARIABLES ?3
COL. #'S: ?2,4,5
* # DECIMALS = 3

          RELKNO   RELACC   LOGTOT
RELKNO    1.000
RELACC   -0.845    1.000
LOGTOT   -0.901    0.610    1.000
```

Circle this observation in Figure A-18. Show by an X where it should actually be plotted.

What suggested these two logarithmic transformations? Explain.

```
>  SCAT

*  VERT. VAR. :  LOGSAL
*  HORIZ. VAR. : LOGTOT

SCATTER PLOT OF STANDARDIZED VALUES OF 'LOGSAL' VS. 'LOGTOT'
```

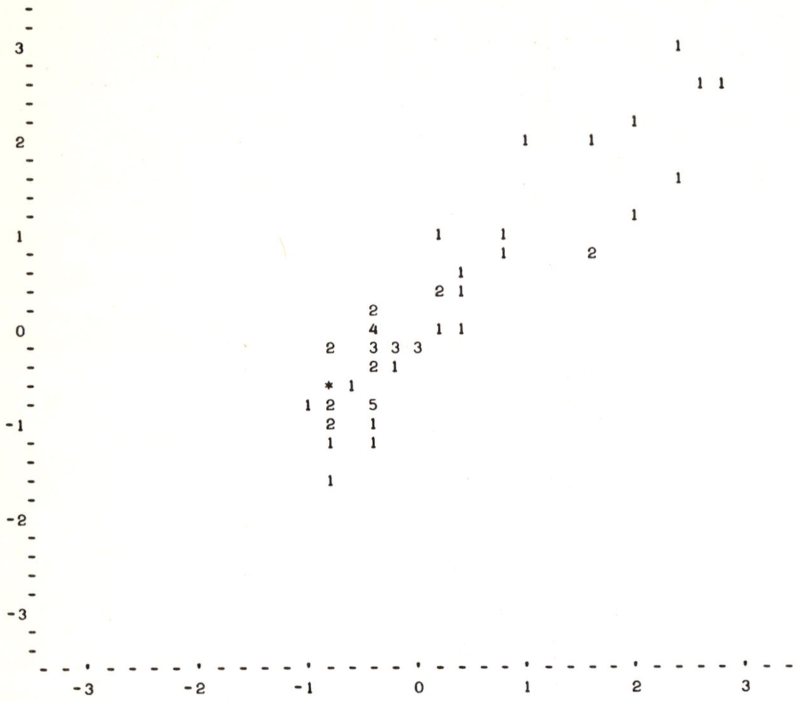

```
     -
   3 -                                              1
     -                                                1 1
     -                                             1
   2 -                               1   1
     -                                                1
     -
   1 -                         1   1
     -                             1       2
     -                           1
     -                         2 1
     -                       2
   0 -                       4   1 1
     -                   2   3 3 3
     -                       2 1
     -                   * 1
     -                     1 2 5
  -1 -                       2 1
     -                       1 1
     -
     -                       1
  -2 -
     -
     -
     -
  -3 -
     -
     -------|-------|-------|-------|-------|-------|-------
          -3      -2      -1       0       1       2       3
```

```
                  MEAN            STD. DEV.
VERT.  VAR.     9.44564            .276618
HORIZ. VAR.     5.94398            .439806
SAMPLE SIZE = 64
```

Figure A-19. Scatter Plot of Logarithmic Values for 'SALARY' vs. 'TOTAL' for M&M Data

```
>  REGR
*  DEP. VAR. = LOGSAL
HOW MANY INDEP. VAR. ? 1
INDEP. VAR.   1 = ? LOGO-TOT
ANALYZING RESIDUALS ...

>  COEF

VARIABLE   B(STD.V)        B         STD.ERROR(B)     T

LOGTOT     0.9195      5.7833E-01    3.1398E-02    18.419
CONSTANT               6.0081E+00    1.8713E-01    32.106

>  SUMM

            MULTIPLE R    R-SQUARE
UNADJUSTED    0.9195       0.8455
  ADJUSTED    0.9181       0.8430
STD. DEV. OF RESIDUALS =      .109607
```

Sketch, with reasonable care, the least squares regression line on Figure A-19.

Which scatter plot, Figure A-18 or Figure A-19, appears closer to what you would expect to see in a sample if the standard linear regression model were appropriate? Why?

In your preferred scatter plot, what model inadequacies appear to remain?

Do you feel that differently-sloped lines are appropriate for different ranges of the independent variable? Explain briefly.

```
> AUTO

* VARIABLE : RESIDU

* MAX ORDER : 16

ORDER   AUTOCORR.

  1      +0.148
  2      +0.036
  3      -0.044
  4      -0.002
  5      -0.242
  6      +0.024
  7      +0.072
  8      -0.068
  9      -0.039
 10      -0.177
 11      -0.095
 12      -0.199
 13      -0.003
 14      -0.013
 15      +0.099
 16      -0.022

S.E. OF EACH COEF. GIVEN RANDOM MODEL=0.125

BOX-PIERCE STATISTIC= 11.9103
EXPECTATION GIVEN RANDOM MODEL= 16
       (OBS.-EXP.)/(STD.DEV.)=-.677292

> REGR
* DEP. VAR. = LOGSAL
HOW MANY INDEP. VAR. ?3
INDEP. VAR.  1 = ?RELKNO
INDEP. VAR.  2 = ?RELACC
INDEP. VAR.  3 = ?LOGTOT
ANALYZING RESIDUALS ...
WARNING: RESIDUAL IN ROW    4 IS +2.64 S.D. UNITS FROM 0
WARNING: AUTO( 5) = -0.26

> COEF

VARIABLE    B(STD.V)     B         STD.ERROR(B)    T

  RELKNO     0.5871    3.5888E+00   1.2825E+00    2.798
  RELACC     0.3717    3.0625E+00   9.4786E-01    3.231
  LOGTOT     1.2215    7.6825E-01   8.9114E-02    8.621
  CONSTANT             2.1171E+00   1.4024E+00    1.510

> SUMM
              MULTIPLE R   R-SQUARE
  UNADJUSTED    0.9319      0.8684
  ADJUSTED      0.9283      0.8618
  STD. DEV. OF RESIDUALS =    .10283
```

Given that these data are cross sectional, did the execution of AUTO make any sense? If yes, explain what is learned. If no, explain why not.

Judging solely from the information that you have at this stage, which regression model of the two thus far tried seems more promising? Why?

264 ☐ **CONVERSATIONAL STATISTICS**

Indicate how you would calculate a point prediction of LOGSAL if you were given values of the three independent variables of the most recent regression, assuming: (1) that you could use IDA; and (2) that you had to make a calculation by hand or with a desk calculator. For the second case, indicate the actual numerical values that you would enter wherever these are available to you now.

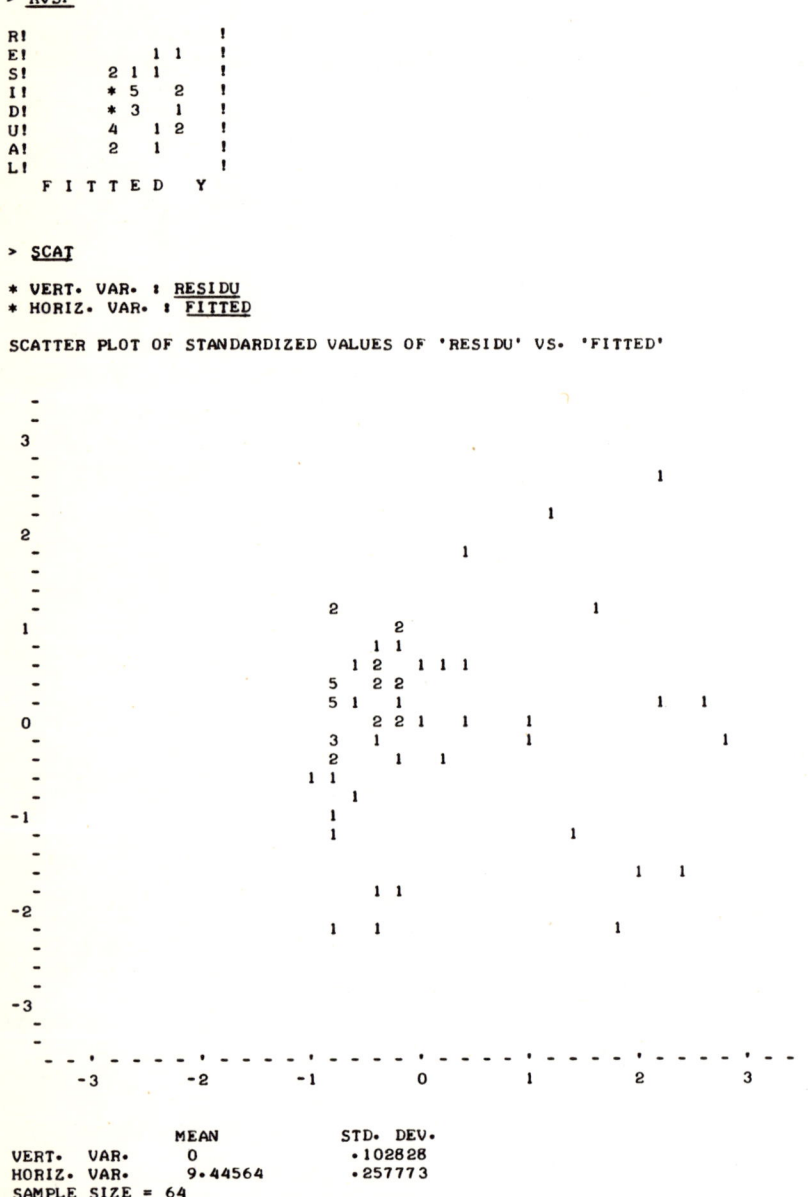

Figure A-20. Scatter Plot of Standardized Values of 'RESIDU' vs. 'FITTED' for Logarithmic Transformation of M&M Data

Draw in the least squares line on the plot above.

Draw in the least squares line on the mini-plot on the left.

Summarize the information on model adequacy given by the scatter plot in Figure A-20 and the mini-plot above it.

Suppose, as suggested in the background discussion, that you were to use the present model as a basis for establishing salary guidelines, given the variables of job evaluation, or, alternatively, to judge whether a given salary is or is not in line with the job evaluation. How would you go about it? What problem would be posed by the one conspicuous symptom of model inadequacy that has turned up?

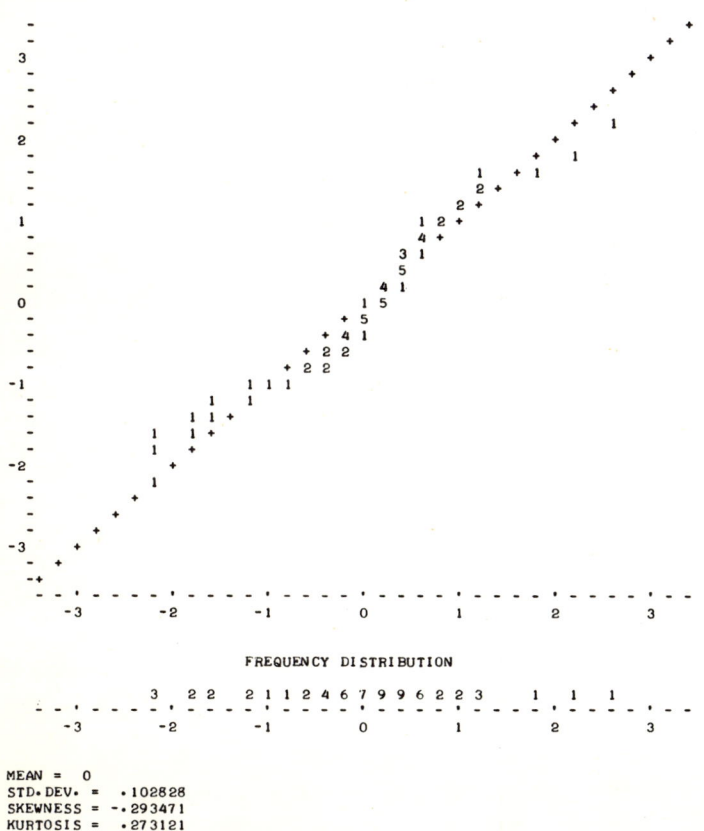

Figure A-21. Normal Probability Plot of Residuals for Logarithmic Transformation for M&M Data

266 □ CONVERSATIONAL STATISTICS

```
> HIST

* VARIABLE : RESIDU

MIN. OBS. =-.234215    MAX. OBS. = .271477
MEAN = 0    STD. DEV. = .102828
SAMPLE SIZE= 64

* MIDPOINT: 0
* WIDTH OF INTERVAL: .02

HISTOGRAM

ABS. FREQ.
                              **
    -                         **
    -                        ***
    -                       *****
  5                         *****
    -                       *****
    -              *      *******   *
    -              *  * * **********
    -            * ***** *********** * * *
      -.'----'----'----'----'----'----'----'----'----'----'-
         -400E-03  -200E-03  0000E+00  2000E-04  4000E-04
                              RESIDU

MEAN = 0
STD. DEV. = .102828
SAMPLE SIZE = 64

WANT CDF ALSO? N
```

Figure A-22. Histogram of Residuals for M&M Data

Summarize the information about model adequacy contained in Figures A-21 and A-22. Do you see any possible conflict between the plots and the numerical information? Explain.

Would you expect normality of disturbances from a regression model if the constant scatter specification were not met? Explain briefly.

```
> SCAT

* VERT. VAR. : RESIDU
* HORIZ. VAR. : CLASS

SCATTER PLOT OF STANDARDIZED VALUES OF 'RESIDU' VS. 'CLASS '

    -
    -
  3 -
    -
    -                          1
    -
    -                          1
  2 -
    -                          1
    -
    -                     2    1
  1 -              1      1
    -                     1  1
    -                        1       4          1
    -              4    4    1
    -              2    3  1 1       1          1
  0 -                     1   1      5
    -                     2  1 2     1
    -                     2          1          1
    -                          2
    -                                1
 -1 -                          1
    -                                2
    -
    -              1                 1
    -                     1          1
 -2 -
    -              1                 2
    -
    -
 -3 -
    -
    -
      - -.'----'----'----'----'----'----'----'----'----'- -
         -3      -2       -1       0        1       2       3

                    MEAN           STD. DEV.
VERT. VAR.          0              .102828
HORIZ. VAR.         4.78125        2.23585
SAMPLE SIZE = 64
```

What interesting practical conclusion about the salary structure at M&M Industries is suggested by Figure A-23. HINT: figure out what each column of observations refers to.

Figure A-23. Scatter Plot of Residuals vs Class for M&M Data

APPENDIX B: COMPUTER NOTES

This book makes intensive use of a system of interactive computer programs called IDA (INTERACTIVE DATA ANALYSIS), which has been implemented on the Hewlett-Packard 2000F time-sharing computer at the Graduate School of Business, University of Chicago, and at certain other business schools that have computers of the HP 2000 series. Although the book can be used without access to the computer programs, the reader can enhance his grasp of the subject matter and gain substantial experience in statistical analysis if he can use IDA or a similar computational package.

The use of IDA requires no previous experience in programming. A few simple procedures and concepts are all that you will need, and most of these are introduced in this appendix, which has been prepared for students at the University of Chicago but which would apply with minor modifications at other HP installations. You should study this material carefully during your study of the early part of the book. Then you will be prepared for the additional instructions in the use of IDA that are woven into the text itself.

There is one caution, however, for those without previous experience with computers: a computer is *very* literal-minded. If, for example, you are told to type "RUN-9998" and you type instead "RUN 9998", the result may be very different. If, therefore, you obtain a result different from the one sought, check back carefully to see if you have followed directions precisely. Assume, contrary to your instinctive feeling, that *you* have erred. You are entitled to question the computer only *after* you have made this careful check; and if you do, be sure to save the exact sequence of operations, as recorded in the printout, that led to the questionable result. It is hard for someone to help you without a detailed record of your problem.

Some of the instructions that follow have been abridged from the Hewlett-Packard manual *2000F: A Guide to Time-Shared Basic*, (HP 02000-90044). If you are interested in capabilities of the computer beyond the use of IDA, you will find this manual helpful for additional study.

The HP 2000F time-shared computer system consists of a central computer, a system of peripheral devices at the computer site, and a number of independent terminals by which users communicate with the computer. Some of these terminals, also called ports, may be connected directly to the computer through a multiplexer. In addition there are ports that can be accessed by telephone from remote locations; a suitable terminal and acoustical coupler are needed.

Since the system has 32 ports, up to 32 terminals can be supported simultaneously. The system is designed so that no user should encounter more than a few seconds delay between entering a command and receiving a response from the system, even when all

32 ports are in use. Occasionally, however, when many users are simultaneously demanding heavy mathematical computations, the lag in response may be more substantial.

If you are using a teletype, here is how you make a connection with the computer:

1. Turn the control knob to LINE.

2. Turn on coupler power.

3. If coupler has a duplex switch, be sure that it is set to FULL or FULL/UP.

4. If coupler has a line switch, set it to ON-LINE.

5. Call the computer number.

6. When the computer answers with a high pitched tone, place the handset in the coupler. (Be sure to check that the handset is inserted in the correct position, which should be marked on the coupler.)

If, however, the teletype is one of the hard-wired terminals, you simply turn the teletype control knob to the LINE position.

If you are using a GE Terminet or other 30 character-per-second terminal, the telephone connection is made similarly to that described above for the teletype except that the telephone number is different.

If you are using a terminal with a half-duplex coupler, see the special instruction in the HP BASIC Manual. The IBM 2741, also requires a different number and you should consult the references in the manual about this terminal.

Logging In

You need an identification code and password to log in. The ID code is a single letter followed by a three digit number. The password consists of one to six regular or control characters. You type a control character by pressing the letter in question and the *ctrl* key simultaneously. The control character is transmitted to the computer, but it does not print out. For example, if the password were $SE^CC^CR^CE^CT$, it would be printed out as ST. The use of control characters in the password serves to keep the password a secret.

As an example, suppose that your ID code were V179 and your password $SE^CC^CR^CE^CT$. Then you would log in by typing HELLO-V179, $SE^CC^CR^CE^CT$. This would actually print out as HELLO-V179,ST. (Note the hyphen and the comma: both are essential!) Then type a carriage return by pressing the *return* key. If you have logged in correctly, the system will cause the terminal to execute one or more linefeeds and possibly to print out a message of interest to system users. After the final linefeed, you are ready to go.

If you are using a GE Terminet or other 30 cps terminal rather than a teletype, you should in addition type a comma followed by "3" (or other digit as explained in the manual) before pressing the *return* key.

HELLO-V179, SE^{C}C^{C}R^{C}E^{C}T, 3 (which prints HELLO-V179,ST,3).

If you make a mistake in logging in, the system will scold you with an error message, and you must try again.

Correction of Typing Mistakes

By typing upper-case "O" (reverse arrow ←) on most terminals, you delete the immediately preceding character; one such character is deleted for each reverse arrow typed. For the GE Terminet, this function is accomplished by pressing the underline key "_".

To delete an entire line, press *control-X*, that is, X^C.

System Commands

After you have successfully logged in, you may wish to issue certain commands to the system to tell the computer what you wish it to do. To put IDA at your disposal, you use the system command GET-, followed by $IDA. Thus: GET-$IDA followed by *return* (that is, pressing the carriage return). Then IDA is at your service. To put IDA to work (to cause it to execute), you then type the system command RUN followed by *return*.

Once you have typed GET-$IDA, *return*, RUN, *return*, you are under the control of IDA rather than the HP system. HP system commands are then unacceptable and IDA commands must be used. Certain other HP system commands will be explained later, but now we introduce IDA by means of an annotated sample session at the terminal.

Sample Session

The following sample session is all you need to get started on IDA. It is recommended that you duplicate it yourself to be sure that you understand the mechanics.

Everything typed into the terminal by the user is underlined. To make room for some typewritten explanations, there are occasional spaces in the printout at points where they would not ordinarily occur.

If some of the explanations given by IDA use terminology that you are not familiar with, you will find that it is explained in the main text. You may later find it useful, however, to glance back at this sample session.

The user logged in as shown below. Remember after typing each line, the carriage return key was pressed.

CONVERSATIONAL STATISTICS

```
HEL-F097,PWGR

SEP 21 1974      11:05 A.M.      PORT # 0


OPEN-FILE1,1
```
Here an HP system command OPEN- was used to create a file named FILE1 which can be used later to store data. The "1" after the comma indicates that the file is one record in length; this means that up to 128 numbers (not just digits) can be stored in the file.

```
GET-$IDA
RUN
IDA

14 JUN 74 VERSION

GOOD MORNING.  NEED HELP ?YES
```
The command "GET-$IDA" puts IDA at the user's service. The command "RUN" causes IDA to start execution. Now you can no longer issue HP system commands. Instead you issue IDA commands following the readiness symbol printed by the terminal: ">". Keep in mind that IDA understands only IDA commands, and that the HP system understands only HP system commands.

Once you are familiar with IDA, you will usually type NO here and start to work.

```
IDA  (AN ACRONYM FOR 'INTERACTIVE DATA ANALYSIS')

IS A SYSTEM OF PROGRAMS CHAINED TO THE CONTROL PROGRAM $IDA.

IN USING IDA, YOU ISSUE A COMMAND FOR THE TASK YOU WANT DONE,
IDA DOES IT, YOU EXAMINE THE RESULTS, AND ON THE BASIS OF THE
RESULTS YOU DECIDE WHICH TASK YOU WISH IDA TO EXECUTE NEXT AND
ISSUE THE APPROPRIATE COMMAND.
WHEN YOU DO NOT WANT TO DO MORE, YOU TYPE THE COMMAND 'QUIT'.

CURRENTLY THERE ARE OVER 100 IDA COMMANDS.

>    IS THE SYMBOL INDICATING THAT IDA IS READY FOR YOU TO ISSUE
     A COMMAND WORD INDICATING WHICH TASK IS TO BE EXECUTED NEXT.

PROMPTS ARE GIVEN BY IDA WHEN MORE INFORMATION IS NEEDED TO EXECUTE
     THE TASK INDICATED BY THE COMMAND.
     THE PROMPTS ARE DETAILED (LEVEL 1), LESS DETAILED (LEVEL 2)
     OR BRIEF (POSSIBLY CRYPTIC) IN LEVEL 3.
     PROMPT LEVEL 1 IS IN EFFECT UNLESS YOU ISSUE THE COMMAND
     CHGP  TO CHANGE THE PROMPT LEVEL.
     IT IS SUGGESTED YOU USE PROMPT LEVEL 1 THE FIRST TIME YOU
     ISSUE A COMMAND IN ORDER TO TAKE ADVANTAGE OF THE
     ADDITIONAL EXPLANATIONS AVAILABLE AT THAT LEVEL.

*   WILL PRECEDE A PROMPT IF FURTHER INFORMATION WILL BE
    FORTHCOMING IF YOU
    (1)  TYPE  ?  OR
    (2)  JUST SIT AND WAIT A BIT.

*   WANT MORE DETAILS ? Y
```
"Y" is an acceptable abbreviation for YES.

```
YOU CAN GET MORE INFORMATION ABOUT IDA
    (1)  FROM 'CONVERSATIONAL STATISTICS' AND ITS 'COMPUTER
         PREFACE', OR
    (2)  IF YOU GET-$IDA, RUN IT, AND
         ISSUE THE IDA COMMANDS:
             EXPL  TO GET AN EXPLANATION FOR A SPECIFIC COMMAND
             INFO  TO GET EXPLANATIONS OF ALL THE COMMANDS IN A
                   GROUP--SUCH AS TRANSFORMATION COMMANDS
         OR, IN SOME CASES,
         ISSUE THE COMMAND AT PROMPT LEVEL 1.

IN ORDER TO ANALYSE DATA WITH IDA,
DATA MUST FIRST BE ENTERED IN THE IDA DATA MATRIX.
YOU CAN THINK OF THE DATA MATRIX AS A TABLE WITH NUMBER OF ROWS
EQUAL TO THE NUMBER OF OBSERVATIONS (QUESTIONNAIRES) AND NUMBER
OF COLUMNS EQUAL TO THE NUMBER OF VARIABLES.

YOU MAY ENTER DATA IN THE IDA DATA MATRIX BY
    (1)  USING DATA FILE(S) AND ONE OF THE FOLLOWING COMMANDS:
         ENTER, ENTS, ENRA, CRSP, OR EOBR;

    (2)  INPUTTING DATA DIRECTLY FROM THE TERMINAL WITH TAPE OR
         KEYBOARD, USING 'ENTER';

    (3)  ENTERING DATA GENERATED BY IDA, USING 'RAND' OR 'INDX'.

AFTER DATA IS ENTERED, YOU MAY EXECUTE OTHER COMMANDS TO:
DESIGNATE VARIABLES FOR ANALYSIS OF CROSS-SECTIONAL AND TIME-SERIES
    DATA BY SIMPLE AND MULTIPLE REGRESSION AND RELATED TECHNIQUES;
TRANSFORM THE DATA AND PLACE THE RESULTS IN THE DATA MATRIX;
ADD OTHER VARIABLES TO THE DATA MATRIX;
DELETE OBSERVATIONS;
RETRIEVE DELETED OBSERVATIONS;
SORT THE DATA INTO ASCENDING ORDER;
SAVE PART OR ALL OF THE DATA MATRIX OR FITTED OR RESIDUAL
    VALUES IN ONE OF YOUR FILES;
EXAMINE THE DATA OR FITTED OR RESIDUAL VALUES BY DISPLAYING
    THEM IN PLOTS OR HISTOGRAMS;
PRINT TABLES OF DATA VALUES AND CROSS TABULATIONS OF FREQUENCIES
    AND OF MEANS;
ANALYSE THE DATA IN VARIOUS WAYS;
COMPUTE AND PRINT OUT SUMMARY AND ONE SAMPLE STATISTICS,
PERFORM OTHER TASKS BY USING THE IDA COMMAND 'NEWC' AND A PROGRAM
    WRITTEN BY YOU TO BE USED WITH IDA.
```

```
OR, YOU CAN USE IDA TO:
CREATE NEW DATA FILES BY SAVING AN EDITED VERSION OF SOME OR ALL
   COLUMNS OF THE IDA DATA MATRIX WITH 'SAVF' OR BY USING 'CRFI'
   FOR LARGER SETS OF DATA;
LIST THE CONTENTS OF FILES WITH 'FILE';
COMPUTE NORMAL PROBABILITES WITH 'GAUS';
SELECT RANDOM SAMPLES WITH 'PSAM'.

YOU CAN NORMALLY ENTER A MAXIMUM OF 100 ROWS (OR OBSERVATIONS)
AND A MAXIMUM OF 19 COLUMNS (OR VARIABLES) OF DATA IN THE IDA
DATA MATRIX, BUT YOU CAN USE THE IDA COMMAND 'RDIM'  TO RE-
DIMENSION THE DATA MATRIX TO MORE ROWS (A MAX. OF 563) AT
THE EXPENSE OF FEWER COLUMNS (A MIN. OF 1).

YOU CAN STOP THE NORMAL EXECUTION OF IDA BY
     (1)  USING C-CONTROL IF IT IS AWAITING INPUT BY YOU, OR,
          OTHERWISE,
     (2)  USING THE 'BRK', 'BREAK', OR 'INTERRUPT' KEY.

IF YOU THEN WISH TO GET BACK TO THE COMMAND LEVEL, TYPE
'RUN-9998', THEN CARRIAGE RETURN
AND IDA WILL RESPOND WITH
                         >
                         THE COMMAND READINESS SYMBOL.

TO STOP USING IDA, TYPE THE IDA COMMAND
                                      QUIT.
TO GET A LIST OF IDA COMMANDS, TYPE THE IDA COMMAND,
                                                   COMM
TO GET ADDITIONAL DETAILS, TYPE THE IDA COMMAND,
                                                INFO

> INFO
* HOW MANY CATEGORIES ?

YOU CAN HAVE HELP ON ANY OR ALL OF THE FOLLOWING :

  1.  GENERAL COMMENTS ABOUT IDA
  2.  DATA DEFINITION
  3.  DATA EDITING
  4.  DATA DISPLAY (PRINT)
  5.  DATA DISPLAY (PLOT)
  6.  TRANSFORMATIONS
  7.  SUMMARY STATISTICS
  8.  ONE SAMPLE STATISTICS
  9.  REGRESSION OR TIME-SERIES ANALYSIS
 10.  MISCELLANEOUS COMMANDS

HOW MANY OF THE ABOVE CATEGORIES DO YOU NEED HELP ?1
WHICH  1 ?  GIVE NUMBERS, SEPARATED BY COMMAS : ?1


GENERAL COMMENTS :

1.  MAXIMUM SIZE OF DATA MATRIX IS NORMALLY 100 BY 19.
    COLUMNS OF THE DATA MATRIX ARE REFERRED TO AS
    VARIABLES;  ROWS, OBSERVATIONS.  UNIVARIATE DATA
    SHOULD BE STORED AS A COLUMN VECTOR.
    IF YOU HAVE MORE THAN 100 ROWS IN YOUR
    MATRIX, YOU MAY RE-DIMEMSION THE SIZE
    BY EXECUTING THE COMMAND 'RDIM'.
2.  COMMAND STRUCTURE :  THE SYSTEM PRINTS THE SYMBOL
    '>' WHEN IT WAITS FOR THE USER TO TYPE A COMMAND
    WORD FOR A TASK.  ONLY THE FIRST 4 CHARACTERS OF
    A COMMAND WORD ARE CHECKED BY THE SYSTEM.  FOR
    EXAMPLE, ONE OF THE COMMANDS AVAILABLE IS 'EXPLAIN'.
    THIS TASK WILL BE EXECUTED WHETHER THE USER TYPES
    'EXPLAIN' OR ANY WORD THAT BEGINS WITH 'EXPL'.
    SOME COMMAND WORDS ARE CONTRACTIONS, SUCH AS 'PARC'
    FOR THE COMPUTATION OF 'PARTIAL CORRELATIONS'.  TO
    OBTAIN THE ENTIRE LIST OF VALID COMMAND WORDS, YOU
    MAY ISSUE THE COMMAND 'LIST'.
3.  PROMPTS :  IN ALMOST ALL CASES, ONCE A COMMAND IS
    ISSUED BY THE USER, IDA WILL NEED ADDITIONAL INFOR-
    MATION BEFORE THE TASK CAN BE EXECUTED.  THE USER
    WILL BE PROMPTED FOR THE INFORMATION.  IDA HAS THREE
    LEVELS OF PROMPTS WHICH THE USER CAN CHOOSE DEPENDING
    ON HIS FAMILIARITY WITH THE SYSTEM.  UNLESS OTHERWISE
    INSTRUCTED BY THE COMMAND 'CHGP' (FOR CHANGING THE
    LEVEL OF PROMPTS), IDA WILL GIVE 1ST LEVEL PROMPTS
    WHICH ARE MEANT TO BE USED BY THE NOVICE -- THESE
    PROMPTS ARE GENERALLY DETAILED AND LENGTHY.  2ND
    LEVEL PROMPTS ARE MORE CONCISE AND ABBREVIATED, AND
    3RD LEVEL PROMPTS ARE VERY BRIEF, POSSIBLY CRYPTIC.
    WHEN A PROMPT IS PRECEDED BY THE SYMBOL '*', THE USER
    WILL AUTOMATICALLY OBTAIN FURTHER EXPLANATION IF HE
    WAITS A CERTAIN AMOUNT OF TIME (USUALLY 30 SECONDS)
    WITHOUT RESPONDING, OR IF HE TYPES ' ? ' OR ANY
    ALPHAMERIC CHARACTERS WHEN NUMERIC INPUT IS CALLED
    FOR.
```

This is the first IDA command issued during the session.

Since the user hesitated to type a response to the query, HOW MANY CATEGORIES?, IDA automatically went ahead to explain the categories for which detailed help is offered.

272 ☐ CONVERSATIONAL STATISTICS

```
   4.  IDA HAS A NUMBER OF BUILT IN CHECKS FOR ERRORS IN
       THE USER'S INPUT.  HOWEVER, ERRORS WILL OCCASIONALLY
       CAUSE YOU TO BE KICKED OUT OF THE SYSTEM IDA.  ALSO
       HITTING THE 'BREAK' KEY DURING EXECUTION WILL SURELY
       GET YOU OUT OF IDA.  IN EITHER CASE, YOU CAN GET BACK
       TO IDA (WITHOUT LOSING YOUR ACTIVE DATA) BY TYPING :
       RUN-9998
       AND YOU'LL BE BACK AT THE IDA COMMAND LEVEL AND CAN
       PROCEED FROM WHERE YOU LEFT OFF.
   5.  ACTIVE DATA : WHEN YOU ENTER YOUR DATA MATRIX, IT
       BECOMES ACTIVE.  ALL COMMANDS WILL REFER TO THIS
       MATRIX.  WHEN YOU DELETE A ROW (BY 'DELO') OR A
       BLOCK OF ROWS (BY 'DELB'), THE ROWS ARE NOT PHYSICALLY
       DELETED.  THEY ONLY BECOME INACTIVE IN SUBSEQUENT
STOP
RUN-9998
IDAE39
```

At the words IN SUBSEQUENT above, the explanation was terminated by pressing the <u>break</u> or <u>interrupt</u> key. As a result, the user is back under HP system control as indicated by the message "STOP". To return to IDA, the command RUN-9998 was typed, as shown. Caution: never use the <u>break</u> or interrupt key when a computation may be underway.

```
> COMM
CURRENT, OR SOON TO BE IMPLEMENTED IDA COMMANDS ARE:

MISCELLANEOUS COMMANDS:
CHGP   CRTS   MISS   QUIT
COMM   EXPL   HELP   INFO   LIST   NEWC   NEWS   PAUS
CALC   CRFI   FILE   GAUS   PSAM
NAME   RDIM   SPAD   SAVF

DATA DEFINITION COMMANDS:
ENTE   ENTS   ENRA
CRSP   INDX   RAND   EOBR   EOBS

DATA EDITING COMMANDS:
CHGO   APPO   APPB   APPV   APPS
DELO   DELB   RECO   RETO   RETB   SELR
DELV   MOVE   MSOR   PSOR   SAVR   SORT
ZERC   ZERR   ZERS

PRINT AND PLOT COMMANDS:
PRTS   PRTO   PRTV   PRTR   PRTF
FPRS   FPRO   FPRV   FPRR   FPRF
FREQ   HIST   NORM   PLTS   SCAT
CTAB   MTAB   MPLS   RVSF

TRANSFORMATION COMMANDS:
CATE   RANK   DIFF   LAGG
ABSO   ADDC   MULC   EXPO   LOGE   LOG1   POWE   STAN
ADDV   DIVI   DOTP   MULV   SUBV

SUMMARY AND ONE SAMPLE STATISTICS:
CORR   COVA   PARC   MEAN
DURB   RUNS   SERC

REGRESSION ANALYSIS:
REGR   FORW   STEP   BACK   ALLS   SUBS   SWEE   SAMP
COEF   SUMM   BCOR   BCOV   SEPR   ANOV

TIME-SERIES ANALYSIS:
AUTO   BOXJ   CROS   PACF   SPEC   STAR

COMMANDS TO BE IMPLEMENTED SOME TIME IN THE FUTURE:
COMP   DUMP   ENLI   PLTC   TRAN   WLSR

EXECUTING SOME OF THE ABOVE COMMANDS AT THE FIRST PROMPT LEVEL
WILL GIVE AN OPTION OF AN EXPLANATION.
IF NOT, THE COMMAND 'EXPLAIN', OR 'EXPL', WILL USUALLY GIVE
A BRIEF EXPLANATION.
'INFO' WILL GIVE EXPLANATIONS FOR ALL COMMANDS WITHIN A GROUP,
E.G., ALL TRANSFORMATIONS.

> NEWS
1AUG74 VERSION

  1AUG74
  USE OF 'NORM' FOR RESIDUALS AFTER 'STEP', 'BACK' OR FORW'
  NOW GIVES PROPER VALUE FOR STD.DEV. OF RESIDUALS (BUG IN
  IDA30 CORRECTED.)

  14JUL74   REMOVED BUG IN >RETB

  13JUN74   NEW COMMANDS:
            'MISS'--TO DECLARE MISSING VALUES IN DATA
            'BOXJ'--BOX-JENKINS ESTIMATION OF ARIMA MODELS
```

NEWS is a useful command for keeping up to date with changes in IDA.

COMPUTER NOTES ☐ 273

```
10JUN74
ENRA  REVISED COMPLETELY. ENTER SELECTED OBSERVATIONS AND
      SELECTED VARIABLES FROM RANDOM ACCESS FILE

NEW COMMANDS AVAILABLE:
EOBR  USING LIST OF OBS. #'S IN IDA DATA COL., ENTER SELECTED
      VARIABLES FROM A RANDOM ACCESS FILE

CRFI  CREATE A RANDOM ACCESS FILE BY INPUTTING DATA FROM TERMINAL

CROS  COMPUTES CROSS CORR
STOP
RUN-9998
IDA98
```

Note to the curious: IDA98 is the IDA subprogram that implements the NEWS command. You can think of that program as remaining in memory even after pressing the <u>break</u> key. RUN-9998 causes execution to begin at line 9998 of IDA98, which in turn is an instruction to chain back to the part of IDA (the main program, called "IDA") that causes the readiness symbol to be issued.

```
> ENTER
WANT EXPLANATION ?N

* MODE OF INPUT : FROM 'FILE' OR 'TERMINAL' ?T

* SAMPLE SIZE (N) = 20
* NO. OF VARIABLES (K) = 2
ENTER ELEMENTS OF MATRIX BY ROWS BELOW :
ROW  1   : ?433,39
ROW  2   : ?483,60
ROW  3   : ?479,42
ROW  4   : ?486,53÷2
ROW  5   : ?494,47
ROW  6   : ?498,51
ROW  7   : ?511,45
ROW  8   : ?534,60
ROW  9   : ?478,39
ROW 10   : ?440,41\
440,41
ROW 11   : ?372,22
ROW 12   : ?381,18
ROW 13   : ?419,27
ROW 14   : ?449,33
ROW 15   : ?511,48
ROW 16   : ?520
??51
ROW 17   : ?477,33
ROW 18   : ?517,46
ROW 19   : ?548,54
ROW 20   : ?629,100

DATA MATRIX HAS 20    OBSERVATIONS ON 2    VARIABLE(S)
COMPUTING MEAN(S) AND STD.DEV.(S)...

YOU MAY GIVE EACH VARIABLE A 1 TO 6 CHARACTER NAME.
YOU ARE STRONGLY URGED TO SUPPLY NAMES FOR YOUR VARIABLES.
IF YOU DO, SEQUENCE PLOTS, TABLES, ETC. WILL BE LABELED.
IF YOU DON'T, YOU WILL HAVE TO CALL YOUR VARIABLES BY COL.#.

WANT TO SUPPLY NAME(S) ?YES
PLEASE GIVE NAME OF:
VAR.  1 ?INCOME
VAR.  2 ?INVEST

> CHGO
GIVE ROW NUMBER I, COLUMN NUMBER J,
AND THE NEW DATA VALUE X(I,J) :
I, J, X(I,J) = ?12,2,17

UPDATING MEAN, STD.DEV....

> PRTS
WANT EXPLANATION ?N
TO PRINT SUBMATRIX BETWEEN ROWS I1 AND I2 INCLUSIVE,
GIVE 2 ROW NUMBERS, SEPARATED BY A COMMAS, FOR I1 AND I2, OR
FIRST ROW, LAST ROW = ?1,20

* DO VARIABLES TO BE PRINTED OCCUPY
   CONSECUTIVE COLUMNS OFDATA MATRIX ? YES
GIVE 2 COL. NUMBERS, SEPARATED BY A COMMA, FOR:
```

The following illustrates the entry of data into IDA.

T indicates data entry directly from the terminal. Note "T" is an acceptable abbreviation for "TERMINAL".

Data matrix will have 20 rows and 2 columns.

Note the comma between the two numbers; a <u>return</u> was typed at the end of each line.
Only two numbers (the number of variables) are entered per row.

Here a typing mistake was corrected by typing a back arrow; the second number is now 52, not 53.

The entire line was deleted with control-x (x^c that is), typed while holding down the control key.
Here is the correctly typed line.

A mistake was made here that was not noticed until after hitting the <u>return</u>. It will be corrected later.

Here <u>return</u> was typed before typing a comma and the second number -- so IDA came back with "??" to ask for the second number, which was then typed in.

Here IDA is computing the mean and standard deviation of the variables. You can print them out by issuing the command MEAN.

If later you forget the names you assigned, type the command NAME in response to the readiness symbol >.

Next the command CHGO (CHANGE OBSERVATION) is used to correct the entry of the 12th observation, 2nd variable, above.

This message means that IDA is now updating the summary measures computed at the time of the original ENTER command.

The command PRTS is used to print out the current data matrix, (or any part of it I choose).

274 □ CONVERSATIONAL STATISTICS

```
FIRST COL., LAST COL. = ?1,2
  ROW         INCOME        INVEST
*   1  *     433.00000      39.00000
*   2  *     483.00000      60.00000
*   3  *     479.00000      42.00000
*   4  *     486.00000      52.00000
*   5  *     494.00000      47.00000
*   6  *     498.00000      51.00000
*   7  *     511.00000      45.00000
*   8  *     534.00000      60.00000
*   9  *     478.00000      39.00000
*  10  *     440.00000      41.00000
*  11  *     372.00000      22.00000
*  12  *     381.00000      17.00000
*  13  *     419.00000      27.00000
*  14  *     449.00000      33.00000
*  15  *     511.00000      48.00000
*  16  *     520.00000      51.00000
*  17  *     477.00000      33.00000
*  18  *     517.00000      46.00000
*  19  *     548.00000      54.00000
*  20  *     629.00000     100.00000
```

*Note: PRTS prints out what you have in the data matrix; it does **not** save it for you. The procedure for saving the data will be illustrated next.*

Note that observation 12 has been corrected.

```
> SAVE
WANT EXPLANATION ?N

NAME OF OUTPUT FILE TO SAVE DATA MATRIX IS ?FILE1
```

Often you will want to save data in your personal library for easy use at a future session at the terminal. This is next illustrated by use of the file FILE1 that was opened at the start of this session. Since there are 40 numbers in the above data matrix, and since the number of observations and the number of variables is automatically saved, 42 numbers will be saved in total. Since the single record of FILE1 will hold up to 128 numbers, there will be plenty of room.

```
WANT ENTIRE ACTIVE DATA MATRIX SAVED ?Y

THE FOLLOWING ARE NOW SAVED IN FILE1 :
 20
 2
THE 20    BY 2    DATA MATRIX
NAMES:
INCOME
INVEST
YOU MAY ADD EXTRA TEXT, ONE LINE AT A TIME, TO DESCRIBE FILE1

WANT TO ADD SOME TEXT ?Y

WHEN YOU ARE DONE, PRESS THE CARRIAGE RETURN ONLY.

INPUT LINE OF LESS THAN 73   CHARACTERS.
DATA OF HAAVELMO, QUOTED BY ZELLNER, OF $ PER CAPITA INVESTMENT

INPUT LINE OF LESS THAN 73   CHARACTERS.
AND INCOME, DEFLATED, U.S., 1922-1941

INPUT LINE OF LESS THAN 73   CHARACTERS.

DESCRIPTIVE TEXT SAVED AT END OF FILE1
```

And room also for some text!

Now the data are in the file. They will stay there until (1) you save other data on FILE1, in which case the file space will be automatically cleared before the new data are saved or (2) you get rid of the file by the HP-system command KILL-FILE1. If you want to see what files you have saved in your personal library, issue the HP system command CAT.
(Note: Never kill a file when you are temporarily out of IDA.)

```
> FILE
NAME OF FILE TO PRINT IS ?FILE1
FIRST ELEMENT OF 'FILE1' IS 20
SECOND ELEMENT IN 'FILE1' IS 2
'FILE1' PROBABLY HAS DATA MATRIX OF 20   ROWS AND 2   COLUMNS.
HOW MANY ROWS DO YOU WANT TO PRINT ?5

ROW  1 :    433     39
ROW  2 :    483     60
ROW  3 :    479     42
ROW  4 :    486     52
ROW  5 :    494     47
```

COMPUTER NOTES ☐ 275

```
* WANT TO CHECK NAMES OR TEXT IN 'FILE1' ? Y

NAMES:
INCOME
INVEST
TEXT:
DATA OF HAAVELMO, QUOTED BY ZELLNER, OF $ PER CAPITA INVESTMENT
AND INCOME, DEFLATED, U.S., 1922-1941
END OF FILE CONTENTS.
```

Next, the ENTER command is used to illustrate the entry of data from a file. The command will clear your IDA worksheet and replace it with data from the file. In this example, however, the data will be entered from FILE1.

```
> ENTER
WANT EXPLANATION ?N

FURTHER USE OF 'ENTE' WILL ERASE
DATA NOW IN FIRST 20   ROWS, 2     COLUMNS.
WANT TO CONTINUE WITH  ENTE?Y

* MODE OF INPUT : FROM 'FILE' OR 'TERMINAL' ?F

NAME OF INPUT FILE IS ?FILE1

* ARE THE FIRST TWO ELEMENTS OF YOUR DATA FILE
  VALUES FOR N AND K (SIZE OF DATA MATRIX)? Y

DATA MATRIX HAS 20    OBSERVATIONS ON 2     VARIABLE(S)

COMPUTING MEAN(S) AND STD.DEV.(S)...

NAMES READ FROM DATA FILE ARE:
INCOME
INVEST

WANT TO SUPPLY NEW NAMES(S) ?N

> PRTS
WANT EXPLANATION ?N
TO PRINT SUBMATRIX BETWEEN ROWS I1 AND I2 INCLUSIVE,
GIVE 2 ROW NUMBERS, SEPARATED BY A COMMAS, FOR I1 AND I2, OR
FIRST ROW, LAST ROW = ?18,20

* DO VARIABLES TO BE PRINTED OCCUPY
  CONSECUTIVE COLUMNS OFDATA MATRIX ? Y
GIVE 2 COL. NUMBERS, SEPARATED BY A COMMA, FOR:

FIRST COL., LAST COL. = ?1,1

   ROW         INCOME

*  18 *       517.00000
*  19 *       548.00000
*  20 *       629.00000


> QUIT

DONE
```

CAUTION: If the file was created by IDA, it is essential that you enter YES or Y in response to this prompt.

This time only a small part of the current data matrix will be requested. Be sure that you understand how this is done.

This command gets you out of IDA and puts you back under HP system control.

Here the HP system informs you that the current execution of IDA is done.

At this point you are back under HP system control. However, as long as the current session continues, and as long as you do not issue either of the HP system commands SCRATCH or GET-XXXXXX, or KILL-XXXXXX, you can get back into IDA with data intact by the following use of RUN-9998.

Next CAT (CATALOG) was used to get a listing of the programs and files (indicated by F) in my personal library. FILE1 is boxed in; its length is shown as one record.

```
CAT
NAME      LENGTH    NAME      LENGTH    NAME      LENGTH    NAME      LENGTH
ARG          419    ARG1        2547    ATEMP        182    CHAPRO F       1
DUNN   F       1    FERTIL F       1    FILE1  F       1    FLTRUN       126
GNP58  F       1    HVR4        1062    INTINF F       4    LABEL         63
POWELL F       1    TEMP   F      10    TEMP1  F       4    UAL1   F       2
UAL2   F       2    UAL3   F       7    WINNO4       286    WLLACE       145
ZFLIST       211
```

To get rid of FILE1, kill it, as shown. The subsequent CAT verifies that FILE1 is gone.

```
KILL-FILE1
CAT
  NAME      LENGTH      NAME     LENGTH     NAME     LENGTH     NAME     LENGTH
  ARG          419    ARG1         2547   ATEMP         182   CHAPRO F      1
  DUNN   F       1    FERTIL F        1   FLTRUN        126   GNP58  F      1
  HVR4        1062    INTINF F        4   LABEL          63   POWELL F      1
  TEMP   F      10    TEMP1  F        4   UAL1   F        2   UAL2   F      2
  UAL3   F       7    WINN04        286   WLLACE        145   ZFLIST      211

TIME
CONSOLE TIME = 32 MINUTES.  TOTAL TIME = 102 MINUTES.
BYE
032 MINUTES OF TERMINAL TIME
JC
```

The last two commands, above, tell how much connect time was used both during the current month and during the current session, and log you off the system. After logging off, hang up phone.

When in doubt as to what to do next in IDA, a good general rule is simply to wait for a half-minute or so, when you may be helped automatically by queries or explanations. If you want to think a bit without being interrupted by IDA's well-meaning attempts to help you, issue the command PAUS.

If you want to abort an IDA command just issued, either hit the *break* key during printout by the terminal or hit *return*, *break* in rapid succession. Then type RUN-9998 to get the readiness symbol for another IDA command. Remember: do not hit *break* if IDA may be in the process of computing, updating, or transforming data. Often, but not always, IDA tells you when this is happening.

If, as in the example of the command ENTER, IDA is requesting input by issuing question marks and you wish to abort the procedure, type C^c followed by *return*. Then type RUN-9998.

Occasionally you may ask IDA to do something impossible, like taking the square root of a negative number. Instead of getting a reprimand from IDA, you will get kicked out of IDA and receive an HP system error message. To recover, type RUN-9998, as usual. Remember always that IDA commands are accepted only when you are in IDA.

You have now covered the relatively simple mechanical prerequisites for the use of IDA. You will be introduced to many of IDA's statistical capabilities, in easy stages, in the main chapters of this book.

SUMMARY

CHAPTER 1: INTRODUCTION

Statistics can be thought of as common sense disciplined by calculation in the study of problems in which uncertainty is present. The potential for good statistical application is great but is as yet largely untapped. The major divisions of statistics are statistical inference and statistical decision theory; some would add statistical description as a third field. There are two major philosophical approaches to statistics: sampling theory and Bayesian.

This book concentrates on an informal development of a broad body of techniques called "regression" and the general statistical ideas behind regression. The aim is to develop some capability of applying these techniques to problems; the hoped-for byproduct is to develop skill in the consumption of statistical studies by others. Quite apart from technical knowledge, however, the intelligent consumer of statistics must learn to use his common sense to avoid being duped by the outrageous misuses of statistics that bombard him.

CHAPTER 2: EFFECTIVE USES OF STATISTICS

Inference pertains to the process by which we acquire knowledge from experience. Statistical inference narrows the scope to situations in which relevant aspects of experience are captured in data. Decision pertains to the process by which we take actions on the basis of knowledge. Statistical decision theory narrows the scope to applications in which the decision process can be given a mathematical description. Although a grand design for statistical decision theory has been developed, most decisions appear too complex for a full implementation of this grand design given the current state of the art of decision theory. We shall rather focus on statistical inference, which often can provide essential background for direct judgmental attacks on tough decisions.

Statistical associations or relationships do not necessarily imply causation. When it is possible to perform randomized experiments, however, unambiguous insight into causation may sometimes be obtained. Randomization permits the formation of two or more groups, which will be given different experimental treatments, in such a way that the groups will be closely comparable in respects other than the treatments so long as the size of the groups is moderately large. Randomization also permits precise assessment of the degree of uncertainty of comparisons of responses of the different treatment groups. Randomization in the formation of treatment groups for experimentation, for instance, can be achieved by assigning a 5-digit random number to each individual in the experiment, sorting the individuals by the order of magnitude of the random numbers, and then constructing the treatment groups consecutively from the sorted list. This can be done by using the subcommand RAN1 of the IDA command PSAM. The computer, however, does not store a table of random numbers internally; rather it generates pseudorandom numbers, which are used for the same purposes as random numbers. Failure to apply randomization in the formation of treatment groups can lead to disastrously misleading conclusions. The subject of

GLOSSARY

Sampling theory refers to the broad stategy in which attention is focussed on *sampling distributions*; inferences are drawn directly therefrom.

In the *Bayesian* approach, inferences are made by application of *Bayes' theorem*. A brief and non-technical discussion is given in Chapter 7.

The key idea of *randomization* is extremely thorough shuffling, which is implemented in practice by the use of *random numbers*. A table of random numbers can be made by a process that turns up a digit 0, 1, 2, 3, 4, 5, 6, 7, 8, or 9 on each trial in such a way that each of the 10 possible digits is equally probable.

Pseudorandom numbers are computer-generated by a deterministic computing scheme or algorithm. These numbers could be exactly predicted if you knew the algorithm, but they behave for statistical purposes essentially as random numbers.

statistical design of experiments encompasses many devices other than randomization; the aim is to increase the precision of inferences. One such device is the randomized block.

Sometimes it is possible to run a statistical experiment without disturbing an ongoing business operation; one interesting approach along these lines is called Evolutionary Operation or EVOP.

In the absence of randomized experimentation, it is much more difficult to learn about cause and effect from statistical studies.

Another major application of statistics is sampling from finite populations. A complete enumeration, or census, may be made if the desired information about the population is contained in a computer data file. Often, however, sampling is the only economical way to obtain the desired information. That is, when cost of obtaining the information is taken into account, it may be preferable to accept sampling error in the resulting estimates than to eliminate sampling error by taking a census. Either a sample or a census must contend with non-sampling error; sometimes the diseconomies of scale of large data-gathering operations are such that samples may be more accurate than a census, because sampling error plus the non-sampling error of the samples may be less than the non-sampling error of the census.

A probability sampling design is often desirable for sampling applications. Random numbers are used to implement such designs. The simplest probability sampling design is simple random sampling. Simple random sampling assures that each possible sample of n elements in a population of N elements has an equal chance of being chosen. This goal can be implemented in IDA either by the subcommand RAN1 of PSAM, already mentioned for formation of treatment groups in experimentation, or by the subcommand RAN2. RAN2 works by choosing the sample one individual at a time, without replacement, from those not yet chosen. Each individual has an equal chance of coming in at each stage unless chosen at an earlier stage. Very small random samples can be extremely non-representative since the sampling error is then very high. As the sample size is increased, the sampling error will typically decline. No sampling method, however, can guarantee representativeness unless the characteristics of the population are completely known in advance, in which event the sample would be superfluous.

There are many probability sampling designs other than simple random sampling. Systematic sampling is implemented by the subcommand SYST of PSAM; a special kind of stratified random sampling is implemented by the subcommand SRAN. In stratified random sampling, the fraction of sample elements is often the same in each stratum, but for certain applications it is very desirable to vary the sampling fraction. In estimating total sales from a sample of automobile dealers, for example, the sampling fraction would be higher for strata in which past sales per dealer were higher. Many other devices are used in probability sampling, especially in complex nationwide investigations such as the Current Population Survey. One such device is cluster sampling or area sampling.

Non-probability sampling methods are widely used. Although they cannot be justified by the theoretical base underlying probability sampling, they may sometimes be useful, especially when they borrow some of the features of probability sampling. One well-known type of non-probability sampling is quota sampling, often used in opinion studies.

GLOSSARY

In the *randomized block* design, the entire group of subjects is first divided judgmentally into smaller, more homogeneous groups called blocks; randomization is used within each block to allocate subjects to the different treatments.

A *finite population* is any finite collection of objects, such as the student body in a college, the fish in a lake, or the farms in Iowa.

A *sample* is any part of a population.

A *census* is a complete enumeration, or 100 percent sample, of a population.

All errors that would be present even in an attempted census are lumped under the heading *non-sampling errors*.

A *probability sample* is a sample in which every individual in the population is (or can be) assigned a known, positive probability of inclusion.

In *simple random sampling*, each possible sample of n from a population of N has an equal chance of selection.

In *systematic sampling* with skip interval k, every kth individual, beginning with a random start in the first k, is chosen. In stratified random sampling, the population is first broken up into subpopulations or *strata*, intended to be relatively homogeneous. Then simple random sampling is applied to each such stratum.

In *cluster sampling*, a population or stratum is subdivided into groups of individuals (often living contiguously in small areas), and these groups or clusters are sampled at random.

Quota-sampling assigns quotas of respondents with specified characteristics, but fullfilment of the quotas is by interviewer judgment rather than random selection.

In all sampling investigations, failure to obtain information from individuals designated by the sampling method is potentially a serious problem, since those who do not respond are often systematically different from those who do.

No rule of thumb can be given as to the sample size needed for satisfactory accuracy in sampling. The statistician can usually make a rough assessment of the sampling error entailed in any proposed sampling design, but the user must decide how much it is worth paying to reduce sampling error. In most applications, it is the absolute sample size n rather than the relative sample size n/N that is the prime determinant of sampling accuracy.

CHAPTER 3: OBSERVATION AND MEASUREMENT

Although observation and measurement lie outside the conventional boundaries of statistics, the producer or user of statistical studies can ignore them only at great peril. Writing a questionnaire, for example, is far from a trivial exercise; the best rule for learning is to try, pretest, try, pretest, etc., until an acceptable instrument is obtained.

CHAPTER 4: ANALYSIS OF SEQUENCE

Data often arise in a natural time sequence. To get started on a statistical analysis, it is usually advisable to plot the data in time sequence by the IDA command PLTS in order to see what is going on. The aim is to see what the sequence of observations occurring in the past can contribute to the statistical prediction of observations not yet made on the same underlying process. If it is judged that the past sequence is irrelevant to what will happen in the future, the underlying process generating the data is judged to be random and statistical analysis is much simplified

Visual analysis of the sequence plot is best accomplished by Berkson's interocular traumatic test--giving the data a chance to hit you between the eyes. Departures from randomness include: any tendency for points close together in the time sequence to resemble each other either more or less closely than more distant points; systematic drifts upward or downward; systematic wave-like movements; abrupt shifts of level; variations of the vertical scatter of the data at different segments of the sequence. Berkson's test can be applied in two ways: look at the data at arm's length to get the general visual impression; then look at the data observation-by-observation to see evidences of persistence (or of alternation) of observations close together in the time sequence. One danger for beginners is to confuse the appearance of stationarity (see Chapter 11) with the appearance of randomness. (Although a random process is stationary, a stationary process is not necessarily random)

There are three broad kinds of checks on the visual examination of sequence plots: prior information bearing on the reasonableness of the random specification, numerical conformity measures computed for the data under examination, and prediction to new data. One conformity measure that is easily understood without technical background is the runs count. If there are two kinds of observations (say above and below a certain level) and if there are r observations of one kind and n-r of the other (where n is the sample size), then the expected number of runs for a random process is 2r(n-r)/n + 1. If the actual number of runs is less than this expectation, the data display more persistence or clustering than would be expected on the average from a random process. If the actual number is greater than

GLOSSARY

Non-response refers to the failure to obtain information from individuals designated by the sampling method.

A graph of sample data in time sequence is called a *sequence plot*. The values of the observations are plotted on the vertical axis; the time sequence is plotted on the horizontal axis.

The judgment of *randomness* is tantamount to the judgment that the sequence of the data is irrelevant to analysis. In quality control terminology, a random process is said to be in a state of *statistical control*.

For comparison with more technical treatments: a random process, as the term is used here, is a sequence of *independent and identically distributed random variables*.

Stationarity, to be explained in Chapter 11, is a broader concept than randomness; it includes randomness as a special case.

A *run* is a consecutive string of like observations in a sequence plot.

A *data matrix* is a table of data laid out in n rows and k columns. Each of the rows represents an observation; each of the columns represents a variable or measurement on the observations.

expectation, the data display less persistence or clustering--more alternation--than would be expected on the average from a random process. The RUNS command in IDA makes possible a rule of thumb judgment as to whether the data are behaving compatibly with the specification of a random model.

The IDA command PLTS for obtaining a sequence plot is scaled in standardized units: each observation is expressed as a deviation from the sample mean $\bar{x}$ in units of the sample standard deviation s. This device makes it possible to compare patterns in different sets of data in which the original units may be different. The IDA command MEAN permits you to print out the mean and standard deviation of all variables in the current data matrix.

If the specification of a random model underlying the observed data is not tenable, the standard tools of statistical analysis (to be developed subsequently) are not immediately applicable. One way to alleviate this problem is to transform the data by working with the first differences or changes in the observations rather than with the observations themselves. When the sequence plot of the original data suggests pronounced wave-like movements, it may happen that the sequence plot of the differences will behave randomly, or at least more nearly randomly. The differencing transformation can be implemented in IDA by the command DIFF.

Sometimes there is no natural time sequence underlying the observations, so the procedures of the chapter are not directly applicable.

To get a better intuitive feeling for the behavior of data generated randomly, it is often useful to do simulations, as permitted by the IDA command RAND.

CHAPTER 5: DISTRIBUTIONS

If, on the basis of the approach of Chapter 4, the specification of the random model is made, further attention can be focussed on the pattern of the data after sequence is lost. Then it is helpful to collapse the sequence plot into a histogram, which shows the general pattern or distribution of the data. If, however, the specification of randomness is not made, the collapsing of the sequence plot and subsequent analysis is ordinarily of little interest, and can be seriously misleading. If the sequence plot is wandering around non-randomly, our attempts at prediction should take into account the information of the sequence; this information is lost in the histogram. A histogram of standardized values is given automatically by IDA after, and in alignment with, the sequence plot, by the command PLTS. In this plot, each interval has width of 0.2 standard deviations. To make histograms in non-standardized units, and to tailor the width and centering of intervals to one's own choice, the command HIST is available. In choosing these interval-widths, one should ordinarily select a width that is a relatively small fraction of the sample standard deviation; the danger is more from intervals that are too broad than those that are too narrow, because broad intervals may smooth out important features of the data.

By showing the general pattern of the data, histograms may provide clues as to the underlying conditions that gave rise to the data. For example, fudging of the original data may leave a suspicious trace in the histogram. Even more importantly, histograms convey information about the appropriate probability distribution to specify as a part of the statistical model. The relative frequencies of the histogram can be thought of as first approximations to underlying probabilities.

GLOSSARY

The *arithmetic mean* of a sample of n observations $x_1, x_2, \ldots, x_n$ is the sum of the n observations divided by the sample size n:

$$\sum_{i=1}^{n} x_i/n ,$$

which is usually denoted $\bar{x}$.

The sample *standard deviation* s, a useful measure of dispersion, is defined by the following formula:

$$\sqrt{\sum_{i=1}^{n} (x_i - \bar{x})^2/(n-1)} .$$

If the first-differences of data behave randomly, the original data are non-random and are said to follow a *random walk*.

When a sample is obtained essentially at the same time, with no natural time sequence, the data are said to be *cross sectional*, as opposed to *time series* data.

A statistical *model* is a specification of some or all aspects of the underlying probability structure that gives rise to the data. Thus the *random model* implies that the structure is such that the sequence of the observations conveys no useful information for the purpose of prediction.

A *histogram* is a graph in which possible intervals of values for the observations are graphed on the horizontal axis and the frequencies in each interval are graphed on the vertical axis. The histogram is one way of showing the general pattern or *distribution* of data, without regard to time sequence.

A *probability distribution* for a given process assigns probabilities to various possible sample outcomes.

Probability has two major and logically distinct interpretations: long-run relative frequency and personal or subjective probability. Both are useful to us.

In using sample histograms for data adjudged to have arisen randomly as an aid in assessing the probabilities of an underlying probability distribution, a natural starting point--but only a starting point--is to assess all probabilities as equal to the relative frequencies of the histogram. If we want to refine these rough assessments, we can bring in the judgment that probabilities ought to change smoothly as possible values of the data increase. Procedures are available for direct smoothing of the histogram to express our judgments about the underlying probability distribution. More often, however, we will use the standard statistical tools to accomplish the same purpose.

An alternative graph for showing the information given by a histogram is called the cumulative distribution function or c.d.f. This plot is available in IDA as an option under HIST. A tabular representation of sample frequencies and cumulative frequencies is given by the IDA command FREQ.

CHAPTER 6: NORMAL DISTRIBUTION

The normal distribution represents a theoretical pattern to which data *sometimes* correspond more or less closely. If one thinks of a properly-scaled histogram for a sample of astronomical size for which the normal distribution is an appropriate model, the appearance would be that of a "bell" or, better, a "cocked-hat". This characteristic shape has a precise mathematical description. Probabilities implied by this model can be computed by the IDA command GAUS.

The normal distribution is not the "usual" distribution encountered in data. Under favorable circumstances, however, the normal distribution can give a good approximation to what is going on in the process producing the data. Favorable circumstances are these: each observation is the resultant of a substantial number of random factors that are not observed, that act independently of each other, that are additive in their contribution to the observation actually made, and that are not dominated by the contribution of a single factor among them. Good approximations to normality of data are often found in errors of measurement in careful scientific studies, variations of outputs of carefully controlled industrial processes, and certain biological measurements. A major interest in the normal distribution in statistical work stems from the fact that what is left over after our attempts to capture systematic variation in regression models (see Chapter 8) may well conform to the normal model.

There is a widespread but erroneous notion that normality (a property of distribution) implies randomness (a property of sequence). All four of the following conditions are possible: (1) random process, normal distribution; (2) random process, non-normal distribution; (3) nonrandom (but stationary) process, normal distribution; (4) nonrandom (but stationary) process, non-normal distribution.

If the observations are both random and normal, their first differences will be normal but *non*random (too much alternation, too little clustering or persistence). If the first differences are normal and random, the cumulative levels will be neither normal nor random (they will be said to follow a *random walk*).

In practical analysis of data, it is desirable to form some judgment from the data themselves as to the plausibility of the normal model.

GLOSSARY

Long-run relative frequency is one interpretation of probability. According to this interpretation, probability can be approximated closely by relative frequency in a random time series of astronomical length.

Subjective or personal probability can be vividly explained in terms of the betting odds at which you would be indifferent as to which side of a bet you took on the occurrence of the event in question.

The *cumulative distribution function* or *c.d.f.* displays cumulative frequencies up to and including the right boundary of each interval.

For a *normal distribution*, the probability of an observation within a distance of one standard deviation from the mean is about two-thirds. The corresponding probabilities for distances of two and three standard deviations are about 0.95 and nearly 1.

The *central limit theorem* is a mathematical formalization of the conditions that lead to normality.

After rough allowance for the raggedness occasioned by sampling variation, the sample histogram should be shaped like the normal distribution, but this is not always easy to judge. A more useful graph is a special kind of cumulative distribution called a normal probability plot. Systematic departures from linearity of the sample data on this plot are suggestive of non-normality. The IDA command NORM gives a normal probability plot plus certain conformity measures, including the skewness coefficient and the coefficient of kurtosis. If the points on the normal probability plot tend to follow a bow-shape, which is concave downward, the corresponding histogram will be blunt on the left and have a long tail pointing to the right; the skewness coefficient will then ordinarily be positive. If the points on the normal probability plot tend to follow the shape of an elongated S, the corresponding histogram will tend to have more observations distant two or three standard deviations than would the normal distribution; the corresponding coefficient of kurtosis will then ordinarily be positive.

A *normal probability plot* is a specially scaled c.d.f. that makes it easy to detect departures from normality by eye.

The *skewness coefficient* is sensitive to departures from the symmetry of distribution about the mean that is implied by normality. For a strictly normal distribution, the skewness coefficient is zero. The *coefficient of kurtosis* is sensitive to differences in shape between the actual distribution and a normal distribution with the same standard deviation. For a strictly normal distribution, the coefficient of kurtosis is zero.

If examination of the data suggest substantial non-normality, a transformation of the data to a scale on which the data will be approximately normal is worth exploring. The statistical analysis can be carried out in terms of the transformed units; the translation back to the original units, if necessary, can be carried out at the end. If the data are positive and asymmetric with a long tail pointing to the right, a logarithmic transformation may be helpful. This can be implemented in IDA by the command LOGE (logarithms to base "e") or LOG1 (logarithms to the base 10). Data that are approximately normal after the logarithmic transformation may sometimes be rationalized in terms of unobserved, independent, underlying factors that are multiplicative in their contribution to the observation actually made.

CHAPTER 7: INFERENCE

Statistical inferences can be thought of as probabilistic predictions, expressed as probability distributions for quantities that are uncertain. Point estimates or "best guesses" are summary measures of such distributions. Individual future observations and parameters are illustrations of the typical objects of inference.

If we specify that the process generating sample values $x_1, x_2, \ldots, x_n$ is random and normal, and if we specify that the sample constitutes the bulk of our cogent evidence ("weak" or "diffuse" prior information), then the probability distribution for a single future observation $\tilde{x}$ is approximated by: (1) a normal distribution; (2) with mean equal to the sample mean $\bar{x}$; (3) with standard deviation equal to $s\sqrt{1 + (1/n)}$. The calculations are given by the IDA command REGRession with the specification of 0 independent variables, reflecting the fact that analysis of a single variable can be thought of as a special case of regression analysis. The normal distribution just specified is a slightly optimistic approximation, optimistic in the sense that the exact distribution would be a little more spread out and slightly more fat-tailed than the normal.

Parameters can be thought of as numbers that are closely approximated by measures computed from random samples of astronomical size. Thus the parameter μ, the process mean, is closely approximated by the mean $\bar{x}$ of a very large random sample.

Statistics are sample measures actually observed, such as $\bar{x}$.

Any quantity about which we are uncertain is called a *random variable*. To distinguish random variables from quantities that will be known at the time of analysis, or from quantities that represent possible values of the random variable, the tilde notation is used. Thus we may speak of the probability that the random variable $\tilde{\mu}$ will exhibit a particular value μ.

The corresponding inference about the long-run process mean, the uncertain parameter $\tilde{\mu}$, is approximated by: (1) a normal distribution; (2) with mean equal to the sample mean; (3) with standard deviation equal to $s/\sqrt{n}$. Note that the standard error $s/\sqrt{n}$ is directly proportional to the standard deviation s of the sample and inversely proportional to the square root of the sample size. To cut the standard error in half, other things equal, a quadrupling of the sample size would be required. Note also that the standard error of the long-run mean $\tilde{\mu}$ is smaller than the standard error of a single future observation $\tilde{x}$, reflecting the fact that the mean of a large number of observations reflects an averaging effect.

The standard deviation of an inferential probability distribution is often called a *standard error*.

As a consequence of the approximate normality of the distribution for, say, $\tilde{\mu}$, the probability is about 2/3 that $\tilde{\mu}$ lies within the credible interval, or Bayesian confidence interval, $\bar{x} \pm s/\sqrt{n}$. Similar approximations for other intervals can be made using the command GAUS. The parallel sampling-theory concept is called a confidence interval The confidence interpretation of the interval above is that the method by which the particular interval above was computed would be right about two times in three in the long run.

GLOSSARY

The probability that the object of inference lies within a particular interval of the approximating distribution can be calculated. This probability is a measure of belief that the interval, called a *credible interval* or *Bayesian confidence interval* includes the object of inference. The corresponding sampling-theory concept is called a *confidence interval*.

In the Bayesian framework, the inferential distributions here discussed are called posterior distributions. To express "diffuse" or "weak" prior information, a prior distribution that is nearly constant or flat is assessed. This diffuse prior distribution together with the specification of random and normal observations leads mathematically to the posterior distribution by an application of Bayes' theorem. When the posterior distribution is approximately normal, as here, the mean and standard error provide a point estimate and measure of uncertainty.

In the Bayesian framework, the distributions here discussed are called *posterior distributions* since they are reached after the sample evidence has been analyzed. A formal expression of non-sample or prior evidence is called a *prior distribution*. Given a specification of the statistical model, the prior distribution is converted into the posterior distribution by *Bayes'* theorem.

Random sampling from a finite population, discussed in Chapter 2, is closely related to sampling from a process, as treated in the subsequent chapters beginning with Chapter 4. The connection is that the use of random numbers to select observations from a finite population makes the sample behave like a time series from an underlying random process. Inferences about finite populations differ from the inferences about processes (sometimes called infinite populations) only in that the sample of n constitutes a substantial fraction of the finite population of N for which uncertainty has been removed by direct observation (assuming, of course, no non-sampling error). In inferences about the mean of a finite population, for example, the results parallel those for a process with just one difference: the standard error is $s/\sqrt{n}$ times a finite-population correction $\sqrt{(N-n)/N} = \sqrt{1-(n/N)}$. But if n is small by comparison with N, as is often true in applications, the finite-population correction is very close to 1, so that the standard error formulas are approximately the same. This is the basis for the earlier statement that the absolute sample size n rather than the relative size n/N is the prime determinant of accuracy in sampling from a finite population. If n = N, the finite multiplier is strictly 0, reflecting the fact that there is no sampling error in a census.

The *finite population multiplier* is a factor by which we multiply the standard error appropriate for sampling from a process to get the standard error for sampling from a finite population.

CHAPTER 8: SIMPLE REGRESSION

Regression is one way of exploiting the possibility that the measurement of central concern in inference (the dependent variable) may be related statistically to other measurements (the independent variables). The appropriate graph for exploratory data analysis is the scatter plot (IDA command SCAT) in which the dependent variable is plotted on the vertical axis and the independent variable is plotted on the horizontal axis.

The *dependent variable* in regression is the variable that we are primarily interested in for prediction.

An *independent variable* in regression is a variable that may help us to make better predictions of the *dependent variable*.

The simple linear regression model contemplates that there is some "true" but unknown straight-line relationship between the dependent variable and the independent variable. This relationship, however, is statistical rather than exact. The purpose of this model, as of all statistical models, is to provide a conceptual scheme to help us do a sensible statistical analysis of data. In particular, the regression model of concern in this chapter says that each observation on the dependent variable $\tilde{y}_i$, for i = 1, 2, ..., n, arises from the following expression: $\beta_0 + \beta_1 x_i + \tilde{\epsilon}_i$. Think of the model in these steps: (1) a value x_i of $\tilde{x}_i$ is observed (ordinarily, nothing is said about the

A plot of a dependent variable against an independent variable is called a *scatter plot*.

The expression for the "true" regression line is

$$\beta_0 + \beta_1 x ,$$

which is the equation of a straight line, for which β_0 is the *intercept* and β_1 is the *slope*.

distribution of $\tilde{x}_i$); (2) given $\tilde{x}_i = x_i$, the distribution of $\tilde{y}_i$ is a normal distribution with mean of $\beta_0 + \beta_1 x_i$ and standard deviation σ_ϵ, a constant for all i; (3) then a value y_i of $\tilde{y}_i$ is observed as a drawing from this normal distribution. In applications, all we can ever observe are the points (x_i, y_i) for each of the n observations in the sample. The quantities $\tilde{\beta}_0, \tilde{\beta}_1$, and the disturbances $\tilde{\epsilon}_1, \tilde{\epsilon}_2, \ldots, \tilde{\epsilon}_n$ are uncertain parameters about which inferences may be drawn. Given the model just formulated and the further specification that the evidence of the sample constitutes the bulk of the cogent information, we are led to fitting a sample line $\hat{y} = b_0 + b_1 x$, such that the sum of the squared residuals, $\Sigma \hat{\epsilon}^2$, is minimized. The IDA command REGRession computes the sample line and makes certain automatic diagnostic checks on the adequacy of the underlying model as a base for analysis of the data in question. The user is prompted to designate the names or column numbers of the dependent and independent variable. The sample coefficients and other information are given by the command COEF. The sample correlation coefficient r is a special regression coefficient that is the slope coefficient in a regression of the *standardized* y's on the *standardized* x's, where each observation is standardized by subtracting off its sample mean and dividing by its sample standard deviation. On the scatter plot given by SCAT, both variables are standardized. Therefore the least-squares line has slope r on that plot; its intercept is 0, since the line then passes through (0,0). The last fact corresponds to the fact that the least-squares line in the plot of x_i and y_i in original units must pass through the point $(\bar{x}, \bar{y})$. The sample coefficient and its square, "adjusted" versions of these, and the standard deviation of residuals are given by SUMMary. SUMM refers to "multiple R"; simple correlation coefficients r are special cases of multiple R when there is just one independent variable. The "adjusted" R and R^2 will be explained in Chapter 10.

Diagnostic checks of model adequacy in regression parallel the diagnostic checks already applied to analysis of a single variable, except that in regression, attention is focussed on residuals $\hat{\epsilon}$ from regression, and we now can use SCAT for an additional graph. In particular, we can make sequence plots and runs counts of the residuals, plot the residuals on a normal probability plot, examine histograms of residuals, etc. The specifications of linearity and constant scatter can be checked either by examination of the scatter plot of y versus x or, somewhat more easily, the scatter plot of residuals $\hat{\epsilon}$ versus fitted values $\hat{y}$. A mini-version of the latter plot is available from the command RVSF.

Inferences about an unobserved $\tilde{y}$ given that $\tilde{x} = x$ follow a pattern familiar from Chapter 7. Given a diffuse prior distribution, the distribution of $\tilde{y}$ is approximated by: (1) a normal distribution; (2) with mean equal to $b_0 + b_1 x$; (3) with standard deviation slightly greater than the standard deviation of residuals $s_{\hat{\epsilon}}$, the actual formula depending on n and the distance of x from $\bar{x}$. As in Chapter 7, the approximation is slightly optimistic, the actual dispersion being greater than that of the approximating distribution. The necessary computations are given by the IDA command SEPR.

An inference about the height of the "true" regression line at $\tilde{x} = x$ is also given by SEPR. The standard deviation of this distribution is smaller than that of the distribution for a single $\tilde{y}$, just as the standard deviation of an inference about a long-run mean $\tilde{\mu}$ was seen in Chapter 7 to be smaller than that of an inference about a single observation.

For either of the inferences treated in the previous two paragraphs, the standard error increases as x is farther removed from $\bar{x}$.

GLOSSARY

The *disturbance* $\tilde{\epsilon}_i$ is a drawing from a normal distribution with mean 0 and the same standard deviation σ_ϵ for all i; the sequence of the disturbances is random.

The sample estimate of the "true" regression line is denoted

$$\hat{y} = b_0 + b_1 x ,$$

where b_0 is the sample estimate of $\tilde{\beta}_0$, b_1 is the estimate of $\tilde{\beta}_1$, and the *residuals*

$$\hat{\epsilon}_i = \hat{y}_i - b_0 - b_1 x_i ,$$

for i = 1, 2, ..., n, are estimates of the disturbances $\tilde{\epsilon}_i$. The line is fitted by the method of *least-squares*, which minimizes the sum of squared residuals.

The *standard deviation of residuals*,

$$s_{\hat{\epsilon}} = \sqrt{\frac{\Sigma \hat{\epsilon}^2}{n-2}} ,$$

is a sample estimate of the standard deviation σ_ϵ of the disturbances.

The *sample correlation coefficient* r is the slope coefficient in the least-squares regression with standardized x's and y's.

In regression, it is customary to speak of the regression of y *on* x, that is, of the regression of the dependent variable *on* the independent variable. The corresponding plotting convention is to put y on the vertical axis and x on the horizontal axis of the scatter plot.

Standard errors for the various regression inferences are interpreted just as in the inferences of Chapter 7.

Often there is major interest in inferences about the "true" slope $\tilde{\beta}_1$ of the regression model. The inference again follows the familiar pattern. Given a diffuse prior distribution, the distribution of $\tilde{\beta}_1$ is approximated by: (1) a normal distribution; (2) with mean equal to b_1; (3) with standard error given by $s_{\hat{\epsilon}}$ divided by the square root of the sum of squared deviations of the x's about $\bar{x}$. The needed computations are displayed in COEF, as is the parallel information for inferences about the "true" intercept $\tilde{\beta}_0$.

An inference of frequent importance refers to the correct sign of the "true" regression slope $\tilde{\beta}_1$. If, for example, b_1 is greater than 0, what is the probability that $\tilde{\beta}_1$ is also greater than 0? This question can be answered by the command GAUS; compute the probability of a normal observation greater than $L = 0$ for a distribution with mean b_1 and standard deviation equal to the standard error of the slope coefficient (in IDA output, STD.ERROR(B)).

> **GLOSSARY**
>
> The ratio of an estimate to its standard error is often called *t* (in IDA printout T).

There is a good deal of popular confusion about the purposes of a regression analysis. It may be said, for example, that a good (or poor) correlation was found. Actually, the success of a statistical analysis does not turn on whether high or low correlation was found, since "high" or "low" is highly contextual. Rather, success in regression or other statistical studies should be judged by: (1) Have we found an adequate statistical model--one that does justice to what seems to be going on in the data? (2) If we have found an adequate model, does it have useful applications? In building a regression model, the strategy--seen only partially so far--is to capture systematic influences on the dependent variable in the deterministic part of the regression model, that is, the part of the model aside from the disturbance $\tilde{\epsilon}$; and to formulate the model in such a way that the disturbances will conform to the random, normal specification.

CHAPTER 9: AUTOREGRESSION

In Chapter 4 we saw that sequence plots that suggest persistent, wave-like movements can sometimes be remedied by working with first differences or changes of the observations. An alternative strategy is to treat the immediately preceding observation as an independent variable in a regression. In terms of a scatter diagram, we are looking at a plot of y_t versus y_{t-1}. Thus each $\tilde{y}_t$ is specified to be generated by the scheme $\beta_0 + \beta_1 y_{t-1} + \tilde{\epsilon}_t$, where $\tilde{\epsilon}_t$ is a random, normal disturbance.

> The *autoregression model* postulates that the current observation $\tilde{y}_t$ can be expressed as the dependent variable in a simple regression with the previously observed value y_{t-1} as independent variable.

Ordinary least-squares regression, as discussed in Chapter 8, may be applied for statistical analysis of this model so long as, say, the sample correlation coefficient r is less than 0.7 and the sample size is moderately large, say greater than 30. To set up the analysis in IDA, the y_t variable is carried over to y_{t-1} in a new column of the data matrix by the command LAGG with a gap of 1. Then y_{t-1} is treated like an ordinary independent variable, even though it is actually the preceding value of y_t.

The correlation coefficient r associated with ordinary least-squares analysis is the sample autocorrelation coefficient. Since it is based on the scatterplot between y_t and the immediately preceding observation y_{t-1}, it is called the first autocorrelation coefficient. If the process generating $\tilde{y}_t$ is random and normal, the expected value of the sample autocorrelation coefficient is 0 and the standard deviation is approximately $1/\sqrt{n}$. Hence the sample autocorrelation coefficient can serve as another conformity measure or diagnostic check for the random, normal model. Large positive values suggest systematic persistence or clustering; large negative values suggest systematic alternation.

> The correlation coefficient between y_t and y_{t-1} is called the *first autocorrelation coefficient*.

The correlation coefficient between y_t and y_{t-2} --between y_t and its second lagged value--is called the second autocorrelation coefficient. For computation, set up y_{t-2} by using LAGG with a gap of 2. Similarly, there is a third, fourth, etc., autocorrelation coefficient. If the process generating $\tilde{y}_t$ is random and normal, the expected value of all these autocorrelation coefficients is 0 and the standard deviation is approximately $1/\sqrt{n}$, just as for the first autocorrelation coefficient. Thus these coefficients also can serve as a diagnostic check for the random, normal model.

It is useful for purposes of diagnostic checking to compute the sample autocorrelations up to some maximum lag or gap. The IDA command AUTO accomplishes this task. At present the maximum gap is 20; it is usually not well to look at lags more distant than n/4 or n/5. The Box-Pierce statistic is computed from the sample autocorrelation coefficients to provide an overall conformity measure based on autocorrelations. For a random, normal process, the Box-Pierce statistic has an expected value equal to the maximum gap for which sample autocorrelations are computed. The larger the value above this expectation, the stronger the evidence against the random, normal specification. If the Box-Pierce statistic is computed for residuals from a simple autoregression analysis, the expected value should be reduced by one, so it is equal to the maximum gap minus one. IDA, however, always shows the expectation for application to y_t rather than to $\hat{\tilde{\epsilon}}_t$.

Autoregression serves both as a possible model for the analysis of data and, through the associated autocorrelation coefficients, as a diagnostic check for randomness and normality, either for the original model or for residuals from regression or autoregression.

GLOSSARY

The correlation coefficient between y_t and y_{t-i} is called the *ith autocorrelation coefficient*.

The *sample autocorrelation function* is a listing of sample autocorrelation coefficients.

The *Box-Pierce statistic* is another diagnostic check for randomness and normality. If r_i denotes the ith sample autocorrelation coefficient, then the Box-Pierce statistic is

$$n \sum_{i=1}^{G} r_i^2 ,$$

where G is the maximum gap or lag.

CHAPTER 10: MULTIPLE REGRESSION

The multiple regression model is a simple extension of the regression model studied in Chapter 8 to applications in which there are two or more independent variables to be exploited in making predictions about a dependent variable $\tilde{y}$. What is entailed in the model can be seen by the case in which there are two independent variables x_1 and x_2. The model specifies that the $\tilde{y}$'s are generated according to the scheme $\beta_0 + \beta_1 x_1 + \beta_2 x_2 + \tilde{\epsilon}$. With the modification entailed by the need to insert the term for the second independent variable x_2, the summary of the simple regression model in Chapter 8 applies here. Instead of a regression line, we have a regression plane. A compression to a two-dimensional graphical representation can be obtained by plotting y versus the value predicted by regression if the parameters are known: $\beta_0 + \beta_1 x_1 + \beta_2 x_2$. β_1, for example, represents the change of the regression function for a unit increase of x_1 at any given value of x_2. Thus we can say that the regression model permits estimation of the effect on $\tilde{y}$ of a unit increase in x_1 after allowance for the variation of x_2. The method of least squares leads to the fitted sample plane $\hat{y} = b_0 + b_1 x_1 + b_2 x_2$ in such a way that the sum of the squared residuals $\hat{\epsilon}^2 = y - \hat{y}$ is minimized. The computations are, of course, based on the n observed triples (x_1, x_2, y). The basic computations, with automatic screening for possibly poor diagnostics, are accomplished by REGR after the user has specified the two independent variables to be used. The basic outputs of the regression are printed out by COEF and SUMM. The standard user diagnostic

The equation of the "true" regression plane when there are two independent variables is

$$\beta_0 + \beta_1 x_1 + \beta_2 x_2 ,$$

where β_0 is the *intercept*, β_1 is the *slope* in the x_1 direction, and β_2 is the *slope* in the x_2 direction. Sometimes β_1 and β_2 are spoken of as the "true" *regression coefficients*.

Sample estimates of $\tilde{\beta}_0, \tilde{\beta}_1,$ and $\tilde{\beta}_2$ are denoted $b_0, b_1,$ and $b_2,$ and are obtained by the method of least-squares. If the sample regression is denoted $\hat{y} = b_0 + b_1 x_1 + b_2 x_2$, then the *residuals* are defined by $\hat{\epsilon} = y - \hat{y}$. As in simple regression, each residual $\hat{\epsilon}$ can be thought of as an estimate of the corresponding *disturbance* $\tilde{\epsilon}$, which is specified to be random and normal.

checks are applied as in simple regression. Thus PLTS, RUNS, AUTO, and NORM are applied to residuals from regression, and SCAT can be used for residuals versus fitted values.

Interences about individual $\tilde{y}$'s given x_1 and x_2 are interpreted as in simple regression and also computed using SEPR. The same is true for the inferences about the height of the "true" regression line at x_1 and x_2.

Inferences about the "true" regression coefficients $\tilde{\beta}_0$, $\tilde{\beta}_1$, and $\tilde{\beta}_2$ can be derived from the point estimates and accompanying standard errors printed out by COEF. One new wrinkle is that, if we think about these inferences jointly rather than separately, our inference about any two slope coefficients will be correlated; a two-way probability distribution expresses the inference. The appropriate correlation coefficients are printed out by the IDA command BCOR. The correlations may sometimes be so high that it may be impossible to say very much about the statistical effect of any one independent variable.

> In addition to *standard errors* of individual regression coefficients, we may be interested in the correlation between our inferences about pairs of regression coefficients.

Correlations in our inferences about the effects of different independent variables arise because, in most applications outside of randomized experiments, the independent variables of the regression model are themselves more or less correlated. This state of affairs, usually called multicollinearity, does not constitute an intrinsic violation of the multiple regression model; the "independent" variables are not specified to be statistically uncorrelated, and the effects of correlations can be allowed for in analysis by BCOR as explained above. A high degree of multicollinearity does, however, raise some question about the appropriateness of the particular independent variables selected for tentative inclusion in the model, and it may, in the extreme, raise numerical problems in actual computation of regression estimates.

> Correlations among pairs of independent variables in multiple regression are referred to as *multicollinearity*. Correlation coefficients for variables in the data matrix are obtainable by the IDA command CORR.

In building multiple regression models, there is always the problem of deciding which possible independent variables to choose for inclusion in the model from among those that are actually measured or that can be constructed by transformation from those that are measured. The possibility of transformation arises also for the dependent variable. Just as in the simple transformation to achieve a better conformity to normality in Chapter 6, the aim here is to find transformations that will bring the data into closer conformity with the specifications of the regression model. We are guided by prior knowledge and diagnostic checks on the data in trying to achieve closer conformity to randomness and normality of residuals, and linearity and constant scatter, as required by the model specification. We judge our success in model building partly by the diagnostic checks and partly by the smallness of the standard deviation of residuals. Sometimes we achieve rough-and-ready simplification of the model by using the SWEEP command to eliminate unpromising independent variables, as judged by small absolute values of T for the regression coefficients associated with these variables.

> *Transformations* to achieve closer conformity to model specification must often be considered in building regression models.

There is thus room for a certain amount of disciplined and restrained trial-and-error or data-dredging in the search for a good regression model. At the same time, there is danger that by persistent data-dredging we may succeed in overstressing purely chance features of the set of data we are exploring. One safeguard, mentioned earlier, is to build the model on a random sample of the data and then to check to see how well it fits the balance of the data.

> *Data-dredging* is exploration of a body of data with the aim of finding a good regression model to fit the data.

IDA has several commands to facilitate data-dredging. Besides SWEEP and commands for deleting observations and blocks of observations, there are commands for automated search, including ALLS, STEP, SUBS, BACK, and FORW. These commands should be used only with great restraint, lest the search for a good model degenerate into a mindless sledgehammer approach.

The multiple correlation coefficient associated with a particular analysis, denoted by R, is the ordinary correlation coefficient between actual and fitted y--y and $\hat{y}$. Although often used as a measure of fit for a multiple regression model, it has important limitations including the fact that no allowance is made for the number of independent variables introduced. This shortcoming can be overcome by the use of the adjusted multiple correlation coefficient, but other limitations are not so easily overcome. For example, identical models from the point of view of prediction can have different R's, and a clearly superior model can have a smaller R.

CHAPTER 11: TIME SERIES

In this chapter we draw together and develop in a coordinated way the various tools for time series analysis already seen piecemeal in earlier chapters. We concentrate on ways of exploiting the past history of $\tilde{y}_t$ in order to draw inferences about future values of the series or about the underlying process; there is also the possibility of using "leading indicators" such as $\tilde{x}_t$.

A good starting point, as first seen in Chapter 4, is a sequence plot of the series itself and the accompanying diagnostic checks for randomness: the runs count (RUNS) introduced in Chapter 4 and the sample autocorrelation function introduced in Chapter 9. Now, however, we are looking not only for signs of random behavior but, more modestly, signs of stationary behavior. A random process is stationary, but a stationary process is not necessarily random A stationary process, like a random process, tends to vary around a constant level and to display roughly constant vertical scatter through time, without pronounced "trends" or abrupt shifts of level. Unlike a random process, however, a stationary process can display autocorrelation and wave-like movements, even in large samples. Signs of non-stationarity include long-continued movements in one direction or another, non-constant vertical scatter, and an autocorrelation function for which the higher autocorrelations die out only slowly. Still another sign of non-stationarity consists of a visual impression that a low-order polynomial function of time, such as a quadratic, gives a good account of the behavior of the sequence plot, viewed from arms's length.

Many time series of practical interest in business and economics show grossly non-stationary as well as non-random behavior in their sequence plots. If there are apparent strong trends or long wave-like movements, the transformation of first differencing, illustrated by the Dow-Jones Index in Chapter 4, is likely to move the analysis in the right direction. First-differencing may not, in most instances, be as successful in going all the way to a random series as it was in Chapter 4. Another common manifestation of non-stationarity is increasing vertical scatter on the sequence plot when the series reaches higher levels. When this occurs, a logarithmic transformation prior to differencing is often advisable.

The sequence plot and the autocorrelation function can often point to periodic regularities associated with seasons (or days of the week in daily time series). This may happen even in differenced series. Since seasonal variation is periodic or nearly so, it is incompatible with

GLOSSARY

Stepwise regression computations permit an automated search for an appropriate regression model. They should be used with caution and restraint to avoid overfitting.

The *multiple correlation coefficient* is the ordinary correlation coefficient between the dependent variable and the values fitted by the regression model. The *adjusted multiple correlation coefficient* makes an allowance for the number of independent variables involved in the model.

Time-series analysis can exploit for prediction either past values of the time series of central interest or "*leading indicators*", or both.

Roughly speaking, a *stationary process* looks the same statistically wherever you observe it. In particular: (1) the data tend to vary about a constant level; and (2) successive snapshots of sub-sequences of the sequence plot tend to resemble each other more or less like the resemblance between small plots of land in a virgin meadowland, successive snapshots of the cumulus clouds in the sky on a summer's day, or the chocolate chip cookies made in a good bakery.

A *logarithmic* (or similar) transformation followed by *differencing* often offers a good start toward stationarity when the initial series is evidently non-stationary.

Seasonal variation refers to periodic or near-periodic behavior of a time series that is associated with seasons of the year.

stationarity. Sometimes differencing the data at the seasonal gap or lag (12 for monthly data) will remove this source of non-stationarity.

When, after possible transformation and differencing, stationarity appears to have been achieved, the analysis enters a new phase. If the series is not only stationary but random, as in the Dow-Jones differences, we are done. If it is stationary but non-random, we may be able to obtain a good model by judicious use of autoregression for a number of lagged values. A few common sense hints: (1) keep the number of lagged variables (independent variables) reasonably small; (2) give preference to recent lags, expecially the first two; (3) if there is a substantial autocorrelation at the seasonal lag (even if seasonal differencing has been used), put that lag in; (4) use the SWEEP command to eliminate variables with relatively low T values in COEF.

If a promising model is obtained in this way, apply the usual diagnostic checks for randomness and normality of residuals. If these give a clean bill of health, the model may be applied for prediction, using SEPR and carefully unscrambling the differences and transformation that may have been applied. The predictions can then be applied to the original numbers of interest in application.

Often still further improvements in model specification for time series analysis can be obtained by using methods of estimation associated with the names of Box and Jenkins. A major advantage of these methods is that they can often produce more parsimonious models--models including a small number of parameters. Estimation of these models cannot be carried through in terms of ordinary least-squares, but the command BOXJ permits estimation for those familiar with the special concepts.

Sometimes it is recommended that time series data be fitted by trend analysis, in which time and possibly the square of time are treated as independent variables in regression. This approach is usually worse than useless unless combined with the steps outlined above; when combined with those steps, the time variables seldom contribute to the model in the fitting of business and economic data. Even when they appear to do so, however, they seldom can be exploited for predictive purposes. If the use of time variables for trend analysis appears to improve the model, it is worth looking for a catch. Time-trend variables can be created in IDA by use of INDX.

No special "deseasonalization" is needed with the approach to time series outlined in this chapter.

GLOSSARY

When stationarity is obtained by appropriate transformations, systematic use of *autoregression* may yield a satisfactory model.

Trend analysis, as traditionally exposited, consists of the use of time and possibly time squared as independent variables in a regression. It is more useful as a diagnostic check than a serious attempt at model-building and estimation.

290 □ CONVERSATIONAL STATISTICS